Cryptographic Boolean Functions and Applications

Cryptographic Boolean Functions and Applications

Thomas W. Cusick and Pantelimon Stănică

AMSTERDAM • BOSTON • HEIDELBERG • LONDON
NEW YORK • OXFORD • PARIS • SAN DIEGO
SAN FRANCISCO • SINGAPORE • SYDNEY • TOKYO

Academic Press is an imprint of Elsevier

Academic Press is an imprint of Elsevier
525 B Street, Suite 1900, San Diego, CA 92101-4495, USA
Linacre House, Jordan Hill, Oxford OX2 8DP, UK

First edition 2009

British Library Cataloguing in Publication Data
A catalogue record for this book is available from the British Library

Library of Congress Cataloging-in-Publication Data
A catalog record for this book is available from the Library of Congress

ISBN: 978-0-1237-4890-4

For information on all Academic Press publications visit our web site at elsevierdirect.com

Printed and bound by CPI Group (UK) Ltd, Croydon, CR0 4YY

Transferred to Digital Print 2011

To my wife Beverly and my children Alan and Laura – T.W.C.
To my wife Nicoleta Gabriela and my daughter Maria Andreea – P.S.

Contents

Preface

Boolean functions have been an object of study in cryptography for over 50 years, beginning with their use in linear feedback shift registers. It was in the late 1940s that Shannon published the concepts of confusion and diffusion as fundamental concepts for achieving security in cryptosystems. Confusion is reflected in the nonlinearity of parts of the cryptosystem; linear systems are generally easy to break. Diffusion is achieved by ensuring that a small change in the input is spread out to make a large change in the output. Boolean functions can easily provide both confusion and diffusion. One goal of this book is to show how to choose good Boolean functions for this purpose.

The book is designed to serve as a reference for various applications of Boolean functions in modern cryptography, which we can say is about 35 years old. The relevant material in the literature is scattered over hundreds of journal articles, conference proceedings, books, reports and notes (some of them only available online). Until this book, there has been no attempt to gather the gist of this material in one place. Our goal is to present the major concepts and associated theorems, with proofs, except in those cases where the proofs would be too long or too technical. The book is expository and we have attempted to be accurate in assigning credit for the many results quoted here. There is some original research in the book, and we have made quite a few corrections and improvements to earlier work.

The bibliography is extensive, but is not intended to be all-inclusive. In particular, it is common in cryptography for research which has been first published in the proceedings of a conference ('conference version') to be published later (perhaps with refinements or improvements) in a journal ('journal version'). Many authors seem to fear that publishing only a conference version will lead to their results not being as widely read or referred to. In some cases where a more complete journal version of an article is available, we have not listed the earlier, less polished, conference version. Of course this justifies the fear that some of the authors had! We have not hesitated to give online references where these are the only possible ones. We believe that the problem of long-term archiving of online materials will be solved.

We have tried to avoid any errors, but in a book of this size it is likely that some remain. The authors at first thought to resolve this issue by having each one blame the other for any mistakes, but then decided instead to request that readers notify at least one of us of any errors. We will then correct them in later editions, or via a posting of online errata.

We would like to thank our institutions, the State University of New York at Buffalo (T.W.C.), and Naval Postgraduate School and Institute of Mathematics of the Romanian

Academy (P.S.), for their excellent working conditions, and for allowing us to take some time off, while this book was written.

The authors express their gratitude to the following people who undertook the task of going through the various parts of this book in its incipient form: Anna Bernasconi, David Canright, Claude Carlet, Bruno Codenotti, Ed Dawson, Hal Fredricksen, Subhamoy Maitra, and Yuliang Zheng. The second author also thanks the group of students in the 'Cryptographic Boolean Functions' course at Naval Postgraduate School who diligently went through the manuscript in the Fall of 2008.

Thomas W. Cusick
Buffalo, NY

Pantelimon Stănică
Monterey, CA

September 2008

CHAPTER

A bit of history

1

1.1 GEORGE BOOLE (1815–1864)

FIGURE 1.1

Picture reprinted from *MacTutor History of Mathematics*: http://www-groups.dcs.st-andrews.ac.uk/~history/Mathematicians/Boole.html

The definitive biography of Boole is *George Boole: His Life and Work*, by Desmond MacHale (Boole Press, 1985) [286]. We shall be using both this book [286] and the biography written by O'Connor and Robertson for the MacTutor History of Mathematics archive [351].

George Boole, the son of a lower class tradesman, was born in Lincoln, England, at the end of November 1815. His father gave him his first mathematics lessons and instilled in him the love of learning. A family friend (a local bookseller) helped teach him basic Latin. Boole was translating Latin poetry by the age of 12. By 14, the adolescent Boole was fluent in German, Italian and French, as well. He especially liked novels and poetry.

His powers in higher mathematics did not show until he was 17 years old (he read his first advanced mathematics book, namely Lacroix's *Differential and Integral Calculus*). Because his father's business failed, he was forced to work to support his family. At 16 he became an assistant master in a private school at Doncaster, and before he was 20 years old he opened his own school.

In 1838, Boole was offered to take over Hall's Academy in Waddington, after its founder, Robert Hall, died. His family moved to Waddington and helped him run the school. Using mathematical journals borrowed from the local Mechanic's Institute, Boole read *Principia* of Isaac Newton and the works of French mathematicians Pierre-Simon Laplace (1749–1827) and Joseph Louis Lagrange (1736–1813). After learning what these authors previously wrote, Boole, at 24, published his first paper (*Researches on the theory of analytical transformations*) in the *Cambridge Mathematical Journal* (CMJ). It sparked a friendship between George Boole and the editor of CMJ, Duncan F. Gregory, which lasted until the premature death of Gregory in 1844. Gregory influenced Boole to study algebra. Because of his family's financial situation, Boole was unable to take Gregory's advice and audit courses at Cambridge. In fact, in the summer of 1840 he opened a boarding school in Lincoln and again the whole family moved back with him.

After his father died, Boole took up, in 1849, a Mathematics Professorship position at Queen's College in Cork, where he remained and taught for the rest of his life. There, he met a niece of Sir George Everest (of Everest mountain fame), by the name of Mary Everest. She was 17 years younger than him, but they became friends instantly. George began to give Mary lessons on the differential calculus, and in 1855, after her father died, Mary married George Boole. They were quite happy together and five daughters were born: Mary Ellen (b. 1856), Margaret (b. 1858), Alicia (later Alicia Stott) (b. 1860), Lucy Everest (b. 1862), and Ethel Lilian (b. 1864).

The works of Boole are contained in about 50 articles and a few other publications. A list of Boole's memoirs and papers, on logical and mathematical topics, is found in the *Catalogue of Scientific Memoirs* published by the Royal Society, and in a volume on differential equations (edited by I. Todhunter). Boole wrote 22 articles in the *Cambridge Mathematical Journal* and its successor, the *Cambridge and Dublin Mathematical Journal*, 16 papers in the *Philosophical Magazine*, six memoirs in the *Philosophical Transactions (The Royal Society)*, and a few others in the *Transactions of the Royal Society of Edinburgh and of the Royal Irish Academy*, in the *Bulletin de l'Academie de St-Petersbourg* (in 1862, under the pseudonym G. Boldt), and in *Crelle's Journal*, and a paper on the mathematical basis of logic published in the *Mechanics Magazine* (1848).

In 1844, the Royal Society gave him a medal for his contributions to analysis, because of his work on using algebra and calculus to analyze infinitely small and large figures.

Calculus of reasoning, which Boole was preoccupied with, found its way into his 1847 work, *The Mathematical Analysis of Logic*, that expanded on the work of the German mathematician Gottfried Wilhelm Leibniz (1646–1716) and pushed the idea that logic was a mathematical discipline, rather than philosophy. This paper won him the admiration of the distinguished logician Augustus de Morgan, and a place among the faculty of Ireland's Queen's College.

In 1854, Boole published *An Investigation into the Laws of Thought, on Which are Founded the Mathematical Theories of Logic and Probabilities*, which is perhaps his most important work. Boole approached logic in a new way, reducing it to simple algebra, incorporating logic into mathematics, and laying the foundations of the now famous binary approach. Logical expressions are now represented using a mathematical form called in his honor *Boolean Algebra*.

Boole's genius was recognized and he received honorary degrees from the universities of Dublin and Oxford and was elected a Fellow of the Royal Society in 1857. Since his work eventually led people to land on the Moon, it is only natural that Boole is the name of a lunar crater.

One day in 1864, Boole was walking from his home to the college and was caught in a rain storm. He lectured in wet clothes and caught a cold. Because of that, it is unfortunate for mathematics that he died when he was just 49 years old.

1.2 CLAUDE ELWOOD SHANNON (1916–2001)

FIGURE 1.2

Picture reprinted from the Notices of AMS paper by Golomb et al., available at http://www.ams.org/notices/200201/fea-shannonpdf (courtesy of the MIT Museum).

Two good biographies of Shannon were written by Sloane and Wyner, and by Liversidge in the book edited by Sloane and Wyner containing Shannon's collected papers [419].

Boole's work on mathematical logic was criticized and/or ignored by his contemporaries, except for an American logician, Charles Sanders Peirce (1839–1914), who gave a speech at the American Academy of Arts and Sciences, describing Boole's ideas. Peirce spent more than 20 years working on these ideas and their applications in electronic circuitry; ultimately, he designed a theoretical electrical logic circuit.

Unfortunately, Boolean algebra and Peirce's work remained mostly unknown and unused for (too) many years, until the 1940s, when a young student by the name of Claude Elwood Shannon picked up Boole's and Peirce's works and recognized their relevance to electronics design.

Claude E. Shannon was born in Petoskey, Michigan, on April 30, 1916. His father was a businessman and, for a period, Judge of Probate. His mother was a language teacher and for a number of years Principal of Gaylord High School, in Gaylord, Michigan. Shannon remained in Gaylord until he was 16 when he graduated from high school. He showed an inclination for science and mathematics and kept himself busy by building model planes, a radio-controlled model boat, and a telegraph system to a friend's house half a mile away [419].

Following his sister, in 1932 he entered University of Michigan (UM), where he was introduced to the work of George Boole. Shannon graduated from UM in 1936 with dual Bachelor's Degrees of Science in Electrical Engineering and Science in Mathematics. Right away, accepting a research assistant position at Massachusetts Institute of Technology to support himself, he began his graduate studies. He graduated in 1940 with a Master's Degree in Electrical Engineering and a PhD in Mathematics. His Master's thesis *A Symbolic Analysis of Relay and Switching Circuits* is a (successful) attempt to use Boole's algebra to analyze relay switching circuits, while his doctoral thesis deals with population genetics. A version of his Master's thesis was published in *Transactions of the American Institute of Electrical Engineers* (1940), and earned him the Alfred Noble (American Institute of Engineers) Award.

After spending a year at the Institute for Advanced Study, in 1941 Shannon joined AT&T Bell Telephones in New Jersey as a research mathematician to work on fire-control systems and cryptography. He remained affiliated with Bell Laboratories until 1972, but took up other various positions (MIT; Center for the Study of the Behavioral Sciences in Palo Alto; Institute for Advanced Study in Princeton, Visiting Fellow at All Souls College, Oxford; University of California; the IEEE; and the Royal Society).

In 1949 Shannon married Mary Elizabeth Moore and they had three sons and one daughter: Robert, James, Andrew Moore, and Margarita.

In one of his most important works, *A Mathematical Theory of Communication* [417], Shannon founded the subject of information theory and he proposed a linear schematic model of a communications system. This was a revolutionary idea as there was no longer any need for electromagnetic waves to be sent down a wire. One could communicate instead, by sending sequences of 0 and 1 bits. In the next year, he wrote another fundamental paper, *Communication Theory of Secrecy Systems* [418], which is the first analysis of cryptography. It was based on classified work on secrecy systems undertaken by Shannon in the final year of World War II.

Shannon died in 2001 after a long struggle with Alzheimer's disease.

CHAPTER

Fourier analysis of Boolean functions

2

2.1 BASIC DEFINITIONS ON BOOLEAN FUNCTIONS

Take what you need; act as you must, and you will obtain that for which you wish!
René Descartes (1596–1650)

The purpose of this chapter is to make some preliminary definitions on Boolean functions and introduce one of the most important tools in cryptography, namely the Walsh transform (also called Hadamard transform), which is the characteristic 2 case of the discrete Fourier transform. The use of the Walsh transform makes the computation of nonlinearity, and many of the cryptographic properties of a Boolean function, a very easy and enjoyable task. Various other topics needed in the subsequent chapters will be introduced here.

Let $\mathbb{V}_n$ be the vector space of dimension n over the two-element field $\mathbb{F}_2$. For two vectors in $\mathbb{V}_n$, say $\mathbf{a} = (a_1,\dots,a_n)$ and $\mathbf{b} = (b_1,\dots,b_n)$, we define the scalar product $\mathbf{a}\cdot\mathbf{b} = a_1b_1 \oplus \cdots \oplus a_nb_n$, where the multiplication and addition $\oplus$ (called *xor*) are over $\mathbb{F}_2$ (it should not be confused with the direct product of vector spaces, but that is going to be clear from context). We also define the operation $\star$ by $\mathbf{a}\star\mathbf{b} = (a_1b_1,\dots,a_nb_n)$. Given a vector $\mathbf{a}$, it is always a row vector unless context obviously requires it to be a column vector. For example, in Theorem 2.14 below, $\mathbf{w}$ sometimes denotes a row vector and sometimes denotes a column vector.

Definition 2.1. *A* Boolean function f *in* n *variables is a map from* $\mathbb{V}_n$ *to* $\mathbb{F}_2$. *The* (0, 1)-sequence *defined by* $(f(\mathbf{v}_0), f(\mathbf{v}_1),\dots,f(\mathbf{v}_{2^n-1}))$ *is called the* truth table *of* f, *where* $\mathbf{v}_0 = (0,\dots,0,0), \mathbf{v}_1 = (0,\dots,0,1),\dots,\mathbf{v}_{2^n-1} = (1,\dots,1,1)$, *ordered by lexicographical order. (Oftentimes, when there is a danger of confusion because of a similar notation, we shall write* α_i *instead of* $\mathbf{v}_i$*— we shall point it out, nonetheless.) The* $(1,-1)$-sequence *of* f *(or simply sequence) is defined by* $\left((-1)^{f(\mathbf{v}_0)},\dots,(-1)^{f(\mathbf{v}_{2^n-1})}\right)$. *The algebra of all Boolean functions on* $\mathbb{V}_n$ *will be denoted by* $\mathcal{B}_n$.

Obviously, $\mathbf{v}_i = \alpha_i = \mathbf{b}(i)$, where $\mathbf{b}(i)$ is the binary representation of i, $0 \le i \le 2^n - 1$, written as a vector of length 2^n. We shall be using this notation in Chapter 8.

A Boolean function on $\mathbb{V}_n$ can be expressed as a polynomial in

$$\mathbb{F}_2[x_1,\dots,x_n]/(x_1^2 - x_1,\dots,x_n^2 - x_n),$$

the *algebraic normal form* (ANF for short), that is,

$$f(\mathbf{x}) = \sum_{\mathbf{a} \in \mathbb{V}_n} c_{\mathbf{a}} x_1^{a_1} \cdots x_n^{a_n},$$

where $c_{\mathbf{a}} \in \mathbb{F}_2$ and $\mathbf{a} = (a_1, \ldots, a_n)$. Moreover, $c_{\mathbf{a}} = \sum_{\mathbf{x} \leq \mathbf{a}} f(\mathbf{x})$, where $\mathbf{x} \leq \mathbf{a}$ means that $x_i \leq a_i$, for all $1 \leq i \leq n$ (we sometimes say that $\mathbf{a}$ *covers* $\mathbf{x}$). (We refer to the work of McWilliams and Sloane [287, Chapter 13, pp. 370–373] for a method to derive the ANF from the truth table of a Boolean function.) The number of variables in the highest order monomial with nonzero coefficient is called the *algebraic degree*, or simply the degree of f. (The set of all Boolean functions on $\mathbb{V}_n$ of degree $\leq r$ is called the rth order binary Reed–Muller (RM) code, denoted by $\mathcal{R}(r,n)$.) A Boolean function is said to be *homogeneous* if its ANF contains terms of the same degree only. The logical negation or *complement* of a Boolean function f is $\bar{f} = f \oplus 1$ (sometimes denoted by $C(f)$ to avoid awkward expressions).

Let f be a function on $\mathbb{V}_n$ and U be a subspace of $\mathbb{V}_n$. The *restriction* of f to U, denoted by f_U, is a function on U, defined by the rule

$$f_U(\mathbf{u}) = f(\mathbf{u}) \quad \text{for every } \mathbf{u} \in U.$$

The dimension of a vector subspace U of $\mathbb{V}_n$ is denoted by $\dim(U)$.

An *affine function* $\ell_{\mathbf{a},c}$ on $\mathbb{V}_n$ is a function that takes the form

$$\ell_{\mathbf{a},c}(\mathbf{x}) = \mathbf{a} \cdot \mathbf{x} \oplus c = a_1 x_1 \oplus \cdots \oplus a_n x_n \oplus c,$$

where $\mathbf{a} = (a_1 \ldots, a_n) \in \mathbb{V}_n, c \in \mathbb{F}_2$. If $c = 0$, then $\ell_{\mathbf{a},0} (= \ell_{\mathbf{a}})$ is a *linear* function .

To each Boolean function $f : \mathbb{V}_n \to \mathbb{F}_2$ we associate its character form, or *sign function*, denoted by $\hat{f} : \mathbb{V}_n \to \mathbb{R}^* \subseteq \mathbb{C}^*$ and defined by

$$\hat{f}(\mathbf{x}) = (-1)^{f(\mathbf{x})}.$$

The behavior of the sign function on the sum and products of Boolean functions is displayed in the next proposition.

Proposition 2.2. *If f, g are Boolean functions on $\mathbb{V}_n$, then the following statements hold:*

1. $\widehat{f \oplus g} = \hat{f}\hat{g}$.
2. $2\widehat{fg} = 1 + \hat{f} + \hat{g} - \hat{f}\hat{g}$.

Proof. The first claim is straightforward. The second claim follows from the observation $\hat{f}(\mathbf{x}) = 1 - 2f(\mathbf{x})$. ∎

Definition 2.3. *The* Hamming weight *of a vector $\mathbf{x} \in \mathbb{V}_n$, denoted by $wt(\mathbf{x})$, is the number of 1s in the vector $\mathbf{x}$. For a Boolean function on $\mathbb{V}_n$, let $\Omega_f = \{\mathbf{x} \in \mathbb{V}_n :$*

$f(\mathbf{x}) = 1\}$ *be the* support *of* f. *The Hamming weight of a function* f *is the Hamming weight of its truth table, that is, the cardinality of* $f^{-1}(1)$, *or equivalently* $wt(f) = |\Omega_f|$. *The* Hamming distance *between two functions* $f, g : \mathbb{V}_n \to \mathbb{F}_2$, *denoted by* $d(f, g)$ *is defined as*

$$d(f,g) = wt(f \oplus g).$$

The nonlinearity *of a function* f, *denoted by* $\mathcal{N}_f$, *is defined as*

$$\mathcal{N}_f = \min_{\phi \in \mathcal{A}_n} d(f, \phi),$$

where $\mathcal{A}_n$ *is the class of all affine functions on* $\mathbb{V}_n$. *A function of* n *variables is called* balanced *if its weight is exactly* 2^{n-1}.

It is a straightforward exercise to prove the next lemma and we leave this to the interested reader.

Lemma 2.4. *The Hamming weight and distance satisfy the following properties:*

1. $wt(\mathbf{x} \oplus \mathbf{y}) = wt(\mathbf{x}) + wt(\mathbf{y}) - 2wt(\mathbf{x} \star \mathbf{y})$;
2. $d(f,g) = |\{\mathbf{x} \in \mathbb{V}_n : f(\mathbf{x}) \neq g(\mathbf{x})\}|$;
3. $d(f,g) + d(g,h) \geq d(f,h)$;
4. $d(f, \bar{g}) = 2^n - d(f,g)$;
5. $d(f,g) = 2^{n-1} - \frac{1}{2}\sum_{\mathbf{x}} \hat{f}(\mathbf{x}) \cdot \hat{g}(\mathbf{x})$.

Throughout the book, we use the Vinogradov symbols $\gg$, $\ll$ and $\asymp$, and the Landau symbols O and o with their usual meanings. We recall that $F \ll G, G \gg F$ and $F = O(G)$ are all equivalent and mean that $|F(x)| \leq c|G(x)|$ holds with some positive constant c, and x sufficiently large, while $F \asymp G$ means that both $F \ll G$ and $G \ll F$ hold. Further, $F = o(G)$ means that for any $\epsilon > 0$, $|(F/G)(x)| < \epsilon$ as x becomes sufficiently large. We write $\log x$ for the natural logarithm of x, and $\Pr(Z)$ for the probability of some random variable Z.

2.2 WALSH TRANSFORM

We recommend the articles of Bernasconi et al. [30] and Pommerening [367], and the references contained therein, for more on this topic.

Definition 2.5. *The* Walsh transform *of a function* f *on* $\mathbb{V}_n$ *(with the values of* f *taken to be real numbers* 0 *and* 1*) is the map* $W(f) : \mathbb{V}_n \to \mathbb{R}$, *defined by*

$$W(f)(\mathbf{w}) = \sum_{\mathbf{x} \in \mathbb{V}_n} f(\mathbf{x})(-1)^{\mathbf{w} \cdot \mathbf{x}}, \tag{2.1}$$

which defines the coefficients of f with respect to the orthonormal basis of the group characters $Q_{\mathbf{x}}(\mathbf{w})=(-1)^{\mathbf{w}\cdot\mathbf{x}}$; *$f$ can be recovered by the* inverse Walsh transform

$$f(\mathbf{x})=2^{-n}\sum_{\mathbf{w}\in\mathbb{V}_n} W(f)(\mathbf{w})(-1)^{\mathbf{w}\cdot\mathbf{x}}. \tag{2.2}$$

The *Walsh spectrum* of f is the list of the 2^n *Walsh coefficients* given by (2.1) as $\mathbf{w}$ varies.

The simplest Boolean functions are the constant functions $\mathtt{0}$ and $\mathbf{1}$. Obviously, $W(\mathtt{0})(\mathbf{u})=0$ and the Walsh coefficients for the function $\mathbf{1}$ are given by the next lemma.

Lemma 2.6. *If* $\mathbf{w}\in\mathbb{V}_n$, *we have*

$$\sum_{\mathbf{u}\in\mathbb{V}_n}(-1)^{\mathbf{u}\cdot\mathbf{w}}=\begin{cases}2^n & \text{if } \mathbf{w}=\mathtt{0},\\ 0 & \text{else.}\end{cases}$$

Proof. First, if $\mathbf{w}=\mathtt{0}$, then all summands are 1. Now, assume $\mathbf{w}\neq\mathtt{0}$, and consider the hyperplanes $H=\{\mathbf{u}\in\mathbb{V}_n : \mathbf{u}\cdot\mathbf{w}=0\}$, $\bar{H}=\{\mathbf{u}\in\mathbb{V}_n : \mathbf{u}\cdot\mathbf{w}=1\}$. Obviously, these hyperplanes generate a partition of $\mathbb{V}_n$. Moreover, for any $\mathbf{u}\in H$, the summand is 1, and for any $\mathbf{u}\in\bar{H}$, the summand is -1. Since the cardinalities of $H,\bar{H}$ are the same, that is 2^{n-1}, we have the lemma. ∎

2.3 AUTOCORRELATION FUNCTION

Definition 2.7. *The* autocorrelation *function $\hat{r}_{\hat{f}}(\mathbf{a})$ is defined as*

$$\hat{r}_{\hat{f}}(\mathbf{a})=\sum_{\mathbf{x}\in\mathbb{V}_n}\hat{f}(\mathbf{x})\cdot\hat{f}(\mathbf{x}\oplus\mathbf{a}). \tag{2.3}$$

We shall write $\hat{r}(\mathbf{a})$ if there is no danger of confusion. Note that $\hat{r}(\mathtt{0})$ equals 2^n. The correlation value *between two Boolean functions g and h is defined by*

$$c(g,h)=1-\frac{d(g,h)}{2^{n-1}}. \tag{2.4}$$

We define the crosscorrelation function between $f,g:\mathbb{V}_n\to\mathbb{F}_2$ by

$$c(\hat{f},\hat{g})(\mathbf{y})=\sum_{\mathbf{x}\in\mathbb{V}_n}\hat{f}(\mathbf{x})\cdot\hat{g}(\mathbf{x}\oplus\mathbf{y}).$$

A very important relation asserts that the inverse transform of the energy spectrum is the autocorrelation function.

Theorem 2.8 (*Wiener-Khintchine*). *A Boolean function on $\mathbb{V}_n$ satisfies*

$$W(\hat{r})(\mathbf{w})=W(\hat{f})^2(\mathbf{w}),\quad \textit{for all } \mathbf{w}\in\mathbb{V}_n. \tag{2.5}$$

Since

$$\Pr(\hat{f}(\mathbf{x}) \neq \hat{f}(\mathbf{x} \oplus \mathbf{a})) = \frac{1}{2} - \frac{\hat{r}(\mathbf{a})}{2^{n+1}},$$

for large values of n, this theorem allows an efficient computation of these probabilities, requiring $O(n2^n)$ operations, instead of $O(2^{2n})$ for a straightforward computation.

The relationship between the Walsh transforms of f and $\hat{f}$ is displayed in the next lemma.

Lemma 2.9. *We have*

$$W(\hat{f})(\mathbf{w}) = -2W(f)(\mathbf{w}) + 2^n \delta(\mathbf{w}), \tag{2.6}$$

or

$$W(f)(\mathbf{w}) = 2^{n-1}\delta(\mathbf{w}) - \frac{1}{2}W(\hat{f})(\mathbf{w}), \tag{2.7}$$

where $\delta(\mathbf{w}) = 1$ *if* $\mathbf{w} = 0$ *and* 0 *otherwise.*

Proof. Starting from the left-hand side of the first equation we obtain

$$\begin{aligned} W(\hat{f})(\mathbf{w}) &= \sum_{\mathbf{x}\in\mathbb{V}_n} (-1)^{f(\mathbf{x})\oplus \mathbf{w}\cdot\mathbf{x}} \\ &= \sum (1-2f(\mathbf{x}))(-1)^{\mathbf{w}\cdot\mathbf{x}} \\ &= \sum_{\mathbf{x}\in\mathbb{V}_n} (-1)^{\mathbf{w}\cdot\mathbf{x}} - 2\sum_{\mathbf{x}\in\mathbb{V}_n} f(\mathbf{x})(-1)^{\mathbf{w}\cdot\mathbf{x}} \\ &= 2^n\delta(\mathbf{w}) - 2W(f)(\mathbf{w}), \end{aligned}$$

by Lemma 2.6. ∎

2.4 WALSH TRANSFORM ON SUBSPACES

One can find a very important equation between $W(f)$ and f restricted to an arbitrary subspace of $\mathbb{V}_n$, which Lechner [272, Theorem 2.6, p. 147] called the *Poisson Summation Formula*.

Theorem 2.10. *Let* $f : \mathbb{V}_n \to \mathbb{R}$ *and* $W(f)$ *be its Walsh transform. Let* S *be an arbitrary subspace of* $\mathbb{V}_n$ *and let* $S^\perp$ *be the dual (annihilator) of* S*, that is,*

$$S^\perp = \{\mathbf{x} \in \mathbb{V}_n : \mathbf{x}\cdot\mathbf{s} = 0 \text{ for all } \mathbf{s} \text{ in } S\}.$$

Then

$$\sum_{\mathbf{u}\in S} W(f)(\mathbf{u}) = 2^{\dim S} \sum_{\mathbf{u}\in S^\perp} f(\mathbf{u}). \tag{2.8}$$

Proof. We have

$$\sum_{\mathbf{u}\in S} W(f)(\mathbf{u}) = \sum_{\mathbf{u}\in S}\left(\sum_{\mathbf{v}\in\mathbb{V}_n} f(\mathbf{v})(-1)^{\mathbf{u}\cdot\mathbf{v}}\right)$$
$$= \sum_{\mathbf{v}\in\mathbb{V}_n} f(\mathbf{v})\left(\sum_{\mathbf{u}\in S}(-1)^{\mathbf{u}\cdot\mathbf{v}}\right)$$
$$= 2^{\dim S}\sum_{\mathbf{v}\in S^\perp} f(\mathbf{v}). \tag{2.9}$$

■

Corollary 2.11. *For any Boolean function $f:\mathbb{V}_n\to\mathbb{F}_2$*

$$\sum_{\mathbf{u}\leq\mathbf{v}} W(f)(\mathbf{u}) = 2^{wt(\mathbf{v})}\sum_{\mathbf{u}\leq\bar{\mathbf{v}}} f(\mathbf{u}), \tag{2.10}$$

where $\mathbf{u}\leq\mathbf{v}$ means that if $u_i = 1$, then $v_i = 1, 1\leq i\leq n$.

2.5 LINEAR TRANSFORMATIONS AND THE SIGN FUNCTION

Lemma 2.12. *If the Boolean function f can be obtained from g by an affine transformation of the input, that is*

$$g(\mathbf{v}) = f(A\mathbf{v}\oplus\mathbf{b}),$$

with A an invertible matrix and $\mathbf{b}\in\mathbb{V}_n$, then the Walsh transforms of f and g are related by

$$W(g)(\mathbf{u}) = \pm W(f)(\mathbf{u}A^{-1}).$$

Proof. First,

$$W(g)(\mathbf{u}) = \sum_{\mathbf{v}}(-1)^{\mathbf{u}\cdot\mathbf{v}}g(\mathbf{v}) = \sum_{\mathbf{v}}(-1)^{\mathbf{u}\cdot\mathbf{v}}f(A\mathbf{v}\oplus\mathbf{b}).$$

By setting $\mathbf{v} = A^{-1}\mathbf{w}\oplus A^{-1}\mathbf{b}$ and $\mathbf{u}' = \mathbf{u}A^{-1}$, we get

$$W(g)(\mathbf{u}) = \sum_{\mathbf{w}}(-1)^{\mathbf{u}\cdot A^{-1}\mathbf{w}}(-1)^{\mathbf{u}\cdot A^{-1}\mathbf{b}}f(\mathbf{w})$$
$$= \pm\sum_{\mathbf{w}}(-1)^{\mathbf{u}'\cdot\mathbf{w}}f(\mathbf{w}) = \pm W(f)(\mathbf{u}'), \tag{2.11}$$

which proves the lemma. ■

Let the linear function $\ell_{\mathbf{a}}(\mathbf{x}) = \mathbf{a}\cdot\mathbf{x}$. We leave the proof of the next theorem to the interested reader.

Theorem 2.13. *The following statements are true:*

1. $W(\widehat{f \oplus 1})(\mathbf{x}) = -W(\hat{f})(\mathbf{x})$;
2. *If* $g(\mathbf{x}) = f(\mathbf{x}) \oplus \ell_{\mathbf{a}}(\mathbf{x})$, *then* $W(\hat{g})(\mathbf{x}) = W(\hat{f})(\mathbf{x} \oplus \mathbf{a})$;
3. *If* $g(\mathbf{x}) = f(\mathbf{x} \oplus \mathbf{a})$, *then* $W(\hat{g})(\mathbf{x}) = (-1)^{\mathbf{a}\cdot\mathbf{x}} W(\hat{f})(\mathbf{x})$;
4. *If* $g(\mathbf{x}) = f(\mathbf{x}A)$, A *is a nonsingular matrix, then* $W(\hat{g})(\mathbf{x}) = W(\hat{f})(\mathbf{x}(A^{-1})^t)$ *(the superscript t takes the transpose of the invertible matrix);*
5. *If* $h(\mathbf{x}) = f(\mathbf{x}) \oplus g(\mathbf{x})$ *on* $\mathbb{V}_n$, *then*

$$W(\hat{k})(\mathbf{x}) = \frac{1}{2^n} \sum_{\mathbf{v} \in \mathbb{V}_n} W(\hat{f})(\mathbf{v}) W(\hat{g})(\mathbf{x} \oplus \mathbf{v}).$$

6. *If* $h(\mathbf{x}, \mathbf{y}) = f(\mathbf{x}) \oplus g(\mathbf{y})$, f *on* $\mathbb{V}_n$ *and* g *on* $\mathbb{V}_m$, *then*

$$W(\hat{h})(\mathbf{x}, \mathbf{y}) = W(\hat{f})(\mathbf{x}) W(\hat{g})(\mathbf{y}).$$

7. *If* $k(\mathbf{x}) = f(\mathbf{x})g(\mathbf{x})$ *and* $h(\mathbf{x}) = f(\mathbf{x}) \oplus g(\mathbf{x})$, *then*

$$W(\hat{h})(\mathbf{x}) = \frac{1}{2}\left(2^n \delta(\mathbf{x}) + W(\hat{f})(\mathbf{x}) + W(\hat{g})(\mathbf{x}) - W(\hat{h})(\mathbf{x})\right),$$

where δ *is the Kronecker delta function* ($\delta(\mathbf{x}) = 1$, *if* $\mathbf{x} = 0$, *and* $\delta(\mathbf{x}) = 0$, *if* $\mathbf{x} \neq 0$).

Take g, h, two Boolean functions in n variables, such that $h(\mathbf{x}) = g(A\mathbf{x} \oplus \mathbf{a})$, where A is an $n \times n$ matrix over $\mathbb{F}_2$. Theorem 2.7 of [272] investigates *domain encodings*. In the Walsh transform for g,

$$W(g)(\mathbf{w}) = \sum_{\mathbf{x} \in \mathbb{V}_n} g(\mathbf{x})(-1)^{\mathbf{w}\cdot\mathbf{x}},$$

we replace $\mathbf{x}$ by $(\mathbf{y} \oplus \mathbf{a})(A^{-1})^t$, $g(\mathbf{x})$ by $h(\mathbf{y})$ and sum over $\mathbf{y}$ instead of $\mathbf{x}$. We obtain (for easy writing we denote the transpose $(A^{-1})^t$ by A^{-t})

$$\begin{aligned} W(g)(\mathbf{w}) &= \sum_{\mathbf{x}} (-1)^{\mathbf{w}\cdot\mathbf{x}} g(\mathbf{x}) = \sum_{\mathbf{y}} (-1)^{\mathbf{w}A^{-t}\cdot(\mathbf{y}\oplus\mathbf{a})} h(\mathbf{y}) \\ &= (-1)^{\mathbf{w}A^{-t}\cdot\mathbf{a}} \sum_{\mathbf{y}} (-1)^{\mathbf{w}A^{-t}\cdot\mathbf{y}} h(\mathbf{y}) = (-1)^{\mathbf{w}A^{-t}\cdot\mathbf{a}} W(h)(A^{-t}\mathbf{w}). \end{aligned}$$

In a similar manner,

$$W(h)(\mathbf{w}) = (-1)^{\mathbf{w}\cdot\mathbf{a}} W(g)(\mathbf{w}A^t).$$

Thus, we have the following more precise version of Lemma 2.12.

Theorem 2.14. *Let h, g be Boolean functions with $h(\mathbf{x}) = g(A\mathbf{x} \oplus \mathbf{a})$, where A is an $n \times n$ invertible matrix over $\mathbb{F}_2$. Then*

$$W(g)(\mathbf{w}) = (-1)^{\mathbf{w}A^{-t}\cdot\mathbf{a}} W(h)(A^{-t}\mathbf{w})$$
$$W(h)(\mathbf{w}) = (-1)^{\mathbf{w}\cdot\mathbf{a}} W(g)(A^{t}\mathbf{w}).$$

Let us introduce the notion of equivalence next, which will be useful throughout the book.

Definition 2.15. *Two Boolean functions g, h on $\mathbb{V}_n$ are called (affinely)* equivalent *if $g(\mathbf{x}) = h(A\mathbf{x} \oplus \mathbf{a}) \oplus (\mathbf{b} \cdot \mathbf{x}) \oplus c$, where $\mathbf{a}, \mathbf{b} \in \mathbb{V}_n, c \in \mathbb{F}_2$ and A is an $n \times n$ nonsingular matrix. If no such transformation exists, then g, h are called* inequivalent.

Oftentimes, in literature, we encounter the notion of equivalence with $\mathbf{b} = 0, c = 0$.

The following result, which belongs to Preneel [368], generalizes some of the previous relations. One can deduce it easily now from Theorems 2.13 and 2.14. The *concatenation* of two bit strings $\mathbf{b}_1, \mathbf{b}_2$ will be denoted by $\mathbf{b}_1 \,||\, \mathbf{b}_2$ (or simply $\mathbf{b}_1\mathbf{b}_2$ if there is no danger of confusion).

Theorem 2.16. *For a Boolean function h on $\mathbb{V}_n$, $\mathbf{a}, \mathbf{b} \in \mathbb{V}_n, c \in \mathbb{F}_2$ and an $n \times n$ nonsingular matrix A, define an equivalent function g by $g(\mathbf{x}) = h(A\mathbf{x} \oplus \mathbf{a}) \oplus (\mathbf{b} \cdot \mathbf{x}) \oplus c$. Then,*

$$W(\hat{g})(\mathbf{w}) = (-1)^c (-1)^{A^{-1}\mathbf{a}\cdot(\mathbf{w}\oplus\mathbf{b})} W(\hat{h})(A^{-t}(\mathbf{w} \oplus \mathbf{b})). \tag{2.12}$$

If g is the concatenation $g_0 \,||\, g_1$, $\mathbf{w} \in \mathbb{V}_n$, $\mathbf{w}^* = (\mathbf{w}, w_{n+1}) \in \mathbb{V}_{n+1}$, where g_0, g_1 are both defined on $\mathbb{V}_n$, then

$$W(\hat{g})(\mathbf{w}^*) = W(\hat{g}_0)(\mathbf{w}) + (-1)^{w_{n+1}} W(\hat{g}_1)(\mathbf{w}).$$

2.6 PARSEVAL EQUATION

From the definition of the Walsh transform we deduce that $W(\hat{f})(\mathbf{u})$ is equal to the number of 0s minus the number of 1s in the binary vector $f \oplus \ell_{\mathbf{u}}$ (recall that $\ell_{\mathbf{u}}$ is the linear function $\ell_{\mathbf{u}}(\mathbf{v}) = \sum_{i=1}^{n} u_i v_i$, where $\mathbf{u} = (u_1, \ldots, u_n), \mathbf{v} = (v_1, \ldots, v_n)$) and so,

$$\begin{aligned} W(\hat{f})(\mathbf{u}) &= 2^n - 2d\left(f, \sum_{i=1}^{n} u_i v_i\right) \quad \text{or} \\ d\left(f, \sum_{i=1}^{n} u_i v_i\right) &= \frac{1}{2}\left(2^n - W(\hat{f})(\mathbf{u})\right). \end{aligned} \tag{2.13}$$

Also,

$$d\left(f, 1 \oplus \sum_{i=1}^{n} u_i v_i\right) = \frac{1}{2}\left(2^n + W(\hat{f})(\mathbf{u})\right).$$

We summarize this in our next theorem.

Theorem 2.17. *The nonlinearity of f is determined by the Walsh transform of f, that is,*

$$\mathcal{N}_f = 2^{n-1} - \frac{1}{2}\max_{\mathbf{u}\in\mathbb{V}_n} |W(\hat{f})(\mathbf{u})|.$$

Lemma 2.18.

$$\sum_{\mathbf{u}\in\mathbb{V}_n} W(\hat{f})(\mathbf{u})W(\hat{f})(\mathbf{u}\oplus\mathbf{v}) = \begin{cases} 2^{2n} & \text{if } \mathbf{v}=0 \\ 0 & \text{if } \mathbf{v}\neq 0. \end{cases} \tag{2.14}$$

Proof. We have

$$\begin{aligned}\sum_{\mathbf{u}\in\mathbb{V}_n} W(\hat{f})(\mathbf{u})W(\hat{f})(\mathbf{u}\oplus\mathbf{v}) &= \sum_{\mathbf{u},\mathbf{w}\in\mathbb{V}_n} (-1)^{\mathbf{u}\cdot\mathbf{w}}\hat{f}(\mathbf{w}) \sum_{\mathbf{x}\in\mathbb{V}_n}(-1)^{(\mathbf{u}\oplus\mathbf{v})\cdot\mathbf{x}}\hat{f}(\mathbf{x}) \\ &= \sum_{\mathbf{w},\mathbf{x}\in\mathbb{V}_n} (-1)^{\mathbf{v}\cdot\mathbf{x}}\hat{f}(\mathbf{w})\hat{f}(\mathbf{x}) \sum_{\mathbf{u}\in\mathbb{V}_n}(-1)^{\mathbf{u}\cdot(\mathbf{w}\oplus\mathbf{x})} \\ &= 2^n \sum_{\mathbf{w}\in\mathbb{V}_n}(-1)^{\mathbf{v}\cdot\mathbf{w}}\hat{f}(\mathbf{w})^2 = 2^n \sum_{\mathbf{w}\in\mathbb{V}_n}(-1)^{\mathbf{v}\cdot\mathbf{w}},\end{aligned} \tag{2.15}$$

since $\hat{f}(\mathbf{w})^2 = 1$ and the lemma is proved. ■

Corollary 2.19 (*Parseval's equation*). *For any Boolean function f in n variables, the following equation holds*

$$\sum_{\mathbf{u}\in\mathbb{V}_n} \left(W(\hat{f})(\mathbf{u})\right)^2 = 2^{2n}. \tag{2.16}$$

Relations (2.13) give a way to find the closest affine function to f (in terms of Hamming distance): $\ell_{\mathbf{u},a_0}(\mathbf{v}) = a_0 \oplus \mathbf{u}\cdot\mathbf{v}$, where $|W(\hat{f})(\mathbf{u})|$ is the largest.

Example 2.20 (*Sağdiçoğlu [392]*). *Define the Boolean function on $\mathbb{V}_3$ by*

$$f(x_1,x_2,x_3) = 1\oplus x_1 \oplus x_2 \oplus x_2x_3 \oplus x_1x_2x_3.$$

Its truth table is 11010011 *and its (ordered) Walsh coefficients are* $-2,2,2,-2,-2,2,-6,$ -2. *Since $|W(\hat{f})(\alpha_6)| = 6$, it follows that the function $l_{\alpha_6,1} = 1\oplus x_1\oplus x_2$ is the nearest affine function to f. One can check easily that $d(f,l_{\alpha_6,1}) = 1$.*

2.7 ASYMPTOTIC RESULTS ON WALSH COEFFICIENTS

We define $\mathbb{V}_\infty$ as the space of infinite sequences over $\mathbb{F}_2$ with all elements zero except for finitely many. Define $\mathcal{B}$ as the algebra of Boolean functions on $\mathbb{V}_\infty$, and recall that $\mathcal{B}_n$

is the algebra of Boolean functions on $\mathbb{V}_n$ (see Definition 2.1). Let the *spectral amplitude* of a Boolean function be defined by

$$S(f) = \max_{\mathbf{v}\in\mathbb{V}_n} \left| W(\hat{f})(\mathbf{v}) \right|$$

Rodier in [382,383] proved the next interesting result.

Theorem 2.21. *For almost every Boolean function f in $\mathcal{B}$, the restriction, say f_n, of f to $\mathbb{V}_n$ satisfies*

$$\sqrt{2\log 2} - \frac{5\log n}{n} \le \frac{S(f_n)}{2^{n/2}\sqrt{n}} \le \sqrt{2\log 2} + \frac{4\log n}{n}.$$

Corollary 2.22. *For almost every Boolean function f in $\mathcal{B}$, the restriction, f_n, of f to $\mathbb{V}_n$ satisfies*

$$\lim_{n\to\infty} \frac{S(f_n)}{2^{n/2}\sqrt{n}} = \sqrt{2\log 2}.$$

2.8 PROBABILITY DISTRIBUTIONS

Here, we let $i(\mathbf{w})$ be the index of the ith nonzero component of a vector $\mathbf{w} \in \mathbb{V}_n$. Suppose that the components of the input $(x_1,\ldots,x_n)$ of a Boolean function $f(\mathbf{x})$ on $\mathbb{V}_n$ are independent binary random variables with probability distributions $\Pr(x_i = 1) = \frac{1}{2} - \epsilon_i$, thus, $\Pr(x_i = 0) = \frac{1}{2} + \epsilon_i$, $i = 1,2,\ldots,n$. The connection between the probability distribution of a Boolean function f and the probability distribution of its arguments can be expressed in terms of Walsh transforms, even under the nonuniformity condition on the input (Miranovich [333]).

Theorem 2.23. *If f is an arbitrary Boolean function on $\mathbb{V}_n$, then*

$$\frac{1}{2} - \Pr(f=1) = \frac{1}{2^{n+1}} \left[W(\hat{f})(0) + \sum_{s=1}^{n} \sum_{\substack{\mathbf{w}\in\mathbb{V}_n \\ wt(\mathbf{w})=s}} 2^s W(\hat{f})(\mathbf{w}) \epsilon_{1(\mathbf{w})} \cdots \epsilon_{n(\mathbf{w})} \right].$$

Proof. Let $\mathbf{a} = (a_1\ldots,a_n)$. We have

$$\Pr(f=1) = \sum_{\substack{\mathbf{a}\in\mathbb{V}_n \\ f(\mathbf{a})=1}} \Pr(x_1 = a_1,\ldots,x_n = a_n)$$

$$= \sum_{\substack{\mathbf{a}\in\mathbb{V}_n \\ f(\mathbf{a})=1}} \Pr(x_1 = a_1)\cdots\Pr(x_n = a_n)$$

$$= \sum_{\substack{\mathbf{a}\in\mathbb{V}_n \\ f(\mathbf{a})=1}} \left(\frac{1}{2}+(-1)^{a_1}\epsilon_1\right)\cdots\left(\frac{1}{2}+(-1)^{a_n}\epsilon_n\right)$$

$$= \frac{1}{2^n}wt(f)+\sum_{\substack{\mathbf{a}\in\mathbb{V}_n \\ f(\mathbf{a})=1}}\sum_{s=1}^{n}\frac{1}{2^{n-s}}\sum_{\substack{\mathbf{w}\in\mathbb{V}_n \\ wt(\mathbf{w})=s}}(-1)^{\mathbf{a}\cdot\mathbf{w}}\epsilon_{1(\mathbf{w})}\cdots\epsilon_{n(\mathbf{w})}$$

$$= \frac{1}{2^n}wt(f)+\sum_{s=1}^{n}\frac{1}{2^{n-s}}\sum_{\substack{\mathbf{w}\in\mathbb{V}_n \\ wt(\mathbf{w})=s}}\sum_{\substack{\mathbf{a}\in\mathbb{V}_n \\ f(\mathbf{a})=1}}(-1)^{\mathbf{a}\cdot\mathbf{w}}\epsilon_{1(\mathbf{w})}\cdots\epsilon_{n(\mathbf{w})}.$$

Using Lemma 2.9 and the fact that $W(\hat{f})(\mathbf{0})=2^n-2wt(f)$, we get the theorem. ■

Corollary 2.24. *If f is a Boolean function, then*

$$\Delta_f(\epsilon) := \max_{\substack{|\epsilon_i|\le\epsilon \\ 1\le i\le n}}\left|\frac{1}{2}-\Pr(f=1)\right|$$

$$= \frac{1}{2^{n+1}}\max_{\mathbf{x}\in\mathbb{V}_n}\left|\sum_{\mathbf{w}\in\mathbb{V}_n}(-1)^{\mathbf{x}\cdot\mathbf{w}}W(\hat{f})(\mathbf{w})(2\epsilon)^{wt(\mathbf{w})}\right|.$$

Recall that a Boolean function f on $\mathbb{V}_n$ is balanced if the truth table contains as many 1s as 0s. Equivalently, f is balanced if $wt(f)=|\Omega_f|=2^{n-1}$. Using the above corollary, we can get easily that f is balanced if and only if $\Delta_f(\epsilon)=o(1)$.

Lemma 2.25. *The following are true:*

(i) *If $\ell_{\mathbf{a},a_0}(\mathbf{x})=\mathbf{a}\cdot\mathbf{x}\oplus a_0$ on $\mathbb{V}_n$, then $\Delta_{\ell_{\mathbf{a},a_0}}(\epsilon)=\frac{1}{2}(2\epsilon)^{wt(\mathbf{a})}$.*
(ii) *If $g(\mathbf{x})=f(\mathbf{x})\oplus 1$, then $\Delta_g(\epsilon)=\Delta_f(\epsilon)$.*
(iii) *If $g(\mathbf{x})=f(\mathbf{x}\oplus\mathbf{a})$, then $\Delta_g(\epsilon)=\Delta_f(\epsilon)$.*

Proof. We follow Miranovich's argument [333]. Claims (i) and (ii) follow from Theorem 2.13. Next, for (iii), since $W(\hat{g})(\mathbf{w})=(-1)^{\mathbf{a}\cdot\mathbf{w}}W(\hat{f})(\mathbf{w})$, by using Corollary 2.24, we get

$$\Delta_g(\epsilon) = \frac{1}{2^{n+1}}\max_{\mathbf{u}\in\mathbb{V}_n}\left|\sum_{\mathbf{w}\in\mathbb{V}_n}(-1)^{\mathbf{u}\cdot\mathbf{w}}(-1)^{\mathbf{a}\cdot\mathbf{w}}W(\hat{f})(\mathbf{w})(2\epsilon)^{wt(\mathbf{w})}\right|$$

$$= \frac{1}{2^{n+1}}\max_{\mathbf{u}\in\mathbb{V}_n}\left|\sum_{\mathbf{w}\in\mathbb{V}_n}(-1)^{(\mathbf{a}\oplus\mathbf{u})\cdot\mathbf{w}}W(\hat{f})(\mathbf{w})(2\epsilon)^{wt(\mathbf{w})}\right|$$

$$= \frac{1}{2^{n+1}}\max_{\mathbf{u}\in\mathbb{V}_n}\left|\sum_{\mathbf{w}\in\mathbb{V}_n}(-1)^{\mathbf{u}\cdot\mathbf{w}}W(\hat{f})(\mathbf{w})(2\epsilon)^{wt(\mathbf{w})}\right| = \Delta_f(\epsilon),$$

and the lemma is proved. ■

2.9 HADAMARD MATRICES AND NONLINEARITY BOUNDS

A *Hadamard matrix* H of order n is an $n \times n$ matrix of ± 1s such that

$$HH^t = nI_n,$$

where H^t is the transpose of H, and I_n is the $n \times n$ identity matrix. In other words, the product of any two distinct rows of H is zero. Since $H^{-1} = (1/n)H^t$, we also have $H^tH = nI_n$, so the columns have the same property. A trivial remark is that for a ± 1-matrix (the entries are ± 1s) the dot product of any row or column by itself is equal to n.

Multiplying any row or column by -1 changes H into another Hadamard matrix. Thus we can change the first row and column into $+1$s. Such a Hadamard matrix is called *normalized*. A well-known conjecture (see Meier and Staffelbach [318]) claims that Hadamard matrices exist only for $n = 1, 2$ or a multiple of 4. (As of 2007 [254,459], the smallest unknown order of a Hadamard matrix is 668.)

The following is a known construction of Hadamard matrices:

$$H_0 = (1);\quad H_1 = \begin{pmatrix} 1 & 1 \\ 1 & -1 \end{pmatrix};\quad H_2 = \begin{pmatrix} 1 & 1 & 1 & 1 \\ 1 & -1 & 1 & -1 \\ 1 & 1 & -1 & -1 \\ 1 & -1 & -1 & 1 \end{pmatrix},\quad \text{and}$$
$$H_n = \begin{pmatrix} H_{n-1} & H_{n-1} \\ H_{n-1} & -H_{n-1} \end{pmatrix}, \tag{2.17}$$

that is, H_n is the Kronecker product $H_n = H_1 \otimes H_{n-1}$. The Kronecker product is not commutative, that is, in general $A \otimes B \neq B \otimes A$. However, it is associative, $(A \otimes B) \otimes C = A \otimes (B \otimes C)$ and distributive, $(A + B) \otimes C = A \otimes C + B \otimes C$. Starting with $H_0 = (1)$ and applying the above procedure we construct $H_1, H_2, \ldots$, which are called the *Sylvester*, or *Sylvester-Hadamard matrices*. We prefer the latter name. Obviously, H_n is a square matrix of order 2^n.

We can express now the Walsh transform in terms of Sylvester–Hadamard matrices H_n, namely

$$W(f) = fH_n,$$

since $(-1)^{\mathbf{u}\cdot\mathbf{v}}$ is the entry on the position $(\mathbf{u}, \mathbf{v}) \in \mathbb{V}_n \times \mathbb{V}_n$, in the matrix H_n. Consequently,

$$f = \frac{1}{2^n} W(f) H_n \quad \text{or} \quad f(\mathbf{u}) = \frac{1}{2^n} \sum_{\mathbf{v} \in \mathbb{V}_n} (-1)^{\mathbf{u}\cdot\mathbf{v}} W(f)(\mathbf{v}). \tag{2.18}$$

Similarly, if ζ is a $(1, -1)$-sequence in $\mathbb{V}_n$, then its Walsh transform is defined by

$$W_\zeta = \zeta H_n. \tag{2.19}$$

Seberry and Zhang [407] (see, also, [412]) proved by induction the next result.

Lemma 2.26. *If the Sylvester-Hadamard matrix H_n is given by*

$$H_n = \begin{pmatrix} \ell_0 \\ \ell_1 \\ \dots \\ \ell_{2^n-1} \end{pmatrix},$$

where ℓ_i is a row of H_n, then ℓ_i is the sequence of $\ell_{\alpha_i}(\mathbf{x}) = \alpha_i \cdot \mathbf{x}$, α_i as in Definition 2.1.

Define $\ell_{i+2^n} := -\ell_i$. As a consequence of the above lemma, we get that all affine sequences are among the rows of $\pm H_n$, that is, $\ell_0, \dots, \ell_{2^{n+1}-1}$.

From Definition 2.1, we deduce that if f, g are Boolean functions on $\mathbb{V}_n$ whose sequences are η_f, η_g, respectively, then

$$d(f,g) = 2^{n-1} - \frac{1}{2}\eta_f \cdot \eta_g. \tag{2.20}$$

Let f be a Boolean function on $\mathbb{V}_n$ whose sequence is $\eta_f = (a_1, \dots, a_n)$ and ϕ_i be an affine function whose sequence is ℓ_i. From the above observation we infer that

$$(\eta_f \cdot \ell_i)^2 = 2^n + 2\sum_{j<k} a_j a_k h_{ij} h_{ik}.$$

Summing up for $i = 1, 2, \dots, 2^n$, we deduce

$$\begin{aligned} \sum_{i=1}^{2^n} (\eta_f \cdot \ell_i)^2 &= 2^{2n} + 2\sum_{i=1}^{2^n}\sum_{j<k} a_j a_k h_{ij} h_{ik} \\ &= 2^{2n} + 2\sum_{j<k} a_j a_k \sum_{i=1}^{2^n} h_{ij} h_{ik} = 2^{2n}, \end{aligned} \tag{2.21}$$

which is a variant of Parseval's equation. Hence, there exists an index i such that

$$(\eta_f \cdot \ell_i)^2 = (\eta_f \cdot \ell_{i+2^n})^2 \geq 2^n,$$

so either $\eta_f \cdot \ell_i \geq 2^{n/2}$ or $\eta_f \cdot \ell_{i+2^n} \geq 2^{n/2}$. Without loss of generality we may assume that $\eta_f \cdot \ell_i \geq 2^{n/2}$. But $d(f,g) = 2^{n-1} - \frac{1}{2}\eta_f \cdot \eta_g$, hence

$$d(f, \phi_i) \leq 2^{n-1} - 2^{\frac{n}{2}-1}.$$

This shows the next crucial result.

Theorem 2.27. *The nonlinearity $\mathcal{N}_f$ satisfies*

$$\mathcal{N}_f \leq 2^{n-1} - 2^{\frac{n}{2}-1}, \tag{2.22}$$

for any Boolean function f on $\mathbb{V}_n$.

It is known that for n even, the bound is attained by the perfect nonlinear functions, the object of Chapter 5. For odd n, the right-hand side of the inequality of Theorem 2.27 is not an integer, so equality is impossible. However, the question of how close $\mathcal{N}_f$ can get to the right-hand side when n is odd is not completely answered to this day. We shall give many results concerning this question in the next chapters.

2.10 FAST WALSH TRANSFORM

The computation of the Walsh transform would require about 2^{2n} operations (additions, subtractions). There is a faster way to obtain the Walsh coefficients by using the *fast Walsh transform*, which is a discrete version of the fast Fourier transform. MacWilliams and Sloane [287, Theorem 5, p. 422] give a description of the fast Walsh transform.

Theorem 2.28. *The Sylvester-Hadamard matrix H_n can be decomposed as*

$$H_n = M_n^{(1)} M_n^{(2)} \cdots M_n^{(n)},$$

where $M_n^{(i)} = I_{2^{n-i}} \otimes H_1 \otimes I_{2^{i-1}}$, and I_m is the $m \times m$ identity matrix.

Proof. By induction on n (see [287, p. 422]). ■

Example 2.29. $M_2^{(1)} = \begin{pmatrix} 1 & 1 & 0 & 0 \\ 1 & -1 & 0 & 0 \\ 0 & 0 & 1 & 1 \\ 0 & 0 & 1 & -1 \end{pmatrix}; M_2^{(2)} = \begin{pmatrix} 1 & 0 & 1 & 0 \\ 0 & 1 & 0 & 1 \\ 1 & 0 & -1 & 0 \\ 0 & 1 & 0 & -1 \end{pmatrix}.$

Remark 2.30. *It is interesting to note that the matrices $M_n^{(\cdot)}$ commute with each other, that is, $M_n^{(i)} M_n^{(j)} = M_n^{(j)} M_n^{(i)}$, for any i, j.*

The sparse matrix method of Theorem 2.28 enables one to compute the Walsh coefficients of the sign function $\hat{f}$ using only $n2^n$ operations [287, p. 421].

There is another recent different approach for the computation of the Walsh transform of a Boolean function f on $\mathbb{V}_n$ (see the the work of Gupta and Sarkar [216]), where the authors use the ANF of f to speed up the computation of the Walsh transform at a set of points. However, for their algorithm to be feasible, it seems that it is necessary for f to be a sparse polynomial and its support Ω_f must contain vectors of 'nearly' n weight.

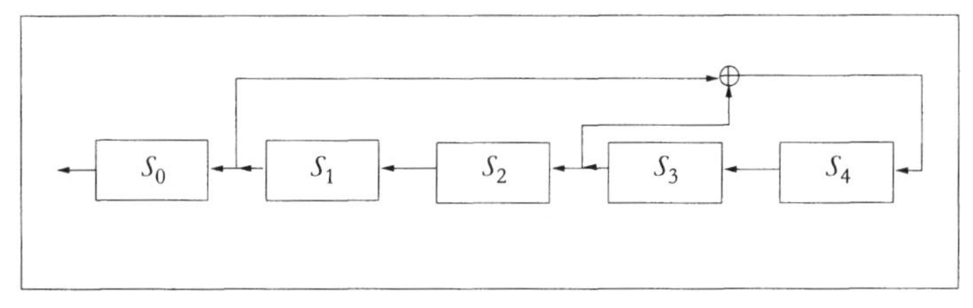

FIGURE 2.1

LFSR with five stages S_i, and feedback bit $S_4 = S_1 \oplus S_3$.

2.11 LFSRs AND LINEAR COMPLEXITY

In digital circuits a *shift register* is a type of sequential logic circuit, mainly for storage of digital data, set up in a linear fashion which has its inputs connected to the outputs in such a way that the data is shifted down the line when the circuit is activated [211,404, 463].

A *linear feedback shift register* (LFSR) is a shift register whose input bit is the output of a linear function of two or more of its previous states (taps). An LFSR of length m consists of m stages numbered $0, 1, \ldots, m-1$, each capable of storing one bit, and a clock controlling data exchange. A vector with entries $s_0, \ldots, s_{m-1}$ would initialize the shift register. At time i, the following operations are performed [445, Section 1.1]:

1. s_i (the content of stage 0) forms part of the output;
2. the content of stage i is shifted to stage $i-1$, for $1 \leq i \leq m-1$;
3. the new content (the *feedback bit*) of the stage $m-1$ would be obtained by xor-ing a subset of the content of the m stages.

The initial input of an LFSR is called a *seed*. Since any register has a finite number of possible states, it must eventually be periodic. However, an LFSR with a well-chosen feedback function and seed can produce a sequence of bits which appears random (good statistical properties) and which has a large period.

LFSRs can be applied in generating pseudorandom numbers, pseudonoise sequences, fast digital counters, whitening sequences, cryptography, etc., and they can be implemented in both software and hardware. We shall use the concept heavily in Chapter 6.

There are many possible configurations, and we present a simple one in Figure 2.1 which starts from an input of all 1s and is very easy to implement in software and hardware. An LFSR of this type will never contain only 0s and would stop if an all 0 binary string is fed into the LFSR. Only some choices of taps (that is, the nonzero coefficients c_i defined below) will generate a maximal sequence with a period of $2^n - 1$ cycles. The depicted LFSR of Figure 2.1 will generate the following sequence if fed the first (left) 4 bits:

$$1,1,1,1,0,1,0,0,1,1,1,0,1,0,0,1,1,1,0,1,0,0,1,1,1,0,1,0,0,1.$$

If the content of the stage S_i is s_i, $0 \le i \le m-1$, then $[s_{m-1}, \ldots, s_1, s_0]$ is called the initial state of the LSFR. From the definition of an LFSR, the output sequence $s_0, s_1, \ldots$ will satisfy the following recursion

$$s_j = \sum_{i=1}^{m} c_i s_{j-i}, \quad j \ge m.$$

The polynomial $C(x) = 1 + c_1 x + \cdots + c_m x^m$ is the *feedback (or connection) polynomial* of the sequence $\{s_j\}_j = \{s_j : j = 0, 1, \ldots\}$. The LFSR is said to be *nonsingular* if $c_m \ne 0$, that is, the degree of its feedback polynomial is m.

In the shown example of Figure 2.1, the constants are $c_1 = 1, c_2 = 0, c_3 = 1, c_4 = 0$, and so, its feedback polynomial is $C(x) = 1 + x + x^3$.

The output sequence of the LFSR can be generated by more than one register. The lowest degree polynomial associated to any such register is called a *minimal polynomial.* Simply put, the polynomial $P(x) = \sum_{i=0}^{d} p_i x^i$ is minimal for $\{s_j\}_j$ if it has the lowest degree such that

$$p_0 s_j + p_1 s_{j+1} + \cdots + p_d s_{j+d} = 0, \quad \text{for all } j \in \mathbb{N}.$$

We define the *linear span* or *linear complexity* $L(\mathbf{s})$ of a binary infinite sequence $\mathbf{s} = \{s_j\}_j$ in the following way [322]:

1. if $\mathbf{s}$ is the zero sequence, then $L(\mathbf{s}) = 0$;
2. if no LFSR generates $\mathbf{s}$, then $L(\mathbf{s}) = \infty$;
3. if there is at least one LFSR that generates $\mathbf{s}$, then $L(\mathbf{s})$ is the length of the shortest LFSR that generates $\mathbf{s}$. Equivalently, $L(\mathbf{s})$ is the degree of the minimal polynomial of $\mathbf{s}$.

For our example of Figure 2.1, we can follow easily the previous algorithm and obtain that the linear complexity is $L = 3$. Of course our LFSR is singular. The periodic sequence 1, 1, 1, 0, 1, 0, 0 can be generated by a four-stage LFSR, where the stages T_i satisfy $T_{3+i} = T_{2+i} + T_i$ $(i = 0, 1, 2, \ldots)$ with input 1, 1, 1. The feedback polynomial $1 + x + x^3$ is the same.

If s^n is a finite binary sequence, then $L(\mathbf{s}^n)$ is the length of the shortest LFSR which has $\mathbf{s}^n$ as its first n terms. The linear complexity satisfies the following properties [322, pp. 198–199].

Theorem 2.31. *Let* $\mathbf{s}, \mathbf{t}$ *be binary sequences. Then*

1. $0 \le L(s^n) \le n$, $n \ge 1$.
2. $L(\mathbf{s}^n) = 0$ *if and only if* $\mathbf{s}^n$ *is the all zero sequence of length* n.
3. $L(\mathbf{s}^n) = n$ *if and only if* $\mathbf{s}^n = 0, 0, \ldots, 0, 1$.
4. *If* $\mathbf{s}$ *is periodic of period* T, *then* $L(\mathbf{s}) \le T$.
5. L *satisfies a triangle-like inequality, that is,* $L(\mathbf{s} \oplus \mathbf{t}) \le L(\mathbf{s}) + L(\mathbf{t})$, *where* $\mathbf{s} \oplus \mathbf{t}$ *is the bitwise xor of* $\mathbf{s}, \mathbf{t}$.

6. *If the feedback polynomial $C(x)$ is irreducible over $\mathbb{F}_2[x]$ and $c_{m-1} \neq 0$, then each of the $2^m - 1$ nonzero states of the associated nonsingular LFSR will produce an output of linear complexity m.*

The reader may consult [129,211,322] for more details on LFSRs.

2.12 THE BERLEKAMP–MASSEY ALGORITHM

The Berlekamp–Massey algorithm was devised to decode Bose–Chaudhuri–Hocquenghem (BCH) codes in 1968–69 [26, Chapter 7], [300]. The algorithm is very efficient for determining the linear complexity of a finite binary sequence $\mathbf{s}^n$ of length n. It will output the minimal polynomial of s in finitely many steps provided we know an upper bound for its degree. We will follow [322] in the description of the algorithm.

Let $G(x) = \sum_{i=0}^{\infty} s_i x^i$ be the generating function of the sequence $\mathbf{s} = \{s_i\}_i$. We assume that n is an upper bound for the degree of the minimal polynomial $P(x)$ of s, and $C(x)$ is the feedback polynomial of an LFSR that generates s. We define the *next discrepancy* d_N by

$$d_N = s_N \oplus \sum_{i=0}^{m-1} c_i s_{N-i-i}.$$

The Berlekamp–Massey algorithm runs according to the following pseudocode:

```
Input: binary sequence s = s_0, s_1, ..., s_{n-1} of length n
Output: linear complexity 0 ≤ L(s^n) ≤ n
begin
  C(x) = 1
  L = 0
  m = −1
  B(x) = 1
  N = 0
  while (N < n)
    d = s_N ⊕ Σ_{i=0}^{m−1} c_i s_{N−1−i} (computes the next discrepancy)
    if (d = 1)
      T(x) = C(x)
      C(x) = C(x) + B(x) · x^{N−m}
      if L ≤ N/2
        L = N + 1 − L
        m = N
        B(x) = T(x)
      end if
    else (N = N + 1)
    end if
  end while
end
```

The Berlekamp–Massey algorithm is based on the next result [300,322].

Theorem 2.32. *Let* $\mathbf{s}^n$ *be a finite binary sequence of linear complexity* $L = L(\mathbf{s}^n)$, *and* $C(x)$ *be the feedback polynomial of an LFSR which generates* $\mathbf{s}^n$. *Then*

(i) *The LFSR of feedback polynomial* $C(x)$ *also generates* $\mathbf{s}^{N+1} = s_0, s_1, \ldots, s_N, s_{N+1}$ *if and only if the next discrepancy* d_N *is 0.*
(ii) *If* $D_n = 0$, *then* $L(\mathbf{s}^{N+1}) = L$.
(iii) *Assume* $d_N = 1$. *Let* m *be the largest integer* $<N$ *such that* $L(\mathbf{s}^m) < L(\mathbf{s}^N)$, *and let* $B(x)$ *be the feedback polynomial of an LFSR that generates* $\mathbf{s}^m$. *Then* $C'(x) = C(x) + B(x) \cdot x^{N-m}$ *is the feedback polynomial of an LFSR of linear complexity* $L' = L$, *if* $L > N/2$, *and* $L' = N + 1 - L$, *if* $L \leq N/2$.

It is interesting to note that if the linear complexity of $\mathbf{s}^n$ is $L \leq n/2$, then there is a unique LFSR that generates $\mathbf{s}^n$ of length L.

The running time of the Berlekamp–Massey algorithm is $O(n^2)$ bit operations, where n is the length of the binary sequence $\mathbf{s}^n$ whose linear complexity is sought. There have been some improvements on the Berlekamp–Massey algorithm, and we cite here Shoup's book [420], which contains such an improvement.

2.13 DE BRUIJN SEQUENCE

The de Bruijn sequence was introduced by the Dutch mathematician de Bruijn in 1946 in [144].

A *k-ary de Bruijn sequence of order n* is a circular k-ary string containing every k-ary string of length n as a substring exactly once, and so it is of length k^n. For $n = 3$ and $k = 2$, the lexicographically smallest de Bruijn sequence is 00010111. If you think of these bits around a circle and travel clockwise you will encounter all of the $k^n = 2^3$ bit strings 000, 001, 010, 011, 100, 101, 110, 111 exactly once. An immediate application of de Bruijn sequences is prompted by this observation: one can encode 24 bits of information into only 8 bits (a byte). Other applications can be found in electronic memory, coding theory, spectral imaging, robot self-location, etc.

We found one example, but that is little evidence that they exist for any value of n, k. However, one can prove (for instance, using graph theoretical methods) that they exist for any value of n, k [375].

Theorem 2.33. *The de Bruijn sequences exist for any positive integers* n, k.

Further, there are a total of $((k-1)!)^{k^{n-1}} \cdot k^{k^{n-1}-n}$ such sequences, a fact discovered by Flye-Sainte Marie (1894) (cf. Fredricksen [177]). If $k = 2$, the formula becomes $2^{2^{n-1}-n}$ and counts the binary de Bruijn sequences of length n.

A *k-ary necklace* is an equivalence class of k-ary strings under rotation. As a representative of such an equivalence class we take the smallest in the lexicographical order. A period n necklace representative with n digits is called a *Lyndon word*. See Table 2.1 for some examples.

Table 2.1 Necklace representative for the binary case $k=2$

$n=2$	0, 01, 1
$n=3$	0, 001, 011, 1
$n=4$	0, 0001, 0011, 01, 0111, 1
$n=5$	0, 00001, 00011, 00101, 00111, 01011, 01111, 1
$n=6$	0, 000001, 000011, 000101, 000111, 001, 001011, 001101, 001111, 01, 010111, 011, 011111, 1
$n=7$	0, 0000001, 0000011, 0000101, 0000111, 0001001, 0001011, 0001101, 0001111, 0010011, 0010101, 0010111, 0011011, 0011101, 0011111, 0101011, 0101111, 0110111, 0111111, 1

The sequence for the number $N_2(n)$ of binary necklaces starts with 3, 4, 6, 8, 14, 20, 36, 60, 108, 188, 352, 632, 1182, ... for $n \geq 2$, and the sequence for the number $L_2(n)$ of Lyndon words starts with 1, 2, 3, 6, 9, 18, 30, 56, 99, 186, 335, 630, 1161, ..., for $n \geq 2$. There are explicit formulas for both of these sequences, for every k, namely

$$L_k(n) = \frac{1}{n}\sum_{d|n}\mu(n/d)k^d; \quad N_k(n) = \frac{1}{n}\sum_{d|n}\phi(n/d)k^d,$$

where $\mu(m)$ is the Möbius function defined by $\mu(1)=1$, $\mu(m)=0$ if m is the product of nondistinct primes (m is not squarefree, that is, m is divisible by the square of a prime), and $\mu(m)=(-1)^k$ if m is the product of k distinct primes; $\phi(m)$ is the Euler's phi function, which is the number of integers between 1 and m that are relatively prime to m.

Fredricksen and Kessler [178] and Fredricksen and Maiorana [179] came up with the following algorithm to build de Bruijn sequences by using Lyndon words.

Theorem 2.34. *For a given positive integer n, the lexicographical concatenation of Lyndon words of length dividing n generates a de Bruijn sequence of span n.*

De Bruijn sequences can be classified by the Hamming weight of the truth tables of the generating functions [177,307,308]. Interestingly enough, the weight classes of the generators are known, but the number of de Bruijn sequences in each weight class is still an unsolved problem at the time of this writing.

To count fixed-density necklaces, we let $N(n_0, n_1, \ldots, n_k)$ be the number of necklaces composed of n_i occurrences of the symbol $0 \leq i \leq k-1$. The density of the necklace is defined by $d = n_1 + \cdots + n_{k-1}$ and $n_0 = n - d$. In [185], Gilbert and Riordan proved that

$$N(n_0, n_1, \ldots, n_{k-1}) = \frac{1}{n}\sum_{j|\gcd(n_0,\ldots,n_{k-1})}\phi(j)\frac{(n/j)!}{(n_0/j)!\cdots(n_{k-1}/j)!}.$$

It is obvious that to get the number of fixed-density necklaces with length n and density d, we sum

$$N_k(n,d) = \sum_{n_1+\cdots+n_{k-1}=d} N(n-d, n_1, \ldots, n_{k-1}).$$

Defined in a similar manner, the fixed-density Lyndon words satisfy

$$L(n_0, n_1, \ldots, n_{k-1}) = \frac{1}{n} \sum_{j \mid \gcd(n_0,\ldots,n_{k-1})} \mu(j) \frac{(n/j)!}{(n_0/j)! \cdots (n_{k-1}/j)!},$$

$$L_k(n,d) = \sum_{n_1+\cdots+n_{k-1}=d} L(n-d, n_1, \ldots, n_{k-1}).$$

When $k = 2$, the formulas take a simple form

$$N_2(n,d) = \frac{1}{n} \sum_{j \mid \gcd(n,d)} \phi(j) \binom{n/j}{d/j}$$

$$L_2(n,d) = \frac{1}{n} \sum_{j \mid \gcd(n,d)} \mu(j) \binom{n/j}{d/j}.$$

An algorithm for generating fixed-density necklaces was developed in [391].

For more on the de Bruijn sequences, we recommend Fredricksen's excellent survey paper [177], and the references therein.

CHAPTER

Avalanche and propagation criteria

3

3.1 INTRODUCTION

The 19th Century, which prides itself upon the invention of steam and evolution, might have a more legitimate title to fame from the discovery of pure mathematics.

Bertrand Russell (1872–1970)

A Boolean function $f(\mathbf{x})$ in n variables is said to satisfy the *Strict Avalanche Criterion* (*SAC* for short) if changing any one of the n bits in the input $\mathbf{x}$ results in the output of the function being changed for exactly half of the 2^{n-1} vectors $\mathbf{x}$ with the changed input bit. For example, the following Boolean function with $n = 3$ is easily seen to satisfy the *SAC*.

Example 3.1 (*A 3-variable function which satisfies the SAC*).

Input	000	001	010	011	100	101	110	111
Output	1	1	1	0	0	1	1	1

The *SAC* is a useful property for a Boolean function in cryptographic applications because satisfying the *SAC* means that a slight change in the input to the function leads to a large change in the output (an avalanche effect), and in fact a large change of a uniform kind (hence the name Strict Avalanche Criterion). This is one aspect of making a Boolean function whose input is difficult to infer from its output, which we shall see is essential in a cryptographic context.

The *SAC* was first defined by Webster and Tavares [457] in a study of ways to design good *S*-boxes (see Chapter 7). Boolean functions can be viewed as pieces in the structure of an *S*-box. *S*-box design using Boolean functions satisfying the *SAC* has been explored in papers of Adams and Tavares [1], Kwangjo Kim [257] and Kim et al. [256], just to mention a few. Since 1990 the *SAC* has mainly been studied in the context of Boolean functions, and that is the point of view we adopt here.

It is clear that Lemma 3.2 below follows from our definition of the *SAC* and so provides an equivalent definition (which already appears in the original paper [457]).

Lemma 3.2. *A Boolean function* $f : \mathbb{V}_n \to \mathbb{F}_2$ *satisfies the* SAC *if and only if the function* $f(\mathbf{x}) \oplus f(\mathbf{x} \oplus \mathbf{a})$ *is balanced for every* $\mathbf{a}$ *in* $\mathbb{V}_n$ *with Hamming weight* 1.

Lemma 3.2 provides a straightforward way of verifying the *SAC* by computation, given the output values of f. For brevity, from now on we will sometimes say that a function that satisfies the *SAC* is an *SAC* function, or simply that the function is *SAC*.

3.2 COUNTING *SAC* FUNCTIONS

As with any criterion of cryptographic significance, it is of interest to count the functions which satisfy the *SAC*. We let S_n denote the number of Boolean functions in n variables which are *SAC* functions. We define $\exp_2(x) = 2^x$, so there are exactly $\exp_2(2^n)$ Boolean functions in n variables. One of the simplest questions we can ask about the number S_n is the size of the ratio L_n defined by $L_n = 2^{-n} \log_2 S_n$.

It is clear that $L_n \leq 1$ and it is natural to conjecture that

$$\lim_{n \to \infty} L_n \text{ exists.} \tag{3.1}$$

Let L denote the limit in (3.1). Cusick [127] conjectured that L exists but only proved that $L_n \geq 1/4$. In the same paper, Cusick also conjectured the result given below in Lemma 3.3, which was proved independently by Cusick and Stănică [130] and Youssef and Tavares [473], using different methods. The proof we give here follows the paper of Youssef and Tavares.

Lemma 3.3. *Given any choice of the values $f(\mathbf{v}_i), 0 \leq i \leq 2^{n-1} - 1$, for a Boolean function in n variables, there exists a choice of the remaining 2^{n-1} values $f(\mathbf{v}_i)$, $2^{n-1} \leq i \leq 2^n - 1$, such that the resulting function $f(\mathbf{x})$ satisfies the* SAC.

Proof. We identify the Boolean function $f(\mathbf{x})$ with its truth table by letting

$$f(\mathbf{x}) = \{f(\mathbf{v}_0), f(\mathbf{v}_1), \ldots, f(\mathbf{v}_{2^n-1})\},$$

where $\mathbf{v}_i = \mathbf{b}(i)$ (see Definition 2.1). The case $n = 1$ of the lemma is trivial, so we assume $n \geq 2$. Given any choice of the values $f(\mathbf{v}_i), 0 \leq i \leq 2^{n-1} - 1$, we define a Boolean function h_{n-1} in $n-1$ variables by

$$h_{n-1} = \{f(\mathbf{v}_0), f(\mathbf{v}_1), \ldots, f(\mathbf{v}_{2^{n-1}-1})\}.$$

We let g_{n-1} denote the Boolean function $x_1 \oplus x_2 \oplus \cdots \oplus x_{n-1} \oplus b$ in $n-1$ variables, where b is fixed and equal to 0 or 1. We shall prove that if we define $f_n(\mathbf{x})$ by the truth table

$$f_n(\mathbf{x}) = \{h_{n-1}(\mathbf{x}), h_{n-1}(\mathbf{x}) \oplus g_{n-1}(\mathbf{x})\},$$

then the function $f_n(\mathbf{x})$ is an *SAC* function. Notice that this proves a result slightly stronger than the lemma, because we show that there exist at least two choices (for $b = 0$ and $b = 1$) of $f_n(\mathbf{v}_i), 2^{n-1} \leq i \leq 2^n - 1$, such that $f_n(\mathbf{x})$ is *SAC*.

Let $\mathbf{a}$ denote one of the vectors in V_n with Hamming weight 1. Let $\mathbf{a}^*$ and $\mathbf{v}_i^*$ be made up of the $n-1$ least significant bits of $\mathbf{a}$ and $\mathbf{v}_i$, respectively. For easy notation we sometimes denote by $C(u)$, instead of $\bar{u}$ the complement of the bit (or word) u. Because of Lemma 3.2, the following two computations prove Lemma 3.3.

Case 1. $\mathbf{a} \neq (1,0,\ldots,0)$

$$\begin{aligned}
\sum_{i=0}^{2^n-1} f_n(\mathbf{v}_i) \oplus f_n(\mathbf{v}_i \oplus \mathbf{a}) &= \sum_{i=0}^{2^{n-1}-1} f_n(\mathbf{v}_i) \oplus f_n(\mathbf{v}_i \oplus \mathbf{a}) \oplus \sum_{i=2^{n-1}}^{2^n-1} f_n(\mathbf{v}_i) \oplus f_n(\mathbf{v}_i \oplus \mathbf{a}) \\
&= \sum_{i=0}^{2^{n-1}-1} h_{n-1}(\mathbf{v}_i^*) \oplus h_{n-1}(\mathbf{v}_i^* \oplus \mathbf{a}^*) \oplus \sum_{i=0}^{2^{n-1}-1} h_{n-1}(\mathbf{v}_i^*) \\
&\quad \oplus h_{n-1}(\mathbf{v}_i^* \oplus \mathbf{a}^*) \oplus g_{n-1}(\mathbf{v}_i^*) \oplus g_{n-1}(\mathbf{v}_i^* \oplus \mathbf{a}^*) \\
&= \sum_{i=0}^{2^{n-1}-1} h_{n-1}(\mathbf{v}_i^*) \oplus h_{n-1}(\mathbf{v}_i^* \oplus \mathbf{a}^*) \oplus C(h_{n-1}(\mathbf{v}_i^*) \oplus h_{n-1}(\mathbf{v}_i^* \oplus \mathbf{a}^*)) \\
&= 2^{n-1}.
\end{aligned}$$

For the second last equality, we used the identity

$$g_{n-1}(\mathbf{v}_i^*) \oplus g_{n-1}(\mathbf{v}_i^* \oplus \mathbf{a}^*) = 1,$$

which holds for any $\mathbf{v}_i^*$ and any $\mathbf{a}^*$ of weight 1.

Case 2. $\mathbf{a} = (1,0,\ldots,0)$

$$\begin{aligned}
\sum_{i=0}^{2^n-1} f_n(\mathbf{v}_i) \oplus f_n(\mathbf{v}_i \oplus \mathbf{a}) &= 2 \sum_{i=0}^{2^{n-1}-1} f_n(\mathbf{v}_i) \oplus f_n(\mathbf{v}_i \oplus \mathbf{a}) \\
&= 2 \sum_{i=0}^{2^{n-1}-1} h_{n-1}(\mathbf{v}_i^*) \oplus h_{n-1}(\mathbf{v}_i^*) \oplus g_{n-1}(\mathbf{v}_i^*) \\
&= 2 \sum_{i=0}^{2^{n-1}-1} g_{n-1}(\mathbf{v}_i^*) = 2^{n-1}.
\end{aligned}$$

We have now proved that $S_n = \exp_2(2^{n-1} + 1)$, so the corollary below follows immediately. ■

Corollary 3.4. *For every* n, $L_n \geq 1/2$.

By a much more complicated argument, D. Biss [39] succeeded in proving (3.1) and evaluated the limit as 1.

Theorem 3.5. *We have* $\lim_{n \to \infty} L_n = 1$.

Proof. See [39]. The proof involves detailed geometric analysis inside cubes of high dimension, among other things. ■

3.3 COUNTING BALANCED *SAC* FUNCTIONS

Boolean functions in cryptographic applications almost always need to be balanced, or nearly so. Therefore it is of interest to see if results like those above can be proved for the number of balanced *SAC* Boolean functions in n variables. We let U_n denote this number. We are also interested in the size of the ratio B_n defined by

$$B_n = 2^{-n} \log_2 U_n.$$

As above, it is clear that $B_n \leq 1$ and it is natural to conjecture that

$$\lim_{n\to\infty} B_n \text{ exists.} \tag{3.2}$$

There is no proof of (3.2), but the following is known from [473, Lemma 6].

Lemma 3.6. *We have* $\liminf_{n\to\infty} B_n \geq 1/2$.

Proof. Let $\mathbf{x}^* = (x_1, \ldots, x_{n-1})$ and let $\mathbf{a}^*$ be any fixed $(n-1)$-vector of odd weight. As in the proof of Lemma 3.3, given any choice of the values $f(\mathbf{v}_i), 0 \leq i \leq 2^{n-1}-1$, we define a Boolean function h_{n-1} in $n-1$ variables by

$$h_{n-1} = \{f(\mathbf{v}_0), f(\mathbf{v}_1), \ldots, f(\mathbf{v}_{2^{n-1}-1})\},$$

but now we impose the extra condition

$$\sum_{wt(\mathbf{v}_i^*)\ \text{odd}} h_{n-1}(\mathbf{v}_i^*) = 2^{n-3}.$$

Since b can be 0 or 1 in the definition of

$$g_{n-1} = x_1 \oplus x_2 \oplus \cdots \oplus x_{n-1} \oplus b$$

and for any vector $\mathbf{a}^*$ of odd weight we have

$$g_{n-1}(\mathbf{x}^*) = g_{n-1}(\mathbf{x}^* \oplus \mathbf{a}^*) \oplus 1,$$

if we define the truth table of a function $F_n(\mathbf{x})$ in n variables by

$$F_n(\mathbf{x}) = \{h_{n-1}(\mathbf{x}^*), h_{n-1}(\mathbf{x}^* \oplus \mathbf{a}^*) \oplus g_{n-1}(\mathbf{x}^*)\}$$

(similar to the definition of $f_n(\mathbf{x})$ in the proof of Lemma 3.3), then by Lemma 3.3, $F_n(\mathbf{x})$ satisfies the *SAC*. The following computation shows that $F_n(\mathbf{x})$ is balanced:

$$\begin{aligned}\sum_{i=1}^{2^n-1} F_n(\mathbf{v}_i) &= \sum_{i=1}^{2^{n-1}-1} h_{n-1}(\mathbf{v}_i^*) \oplus h_{n-1}(\mathbf{v}_i^* \oplus \mathbf{a}^*) \oplus g_{n-1}(\mathbf{v}_i^*) \\ &= \sum_{i=1}^{2^{n-1}-1} h_{n-1}(\mathbf{v}_i^*) \oplus h_{n-1}(\mathbf{v}_i^*) \oplus g_{n-1}(\mathbf{v}_i^* \oplus \mathbf{a}^*)\end{aligned}$$

$$= \sum_{i=1}^{2^{n-1}-1} h_{n-1}(\mathbf{v}_i^*) \oplus C(h_{n-1}(\mathbf{v}_i^*) \oplus g_{n-1}(\mathbf{v}_i^*))$$
$$= \sum_{wt(\mathbf{v}_i^*)\text{ even}} (h_{n-1}(\mathbf{v}_i^*) \oplus C(h_{n-1}(\mathbf{v}_i^*))) + 2 \sum_{wt(\mathbf{v}_i^*)\text{ odd}} h_{n-1}(\mathbf{v}_i^*)$$
$$= 2^{n-2} + 2^{n-2} = 2^{n-1}.$$

Thus we have proved that

$$U_n \geq \binom{2^{n-2}}{2^{n-3}} \exp_2(2^{n-2}+1).$$

Using Stirling's formula $n! \asymp (2\pi n)^{1/2}(n/e)^n$, we find

$$\binom{2^{n-2}}{2^{n-3}} \exp_2(2^{n-2}+1) \asymp 2^{5/2}\pi^{-1/2} \exp_2(2^{n-1} - n/2),$$

which implies $B_n \geq 1/2 - \epsilon(n)$, where $\epsilon(n) \to 0$ as $n \to \infty$. This gives Lemma 3.3. ■

3.4 HIGHER ORDER *SAC*

Now we turn to a generalization of the *SAC* defined by Forré [172], which she named the *SAC of higher order*. A Boolean function $f(\mathbf{x})$ in n variables is said to satisfy the Strict Avalanche Criterion of order k ($SAC(k)$ for short) if fixing any k of the n bits in the input $\mathbf{x}$ results in a Boolean function in the remaining $n-k$ variables which satisfies the *SAC*. For brevity, we may say that such a function is an $SAC(k)$ function, or simply is $SAC(k)$. Thus a function which satisfies the *SAC* as originally defined is an $SAC(0)$ function. The definition of $SAC(k)$ for n-variable functions makes sense only for $0 \leq k \leq n-2$, since the *SAC* is not defined for 1-variable functions. Note that the function in Example 3.1 above is $SAC(1)$.

Forré [172] did not notice that if a function is $SAC(k)$ for $k > 0$, then it also is $SAC(j)$ for any $j = 0, 1, \ldots, k-1$. This was pointed out by Lloyd [279, Theorem 4.3, p. 68], and we give her proof in the next lemma.

Lemma 3.7. *Suppose $f(\mathbf{x})$ is a Boolean function in $n > 2$ variables which satisfies the* SAC *of order k, $1 \leq k \leq n-2$. Then $f(\mathbf{x})$ also satisfies the* SAC *of order j for any $j = 0, 1, \ldots, k-1$.*

Proof. We prove that if f satisfies the *SAC* of order k, then f also satisfies the *SAC* of order $k-1$. Then our lemma will follow by induction. Let g be a function in $n-k+1$ variables obtained by fixing $k-1$ variables in f. We need to prove that g is an *SAC* function, so by Lemma 3.2, it suffices to show

$$S = \sum_{i=0}^{2^{n-k+1}-1} g(\mathbf{v}_i) \oplus g(\mathbf{v}_i \oplus \mathbf{a}) = 2^{n-k} \tag{3.3}$$

for all $\mathbf{a} \in \mathbb{V}_{n-k+1}$ with Hamming weight 1. Without loss of generality, we may take $\mathbf{a} = (0, \ldots, 0, 1)$. Thus $\mathbf{v}_i$ and $\mathbf{v}_i \oplus \mathbf{a}$ have the same first bit, so we may split the above sum S into two sums, one where the first bit of $\mathbf{v}_i$ is 0 and one where that first bit is 1. Let g_0 and g_1 denote the functions obtained from g by fixing the first input bit as 0 or 1, respectively, and let $\mathbf{a}^*$ denote the vector made up of the least $n-k$ significant bits of $\mathbf{a}$. Then we have

$$S = \sum_{i=0}^{2^{n-k}-1} g_0(\mathbf{v}_i) \oplus g_0(\mathbf{v}_i \oplus \mathbf{a}^*) \oplus \sum_{i=0}^{2^{n-k}-1} g_1(\mathbf{v}_i) \oplus g_1(\mathbf{v}_i \oplus \mathbf{a}^*).$$

Both g_0 and g_1 are obtained from f by fixing k variables, so by hypothesis they are both *SAC* functions. Therefore both of the above sums are 2^{n-k-1}, and this proves (3.3). ■

Our next lemma gives the simple result that in testing whether a Boolean function f is $SAC(k)$, we can discard the affine terms (if any) in f.

Lemma 3.8. *If a Boolean function f in n variables satisfies $SAC(k)$ for some k, $0 \le k \le n-2$, then so does $f \oplus g$, where g is any affine function in n variables.*

Proof. This follows immediately from Lemma 3.2. ■

Two fundamental results on $SAC(k)$ functions were given by Preneel et al. [371, Theorems 10 and 11, pp. 169–170]. The key concept used in the proof of these results is the autocorrelation function for a Boolean function $f(\mathbf{x})$ in n variables, which we define for all $\mathbf{a} \in \mathbb{V}_n$ by

$$r_f(\mathbf{a}) = \sum_{i=0}^{2^n-1} f(\mathbf{v}_i) \oplus f(\mathbf{v}_i \oplus \mathbf{a}). \tag{3.4}$$

(Recall that a different but closely related autocorrelation function was defined in Section 2.3.) For example, the autocorrelation function of an affine function f is $r_f(\mathbf{a}) = 2^n f(\mathbf{a})$, which is one of the constants 0 or 2^n. Now we can restate Lemma 3.2 as:

Lemma 3.9. *A Boolean function f in n variables is* SAC *if and only if the autocorrelation function $r_f(\mathbf{a})$ is equal to 2^{n-1} for all $\mathbf{a} \in \mathbb{V}_n$ with Hamming weight 1.*

We need the following lemma for the proof of the results of Preneel et al. [371].

Lemma 3.10. *If f is a Boolean function in $n > 2$ variables and* $\deg(f) = n$, *then $r_f(\mathbf{a})$ does not take on the value 2^{n-1} for any $\mathbf{a} \in \mathbb{V}_n$.*

Corollary 3.11. *If f is a Boolean function in $n > 2$ variables and* $\deg(f) = n$, *then f does not satisfy the* SAC.

Proof. We prove that if $r_f(\mathbf{a}) = 2^{n-1}$ for some $\mathbf{a}$, then the Hamming weight $wt(f)$ is even. This is a contradiction, since clearly $\deg(f) = n$ implies $wt(f)$ is odd. Suppose that $r_f(\mathbf{a}) = 2^{n-1}$; then

$$\begin{aligned} wt(f) &\equiv \sum_{i=0}^{2^n-1} f(\mathbf{v}_i) \equiv \sum_{i=0}^{2^n-1} f(\mathbf{v}_i \oplus \mathbf{a}) \\ &\equiv (1/2)\sum_{i=0}^{2^n-1} f(\mathbf{v}_i) \oplus f(\mathbf{v}_i \oplus \mathbf{a}) \\ &\equiv r_f(\mathbf{a})/2 \equiv 2^{n-2} \pmod 2. \end{aligned}$$

Since $n > 2$, we have $wt(f)$ even and the contradiction follows. The corollary is immediate from Lemma 3.9. ■

Next we prove the two results of Preneel et al. [370, Theorems 10 and 11, pp. 169–170]. The proofs are not given in [370], but do appear in the unpublished thesis [368, Theorem 8.4, p. 238 and Theorem 8.9, p. 249].

Theorem 3.12. *Suppose f is a Boolean function in $n \geq 2$ variables. If f satisfies $SAC(k)$, $0 \leq k \leq n-3$, then $2 \leq \deg(f) \leq n-k-1$. If f satisfies $SAC(n-2)$, then $\deg(f) = 2$.*

Proof. We have $\deg(f) \geq 2$ since a function of degree 1 cannot be *SAC* (for example, by Lemma 3.2). To prove the upper bound on $\deg(f)$ in the second sentence of the theorem, we show that an $SAC(k)$ function f with $0 \leq k \leq n-3$ cannot have any terms of degree $\geq n-k$. Assume on the contrary that f has a term of degree $n-j$ with $0 \leq j \leq k$. If we fix equal to 0 all j variables that do not occur in this term, then we obtain a function g in $n-j$ variables with $\deg(g) = n-j$. Since $n-j > 2$, it follows from Corollary 3.11 that g does not satisfy the *SAC*, which contradicts our assumption that f is $SAC(k)$. Finally, since an $SAC(n-2)$ function is also $SAC(n-3)$, we see that the last sentence of the theorem follows. ■

Theorem 3.13. *Suppose f is a quadratic Boolean function in $n \geq 2$ variables. Then f satisfies $SAC(k)$, $0 \leq k \leq n-2$, if and only if every variable x_i occurs in at least $k+1$ second degree terms in the algebraic normal form of f.*

Proof. For the case $k = 0$ of the theorem, it suffices to show that a quadratic function is *SAC* if and only if every variable occurs in at least one second degree term of the algebraic normal form of f. This condition is plainly necessary, because if a variable x_j is missing we can change input bit j to f without altering any of the outputs of f, which contradicts the definition of *SAC*. The condition is also sufficient, because it implies that given any $x_i, 1 \leq i \leq n$, we can write $f(\mathbf{x}) = x_i g(\mathbf{x}^*) \oplus h(\mathbf{x}^*)$, where $\mathbf{x}^*$ denotes the $(n-1)$-vector $(x_1, \ldots, x_{i-1}, x_{i+1}, \ldots, x_n)$, $g(\mathbf{x}^*)$ is an affine function, and $h(\mathbf{x}^*)$ is a quadratic function. Now we can compute the value of the autocorrelation function at the unit vector $\mathbf{e}_i$ with ith bit equal to 1 and all other bits 0:

$r_f(\mathbf{e}_i) = \sum_{i=0}^{2^n-1} f(\mathbf{v}_i) \oplus f(\mathbf{v}_i \oplus \mathbf{e}_i) = \sum_{i=0}^{2^n-1} g(\mathbf{x}^*) = 2^{n-1}$. By Lemma 3.9, this proves that f satisfies *SAC*.

For the cases in which $k > 0$, the theorem follows from the observation that if k variables are fixed in f, the number of degree 2 terms in which a remaining unfixed variable occurs will be reduced by at most k. ■

Now we turn to the question of counting $SAC(k)$ functions for $k > 0$. The problem of counting the $SAC(k)$ functions for the highest orders $n-2$ and $n-3$ is easier, because of the stringent conditions which these functions must satisfy.

If we define

$$q(x_1, \ldots, x_n) = \sum_{1 \le i < j \le n} x_i x_j,$$

then each variable x_i occurs in exactly $n-1$ terms. It follows immediately from Lemma 3.8 and Theorems 3.12 and 3.13 that any $SAC(n-2)$ Boolean function in n variables has a nonaffine part equal to $q(\mathbf{x})$, so we have a count and a description of all $SAC(n-2)$ functions.

Theorem 3.14. *There are 2^{n+1} $SAC(n-2)$ Boolean functions in n variables; they are exactly the functions $q(\mathbf{x}) \oplus g(\mathbf{x})$, where $g(\mathbf{x})$ is affine.*

This result is given in [371, p. 170]. A longer proof of the count (but not the description) in Theorem 3.14, using only Lemmas 3.2 and 3.7 but not the autocorrelation function, was given by Lloyd [279, Theorem 5.1, pp. 69–74]. Later she refined this work to obtain all of Theorem 3.14 in [283, Theorem 1.8, p. 226].

Note that the function in Example 3.1 above is one of the 16 $SAC(1)$ functions in three variables. Its algebraic normal form is $x_1x_2 \oplus x_1x_3 \oplus x_2x_3 \oplus x_1 \oplus 1$.

Our next theorem makes an interesting connection between $SAC(n-3)$ functions in n variables and the symmetric group S_n of all permutations on $\{1,2,\ldots,n\}$ under composition.

Theorem 3.15. *The set of $SAC(n-3)$ Boolean functions in n variables with no affine terms is in one-to-one correspondence with the set $T(n)$ of all permutations σ in S_n such that $\sigma^2 = I$ (I is the identity permutation).*

Proof. By Theorem 3.12, any $SAC(n-3)$ function f has degree 2. By Theorem 3.13, given such a function f the variable x_1 appears in at least $n-2$ terms x_1x_j. If there are just $n-2$ terms x_1x_j, then there is a unique index $J(1)$ such that the term $x_1x_{J(1)}$ is missing from the algebraic normal form of f. Now $x_{J(1)}$ appears in exactly $n-2$ terms $x_{J(1)}x_j$, and the only missing j in these terms must be $j = 1$. We see that given an $SAC(n-3)$ function f with no affine terms, we can associate with it a unique permutation σ in S_n made up of products of 2-cycles and 1-cycle, as follows: we put the 1-cycles (x_i) in σ if f contains all of the $n-1$ possible terms x_ix_j. We put the 2-cycle (x_ix_j) in σ if f contains only $n-2$ terms x_ix_j, where x_ix_j is the missing term of that form. Clearly, $\sigma^2 = I$, so every $SAC(n-3)$ function with no affine terms corresponds to a unique permutation in $T(n)$. Conversely, any permutation σ in $T(n)$ is a unique product of disjoint 2-cycles $(x_i x_j)$ and 1-cycles (x_k). We associate with σ a unique $SAC(n-3)$ function f with no affine terms, as follows: For every 1-cycle (x_k) in σ, we put all $n-1$ terms $(x_u x_k)$ in f. For every 2-cycle $(x_i x_j)$ in σ, we put the $2(n-2)$ terms $(x_i x_u)$ and $(x_j x_u)$, where u takes on the $n-2$ values $1, \ldots, j-1, j+1, \ldots, n$, in f. The function so defined is $SAC(n-3)$ by Theorem 3.13, so we have the one-to-one correspondence in Theorem 3.15. ■

The set of permutations $T(n)$ defined in Theorem 3.15 was studied by Chowla et al. [112]. Let t_n denote the number of elements in the set $T(n)$. It is not hard to see that

$$t_1 = 1, t_2 = 2, t_n = t_{n-1} + (n-1)t_{n-2} \quad \text{for } n \geq 3. \tag{3.5}$$

Chowla et al. [112] prove (3.5) and then analyze the recursion to obtain the asymptotic formula

$$t_n \asymp (e^{1/4}\sqrt{2})^{-1} e^{\sqrt{n}} (n/e)^{n/2} \quad \text{as } n \to \infty. \tag{3.6}$$

In view of Theorem 3.15, these results give:

Theorem 3.16. *The number of $SAC(n-3)$ Boolean functions in $n \geq 3$ variables is $2^{n+1} t_n$, where the sequence $\{t_i\}_i$ is given by* (3.5). *An asymptotic formula for this number is given by* (3.6).

Lloyd [281, Theorem 4.1, p. 171] used a rather complicated direct counting argument to prove the following exact formula for the number of $SAC(n-3)$ functions.

Theorem 3.17. *The number of $SAC(n-3)$ Boolean functions in $n \geq 3$ variables is*

$$2^{n+1} \sum_{0 \leq i \leq n/2} \frac{n!}{(n-2i)!\, i!\, 2^i}. \tag{3.7}$$

It is not difficult to deduce Equation (3.7) from Theorems 3.12, 3.13 and 3.15. The expression (3.7) is more complicated to compute than the recursion for t_n and it does not seem easy to deduce the nice asymptotic result (3.6) from (3.7).

The only work on estimating the number of $SAC(n-j)$ functions for $j > 3$ seems to be a paper of Cusick [126] giving some complicated upper and lower bounds on the number of Boolean functions which satisfy $SAC(n-4)$. The difficulty with the $SAC(n-j)$ cases with $j > 3$ is that, by Theorem 3.12, functions of degree at least 3 must be considered. The lower bound in [126] is simpler because only quadratic functions are counted.

Now we turn to the problem of counting balanced functions satisfying *SAC* of high order. The easiest case is that of $SAC(n-2)$ functions. Lloyd [283, pp. 226–229] first solved the problem. We give a different proof based on the paper of Gopalakrishnan and

Stinson [212]. From Theorem 3.14, we know that a function $f(\mathbf{x})$ in $n \geq 2$ variables satisfies $SAC(n-2)$ if and only if the algebraic normal form is

$$f(\mathbf{x}) = a_0 \oplus a_1 x_1 \oplus \cdots \oplus a_n x_n \oplus \sum_{1 \leq i < j \leq n} x_i x_j.$$

Since $f(\mathbf{x})$ is balanced if and only if $f(\mathbf{x}) \oplus 1$ is balanced, we may assume without loss of generality that $a_0 = 0$.

Suppose exactly r of the coefficients $a_1, \ldots, a_n$ are equal to one and the rest are equal to zero. Define $S(n,r), 0 \leq r \leq n$, by

$$S(n,r) = \text{number of } \mathbf{x} \in \mathbb{V}_n \text{ such that } f(\mathbf{x}) = 0 \text{ (given } a_0 = 0).$$

The next lemma gives a recurrence relation for $S(n,r)$. It is interesting that this recurrence is the same one satisfied by the binomial coefficients $\binom{n}{r}$.

Lemma 3.18. *For $n \geq 2$ and $0 \leq r \leq n$ we have*

$$S(n,r) = S(n-1,r) + S(n-1,r-1). \tag{3.8}$$

Proof. Renumbering the variables does not affect whether the function is balanced or not, so we may assume

$$f(\mathbf{x}) = x_1 \oplus \cdots \oplus x_r \oplus \sum_{1 \leq i < j \leq n} x_i x_j, \tag{3.9}$$

for some $n, 0 \leq r \leq n$. If $x_{r+1} = 0$, the function $f(\mathbf{x})$ reduces to a function $g_0(\mathbf{x}) = x_1 \oplus \cdots \oplus x_r \oplus \sum_{1 \leq i < j \leq n, i,j \neq r+1} x_i x_j$ in $n-1$ variables and the number of vectors in $\mathbb{V}_{n-1}$ such that $g_0(\mathbf{x}) = 0$ is $S(n-1,r)$. If $x_{r+1} = 1$, the function $f(\mathbf{x})$ reduces to a function

$$\begin{aligned} g_1(\mathbf{x}) &= x_1 \oplus \cdots \oplus x_r \oplus \sum_{1 \leq i \leq n, i \neq r+1} x_i \oplus \sum_{1 \leq i \leq n, i,j \neq r+1} x_i x_j \\ &= x_{r+2} \oplus x_{r+3} \oplus \cdots \oplus x_n \oplus \sum_{1 \leq i \leq n, i,j \neq r+1} x_i x_j \end{aligned}$$

and the number of vectors in $\mathbb{V}_{n-1}$ such that $g_1(\mathbf{x}) = 0$ is $S(n-1, n-r-1)$. This gives

$$S(n,r) = S(n-1,r) + S(n-1,n-r-1). \tag{3.10}$$

Using (3.10) to evaluate $S(n-1,n-r-1)$ we get

$$\begin{aligned} S(n-1,n-r-1) &= S(n-2,n-r-1) + S(n-2,n-1-(n-r-1)-1) \\ &= S(n-2,n-r-1) + S(n-2,r-1) \\ &= S(n-2,r-1) + S(n-2,(n-1)-(r-1)-1) \\ &= S(n-1,r-1). \end{aligned}$$

Substituting this back into (3.10) gives (3.8). ■

The next theorem gives an explicit formula for $S(n,r)$. For this we need the following lemma on binomial coefficients.

Lemma 3.19. *We have the following identities:*

$$\sum_{0\le k\le n, k\equiv 0 \pmod 4} \binom{n}{k} = 2^{n-2} + 2^{(n-2)/2}\cos\left(\frac{n\pi}{4}\right)$$

$$\sum_{0\le k\le n, k\equiv 1 \pmod 4} \binom{n}{k} = 2^{n-2} + 2^{(n-2)/2}\sin\left(\frac{n\pi}{4}\right)$$

$$\sum_{0\le k\le n, k\equiv 3 \pmod 4} \binom{n}{k} = 2^{n-2} - 2^{(n-2)/2}\sin\left(\frac{n\pi}{4}\right)$$

Proof. This is a well-known result; in [212] it is asserted that it goes back at least to the year 1834. ∎

Theorem 3.20. *For $n \ge 2$ and $0 \le r \le n$ we have*

$$S(n,r) = 2^{n-1} - 2^{(n-1)/2}\sin\left(\left(r + \frac{7n-1}{2}\right)\frac{\pi}{2}\right). \tag{3.11}$$

Proof. When $r = 0$, then by symmetry the $SAC(n-2)$ function $f(\mathbf{x})$ in (3.9) has Hamming weight which depends only on the Hamming weight of $\mathbf{x}$. If $wt(\mathbf{x}) = k$, then $wt(f(\mathbf{x})) = \binom{k}{2} \equiv f(\mathbf{x}) \pmod 2$. Since $\binom{k}{2} \equiv 0 \pmod 2$ if and only if $k \equiv 0,1 \pmod 4$, we have

$$S(n,0) = \sum_{0\le k\le n, k\equiv 0,1 \pmod 4} \binom{n}{k}. \tag{3.12}$$

When $r = n$, $wt(f(\mathbf{x}))$ again depends only on $wt(\mathbf{x})$ and if $wt(\mathbf{x}) = k$, then $wt(f(\mathbf{x})) = k + \binom{k}{2} \equiv f(\mathbf{x}) \pmod 2$. Since $k + \binom{k}{2} \equiv 0 \pmod 2$ if and only if $k \equiv 0,3 \pmod 4$, we have

$$S(n,n) = \sum_{0\le k\le n, k\equiv 0,3 \pmod 4} \binom{n}{k}. \tag{3.13}$$

The recurrence (3.8) along with conditions (3.12) and (3.13) completely determines $S(n,r)$ for $n \ge 2, 0 \le r \le n$. The formula (3.11) can be deduced as follows: By Lemma 3.19, (3.12) and (3.13) can be rewritten as

$$S(n,0) = 2^{n-1} + 2^{(n-2)/2}\left(\cos\left(\frac{n\pi}{4}\right) + \sin\left(\frac{n\pi}{4}\right)\right) \tag{3.14}$$

and

$$S(n,n) = 2^{n-1} + 2^{(n-2)/2}\left(\cos\left(\frac{n\pi}{4}\right) - \sin\left(\frac{n\pi}{4}\right)\right). \tag{3.15}$$

By elementary trigonometry, when $r=0$ Equation (3.14) is the same as Equation (3.11); and when $r=n$, Equation (3.15) is the same as Equation (3.11). A tedious computation shows that $S(n,r)$ as given in (3.11) satisfies the recursion (3.8), and this suffices to give the theorem. ■

Now we can prove the result of [283, Theorems 2.4 and 2.5, p. 228] on balanced $SAC(n-2)$ functions.

Theorem 3.21. *If n is even, then there are no balanced $SAC(n-2)$ functions in n variables. If n is odd, then exactly half of the $2^{n+1}SAC(n-2)$ functions in n variables are balanced.*

Proof. We can reduce $f(\mathbf{x})$ to the form (3.9) with no loss of generality and then $f(\mathbf{x})$ is balanced if and only if $S(n,r)=2^{n-1}$. By Theorem 3.20, this is true if and only if

$$\sin\left(\left(r+\frac{7n-1}{2}\right)\frac{\pi}{2}\right)=0. \tag{3.16}$$

Now (3.16) holds if and only if $r+(7n-1)/2$ is an even integer, which is impossible for n even, since then we do not even have an integer. If n is odd, then we get an even integer for exactly half of the $n+1$ values $r=0,1,\ldots,n$, namely even r if $(7n-1)/2$ is even, that is if $n\equiv 3 \pmod 4$; and odd r if $(7n-1)/2$ is odd, that is if $n\equiv 1 \pmod 4$. ■

Lloyd [282, Theorem 4.2, p. 118] also solved the problem of counting balanced $SAC(n-3)$ functions (despite the publication dates, [283] was written before [281] and [282]). We state her result in the next theorem, but omit the complicated proof.

Theorem 3.22. *The number of balanced $SAC(n-3)$ Boolean functions in $n\geq 3$ variables is*

$$2^{n+1}\sum_{0\leq i\leq n/2}\frac{n!}{(n-2i)!\,i!}(2^{-i}-2^{-2i})+2^{n}\sum_{\substack{0\leq i\leq n/2\\ i\not\equiv n \pmod 2}}\frac{n!}{(n-2i)!\,i!}2^{-2i}. \tag{3.17}$$

In contrast to the situation for $SAC(n-2)$ functions as given in Theorem 3.21, Theorems 3.17 and 3.22 imply that for $SAC(n-3)$ functions, the proportion which is not balanced tends to 0 as n tends to ∞. The details are given in Theorem 3.23 below.

We let $|SAC(n-3)|$ denote the number of $SAC(n-3)$ Boolean functions in $n\geq 3$ variables, as given by (3.7), and we let $|BSAC(n-3)|$ denote the number of balanced $SAC(n-3)$ Boolean functions in $n\geq 3$ variables, as given by (3.17).

Theorem 3.23. *We have* $\lim_{n\to\infty}|BSAC(n-3)|/|SAC(n-3)|=1$.

Proof. We define

$$A(a,b)=\sum_{a\leq i\leq b}\frac{n!}{(n-2i)!\,i!\,2^{i}}$$

and

$$B(a,b) = \sum_{a \leq i \leq b} \frac{n!}{(n-2i)!\,i!\,2^{2i}},$$

so the difference $A(0,n/2) - B(0,n/2)$ gives the first summation in (3.17). Obviously the 'tails' $A(3n/8,n/2)$ and $B(3n/8,n/2)$ of these two sums satisfy

$$A(3n/8,n/2) \geq \sum_{3n/8 \leq i \leq n/2} \frac{n!}{(n-2i)!\,i!\,2^{n/2}}$$

and

$$B(3n/8,n/2) \leq \sum_{3n/8 \leq i \leq n/2} \frac{n!}{(n-2i)!\,i!\,2^{3n/4}},$$

so $B(3n/8,n/2)/A(3n/8,n/2) \leq 2^{-n/4}$ and therefore

$$\lim_{n\to\infty} \frac{B(3n/8,n/2)}{A(3n/8,n/2)} = 0. \tag{3.18}$$

A simple calculation shows that for $0 \leq i \leq 3n/8$, the integer

$$\frac{n!}{(n-2i)!\,i!} \tag{3.19}$$

decreases as i decreases; thus by Stirling's formula the largest value of (3.19) in the range $0 \leq i \leq 3n/8$ is asymptotic to

$$\frac{n!}{(n/4)!(3n/8)!} < \left(\frac{6n}{e}\right)^{3n/8}.$$

Hence we get the following estimate for the sum of the low order terms in $B(0,n/2)$:

$$B(0,3n/8) < \frac{3n}{8}\left(\frac{6n}{e}\right)^{3n/8}. \tag{3.20}$$

The last term in $A(0,n/2)$, which is not even the largest one, by Stirling's formula is asymptotic to

$$\frac{n!}{(n/2)!\,2^{n/2}} \asymp 2^{1/2}\left(\frac{n}{e}\right)^{n/2},$$

so this one term is already asymptotically larger than $B(0,3n/8)$ by (3.20). Combining this fact with (3.7), (3.17) and (3.18) we deduce Theorem 3.23. ■

3.5 PROPAGATION CRITERIA

A Boolean function $f(\mathbf{x})$ in n variables is said to satisfy the propagation criterion of degree k ($PC(k)$ for short) if changing any i $(1 \leq i \leq k)$ of the n bits in the input $\mathbf{x}$ results in the output of the function being changed for exactly half of the 2^n vectors $\mathbf{x}$. We may also simply say that $f(\mathbf{x})$ is a $PC(k)$ function. This generalizes the notion of *SAC*, which is clearly identical to $PC(1)$. These criteria were introduced by Preneel et al. [371] and appear in Preneel's thesis [368, 234ff.]. The three-variable function in Example 3.1 satisfies $PC(2)$, but not $PC(3)$.

The propagation criteria are closely connected to properties of the autocorrelation function $r_f(\mathbf{a})$ defined in (3.4), because we have

Lemma 3.24. *A Boolean function $f(\mathbf{x})$ in n variables satisfies $PC(k)$ if and only if all of the given values*

$$r_f(\mathbf{a}) = \sum_{\mathbf{x} \in \mathbb{V}_n} f(\mathbf{x}) \oplus f(\mathbf{x} \oplus \mathbf{a}), \quad 1 \leq wt(\mathbf{a}) \leq k,$$

of the autocorrelation function are equal to 2^{n-1}.

Proof. From the definition of $r_f(\mathbf{a})$ we have

$$\Pr(f(\mathbf{x}) \neq f(\mathbf{x} \oplus \mathbf{a})) = r_f(\mathbf{a})/2^n,$$

so the lemma follows immediately from the definition of $PC(k)$. ■

Note the similarity of Lemma 3.24 to Lemma 4.3 for correlation immunity (if we state Lemma 3.24 using the autocorrelation function $\hat{r}(\mathbf{a})$ for $\hat{f}$ defined in (2.3), then all of the values of $\hat{r}(\mathbf{a}), 1 \leq wt(\mathbf{a}) \leq k$, are zero). As is the case for correlation immunity (see Equation (4.1)), we can interpret $PC(k)$ in terms of information theory. A function f satisfies $PC(k)$ if and only if the information obtained about $f(\mathbf{x})$ given $f(\mathbf{x} \oplus \mathbf{a})$ for any $\mathbf{a}$ with $1 \leq wt(\mathbf{a}) \leq k$ is zero, that is

$$I(f(\mathbf{x}) | f(\mathbf{x} \oplus \mathbf{a})) = 0, \quad 1 \leq wt(\mathbf{a}) \leq k. \tag{3.21}$$

We see from (3.4) that $r_f(\mathbf{a})$ is simply the sum of all of the values of the *directional derivative* $f(\mathbf{x}) \oplus f(\mathbf{x} \oplus \mathbf{a})$ as $\mathbf{x}$ runs through $\mathbb{V}_n$. The next lemma restates Lemma 3.24 in terms of these directional derivatives.

Lemma 3.25. *A Boolean function $f(\mathbf{x})$ in n variables satisfies $PC(k)$ if and only if all of the directional derivatives*

$$f_a(\mathbf{x}) = f(\mathbf{x}) \oplus f(\mathbf{x} \oplus \mathbf{a}), \quad 1 \leq wt(\mathbf{a}) \leq k,$$

are balanced functions.

Proof. The lemma follows from Lemma 3.24 and the definition of $PC(k)$. Alternatively, we can use the fact that $f(\mathbf{x})$ satisfies $PC(k)$ if and only if (3.21) holds. ■

For $PC(k)$ functions, there is an analog of Lemma 3.8 for $SAC(k)$ functions.

Lemma 3.26. *If a Boolean function f in n variables satisfies $PC(k)$ for some k, $1 \leq k \leq n$, then so does $f \oplus g$, where g is any affine function in n variables.*

Proof. This follows immediately from Lemma 3.24. ■

A function $f(\mathbf{x})$ in n variables satisfies $PC(n)$ if and only if $f(\mathbf{x})$ is perfect nonlinear, that is, bent (see Definition 5.15). Such functions are discussed in detail in Chapter 5. We will not repeat the results given there, except to note that a $PC(n)$ function $f(\mathbf{x})$ with $n > 2$ must have deg f even and no larger than $n/2$ (Remark 5.2 and Theorem 5.18).

3.6 HIGHER ORDER $PC(k)$

The definition of higher order *SAC* given in Section 3.4 can be generalized to define higher order $PC(k)$.

Definition 3.27. *A Boolean function $f(\mathbf{x})$ in n variables is said to satisfy the propagation criterion of degree k and order m ($PC(k)$ of order m for short) if $k + m \leq n$ and if fixing any m of the n bits in the input $\mathbf{x}$ results in a Boolean function in the remaining $n - m$ variables which satisfies $PC(k)$. For brevity, we may say that such a function is a $PC(k)$ function of order m, or simply is $PC(k)$ of order m.*

The condition $k + m \leq n$ is imposed because when m bits are fixed, there are only $n - m$ variable bits left which can be changed, as the definition of $PC(k)$ requires. If we allow m bits to be fixed and then subsequently k bits to be changed, this makes sense even if the condition $k + m \leq n$ is removed. Then we can generalize the notion of higher order *SAC* in a different way, by using the interpretation (3.21) (see Lemma 3.25 of $PC(k)$ in terms of information theory):

Definition 3.28. *A Boolean function $f(\mathbf{x})$ in n variables is said to satisfy the extended propagation criterion of degree k and order m ($EPC(k)$ of order m for short) if knowledge of m bits of $\mathbf{x}$ gives no information about $f(\mathbf{x}) \oplus f(\mathbf{x} \oplus \mathbf{a})$ for all $\mathbf{a}$ with $1 \leq wt(\mathbf{a}) \leq k$.*

These definitions were introduced in [371, Section 5]. The quadratic functions satisfying $PC(k)$ were studied in some detail in [370]. It follows from the definitions and Lemma 3.25 that $PC(k)$, $PC(k)$ of order 0 and $EPC(k)$ of order 0 all mean the same thing. Of course we want a function that satisfies $PC(k)$ or $EPC(k)$ of order m to also satisfy the corresponding criterion for all orders $<m$. This is true, as our next lemma states.

Lemma 3.29. *Suppose $f(\mathbf{x})$ is a Boolean function in n variables which satisfies $PC(k)$ or $EPC(k)$ of order $m > 0$. Then $f(\mathbf{x})$ also satisfies $PC(k)$ or $EPC(k)$, respectively, of order j for any $j < m$.*

Proof. The proof of Lemma 3.7, which gives this result for $PC(1) = SAC$, generalizes to prove Lemma 3.29. ■

We can express $EPC(k)$ of order $m > 0$ in terms of correlation immunity (see Section 4.1 of Chapter 4 for definitions):

Lemma 3.30. *A Boolean function $f(\mathbf{x})$ in n variables satisfies $EPC(k)$ of order $m > 0$ if and only if all of the directional derivatives*

$$f_{\mathbf{a}}(\mathbf{x}) = f(\mathbf{x}) \oplus f(\mathbf{x} \oplus \mathbf{a}), \quad 1 \le wt(\mathbf{a}) \le k,$$

are balanced and correlation immune of order m.

Proof. The lemma follows immediately from the definitions of $EPC(k)$ of order m and correlation immunity of order m, by using Lemma 3.25. ■

Lemma 3.31. *Let $f(\mathbf{x})$ be a Boolean function in n variables. If $f(\mathbf{x})$ satisfies $EPC(k)$ of order $m > 0$, then $f(\mathbf{x})$ satisfies $PC(k)$ of order $m > 0$. The converse is true for $k = 1$ and any $m > 0$. The converse is also true for any k if $m = 1$, but is false for $k = m = 2$.*

Proof. The first assertion in the lemma follows from the definitions. To prove the converse for $k = 1$, we observe that if $f(\mathbf{x})$ satisfies $PC(1)$ of order m, then by Lemmas 3.25 and 3.30 $f(\mathbf{x})$ satisfies $EPC(1)$ of order m if and only if

$$\sum_{\substack{\mathbf{x} \in \mathbb{V}_n \\ x_{i(1)} = a_{i(1)}, \ldots, x_{i(m)} = a_{i(m)}}} f(\mathbf{x}) \oplus f(\mathbf{x} \oplus \mathbf{e}_i) = 2^{n-1} \tag{3.22}$$

for each unit vector $\mathbf{e}_i$, $1 \le i \le n$, and for all choices of the variables $x_{i(1)}, \ldots, x_{i(m)}$ and all choices of the fixed values $a_{i(1)}, \ldots, a_{i(m)}$. If none of the indices $i(j)$ equals i, then (3.22) is exactly the condition for $PC(1)$ of order m, according to Lemma 3.25. If the index $i(j)$ is equal to i, then (3.22) is exactly the condition for $PC(1)$ of order $m - 1$, and this is true by our assumption that $f(\mathbf{x})$ satisfies $PC(1)$ of order m.

Next, to prove the converse for any k if $m = 1$, we use an argument similar to the one above based on (3.22) (but simpler, since only one variable x_i is fixed).

Finally, to prove that the converse is false for $k = m = 2$, we show that the function

$$q_n = q_n(x_1, \ldots, x_n) = \sum_{1 \le i < j \le n} x_i x_j \tag{3.23}$$

for $n \ge 4$ satisfies $PC(2)$ of order 2 but does not satisfy $EPC(2)$ of order 2. To verify the former condition, we need to examine the directional derivatives of q_n when any two bits are fixed. By symmetry, we may suppose that the fixed bits are x_{n-1} and x_n, in which case the directional derivative is a function $g(x_1, \ldots, x_{n-2})$ of the form

$$g = q_{n-2} \oplus h(x_1, \ldots, x_{n-2}), \tag{3.24}$$

where h is an affine function. Now all of the directional derivatives

$$g_{\mathbf{a}}(\mathbf{x}) = g(\mathbf{x}) \oplus g(\mathbf{x} \oplus \mathbf{a}), \quad 1 \le wt(\mathbf{a}) \le 2,$$

are nonconstant affine functions and therefore are balanced; so by Lemma 3.25, q_n satisfies $PC(2)$ of order 2. However, q_n does not satisfy $EPC(2)$ of order 2, because if $\mathbf{a}$ of weight 2 has a 1 in positions i and j, then the directional derivative of q_n for $\mathbf{a}$ is

$$q_n(\mathbf{x}) \oplus q_n(\mathbf{x} \oplus \mathbf{a}) = x_i \oplus x_j \oplus 1;$$

this function is correlation immune of order 1 but not of order 2, so by Lemma 3.30 q_n does not satisfy $EPC(2)$ of order 2. ■

3.7 CONSTRUCTION OF $SAC(k)$ AND $PC(k)$ FUNCTIONS

We have already seen a straightforward method of constructing *SAC* functions with n variables in the proof of Lemma 3.3, which shows that in fact the first 2^{n-1} bits in the truth table of such a function can be assigned arbitrarily. The modification of this method given in the proof of Lemma 3.6 allows us to construct balanced *SAC* functions in n variables. These constructions are not useful in situations (such as the design of Boolean functions for use in various cryptosystems) where functions with multiple properties (such as *SAC*, balance, high nonlinearity, etc.) are needed. In such situations one must take into account inevitable tradeoffs between the desired properties (see, for example, Sections 4.6 and 4.9). We do not consider any constructions involving tradeoffs in this section, but confine ourselves to simpler constructions.

In Section 5.15 we discuss constructions which use bent functions to produce functions which are balanced, satisfy the *SAC*, and have high nonlinearity. The bent functions in n variables themselves are exactly those functions which satisfy $PC(n)$, because of Lemma 3.25 and Theorem 5.12(vii). For the rest of this section, we confine ourselves to construction of functions which satisfy $SAC(m) = PC(1)$ of order m for $m > 0$ or $PC(k)$ of order m for $k > 1, m \geq 1$. The special case where these functions are quadratic is analyzed in [371] and [370]. Construction of such functions of degree greater than 2 was first discussed by Kurosawa and Satoh [265]. Their work was generalized by Carlet in [72] and [73]; we follow Carlet's approach here.

We define an *MM* function (short for Maiorana-McFarland; as is explained in Section 5.8, the eponymous pair independently used a particular kind of *MM* function to construct a class of bent functions) to be any Boolean function $f(\mathbf{x},\mathbf{y})$ in n variables of the form

$$f(\mathbf{x},\mathbf{y}) = \phi(\mathbf{x}) \cdot \mathbf{y} \oplus g(\mathbf{x}), \tag{3.25}$$

where $\mathbf{x} = (x_1, \ldots, x_s)$, $\mathbf{y} = (y_1, \ldots, y_t)$, with s and t positive integers such that $n = s + t$; ϕ is any mapping from $\mathbb{V}_s$ to $\mathbb{V}_t$; g is any Boolean function on $\mathbb{V}_s$; and '$\cdot$' is the usual dot product on $\mathbb{V}_t$. We write

$$\phi(\mathbf{x}) = (\phi_1(\mathbf{x}), \ldots, \phi_t(\mathbf{x})), \tag{3.26}$$

where the $\phi_i(\mathbf{x})$ are Boolean functions which we call the coordinate functions of ϕ.

Theorem 3.32. *Suppose* (3.25) *and* (3.26) *define an MM function* $f(\mathbf{x},\mathbf{y})$ *in* $n = s + t$ *variables such that*

(i) *for each* j, $1 \leq j \leq k$, *the sum of any* j *of the coordinate functions* $\phi_i(\mathbf{x})$ *is a balanced function;*
(ii) *for every nonzero* $\mathbf{a}$ *in* $\mathbb{V}_s$ *with* $wt(\mathbf{a}) \leq k$, *and for any* $\mathbf{x}$ *in* $\mathbb{V}_s$, *we have* $\phi(\mathbf{x} \oplus \mathbf{a}) \neq \phi(\mathbf{x})$.

Then $f(\mathbf{x},\mathbf{y})$ *satisfies* $PC(k)$.

Proof. For every $\mathbf{a}$ in $\mathbb{V}_s$ and every $\mathbf{b}$ in $\mathbb{V}_t$, we have

$$f(\mathbf{x},\mathbf{y}) \oplus f(\mathbf{x}\oplus\mathbf{a},\mathbf{y}\oplus\mathbf{b}) = (\phi(\mathbf{x}) \oplus \phi(\mathbf{x}\oplus\mathbf{a}))\cdot\mathbf{y} \oplus \phi(\mathbf{x}\oplus\mathbf{a})\cdot\mathbf{b} \oplus g(\mathbf{x}) \oplus g(\mathbf{x}\oplus\mathbf{a}). \tag{3.27}$$

If $\mathbf{a} = 0$ and $1 \leq wt(\mathbf{b}) \leq k$, then the function (3.27) becomes $\phi(\mathbf{x})\cdot\mathbf{b}$, which is balanced by (i). If $1 \leq wt(\mathbf{a}) \leq k$, then for every fixed $\mathbf{x}$, (ii) implies that the function of $\mathbf{y}$ given in (3.27) is a nonconstant affine function, and so is balanced. Hence for any $\mathbf{a},\mathbf{b}$ such that $1 \leq wt(\mathbf{a}) + wt(\mathbf{b}) \leq k$, the function in (3.27) is balanced. Thus $f(\mathbf{x},\mathbf{y})$ satisfies $PC(k)$ by Lemma 3.25. ■

In the next theorem, we use the terminology of Section 4.5, where a function which is balanced and correlation immune of order k is said to be a k-resilient function.

Theorem 3.33. *Suppose* (3.25) *and* (3.26) *define an MM function* $f(\mathbf{x},\mathbf{y})$ *in* $n = s + t$ *variables such that*

(i) *for each* j, $1 \leq j \leq k$, *the sum of any* j *of the coordinate functions* $\phi_i(\mathbf{x})$ *is a* k*-resilient function;*
(ii) *for every nonzero* $\mathbf{a}$ *in* $\mathbb{V}_s$ *with* $wt(\mathbf{a}) \leq k$, *and for any* $\mathbf{x}$ *in* $\mathbb{V}_s$, *we have that at least* $m+1$ *coordinates of the vectors* $\phi(\mathbf{x}\oplus\mathbf{a})$ *and* $\phi(\mathbf{x})$ *are different.*

Then $f(\mathbf{x},\mathbf{y})$ *satisfies* $PC(k)$ *of order* m.

Proof. It is easy to see that any restriction of an MM function (3.25) obtained by fixing some of the input bits in $\mathbf{x}$ and/or $\mathbf{y}$ is still an MM function. Let

$$f'(\mathbf{x}',\mathbf{y}') = \phi'(\mathbf{x}')\cdot\mathbf{y}' \oplus g'(\mathbf{x}')$$

be any such restriction of the MM function $f(\mathbf{x},\mathbf{y})$ in (3.25) obtained by fixing at most m of the input bits in $\mathbf{x}$ and/or $\mathbf{y}$. The sum of any at least 1 and at most k of the coordinate functions $\phi_i'(\mathbf{x})$ is equal to the restriction of the sum of any at least 1 and at most k of the coordinate functions $\phi_i(\mathbf{x})$. Therefore, by condition (i) of Theorem 3.33, the condition (i) of Theorem 3.32 is satisfied by $f'(\mathbf{x},\mathbf{y})$. By condition (ii) of Theorem 3.33, for every nonzero $\mathbf{a}'$ with $wt(\mathbf{a}') \leq k$, and for any $\mathbf{x}'$ in $\mathbb{V}_s$, we have $\phi'(\mathbf{x}'\oplus\mathbf{a}') \neq \phi'(\mathbf{x}')$. Thus the condition (ii) of Theorem 3.32 is also satisfied by $f'(\mathbf{x}',\mathbf{y}')$, and it follows that $f(\mathbf{x},\mathbf{y})$ satisfies $PC(k)$ of order m. ■

In order to use Theorems 3.32 and 3.33 we must be able to find functions which satisfy the conditions in those theorems. Carlet [72,73] showed how to use a deep result of Delsarte [147] on the dual distance of nonlinear codes to construct MM functions (3.25) which satisfy (i), (ii) in Theorem 3.33 with nonlinear mappings $\phi(\mathbf{x})$. If one is satisfied with linear $\phi(\mathbf{x})$ (which would make the functions (3.25) weak from a cryptographic point of view), then the simpler construction of [265] (which does not mention MM functions at all) could be used. In [73, Proposition 11], Carlet gave a refined version of Theorem 3.33 which contains necessary and sufficient conditions for an MM function to satisfy the extended criterion $EPC(k)$ of order m. The proof is a bit longer so we do not give it here.

We remark that MM functions were already used in Camion et al. [53, Proposition 4.2] to construct correlation immune functions of order at least k (see Theorem 4.20).

The functions which satisfy $PC(k)$ for the highest degrees k (namely, $k = n-2$, $n-1$ or n) were characterized by Carlet [73, Proposition 6, p. 38] (the result was stated without proof in the conference version [72, Proposition 1]). We recall that a function $f(\mathbf{x})$ in n variables satisfies $PC(n)$ if and only if $f(\mathbf{x})$ is perfect nonlinear, that is, bent (see Definition 5.15).

Theorem 3.34. *Let $n \geq 4$ be even. The Boolean functions in n variables which satisfy $PC(n-2)$ are the bent functions, so in fact they satisfy $PC(n)$.*

Let $n \geq 3$ be odd. The Boolean functions in n variables which satisfy $PC(n-1)$ are the functions of the form

$$f(x_1,\ldots,x_n) = g(x_1 \oplus x_n,\ldots,x_{n-1} \oplus x_n) \oplus h(x_1,\ldots,x_n), \tag{3.28}$$

where g is any bent function in $n-1$ variables and h is any affine function in n variables. The functions which satisfy $P(n-2)$ are those of form (3.28) and of the forms

$$g(x_1 \oplus x_n,\ldots,x_{i-1} \oplus x_n, x_i, x_{i+1} \oplus x_n,\ldots,x_{n-1} \oplus x_n) \oplus h(x_1,\ldots,x_n) \tag{3.29}$$

or

$$g(x_1 \oplus x_{n-1},\ldots,x_{n-2} \oplus x_{n-1}, x_n) \oplus h(x_1,\ldots,x_n), \tag{3.30}$$

where g and h are as in (3.28).

Equivalently, for odd $n \geq 3$, the functions which satisfy $PC(n-2)$ are the functions satisfying the conditions: there exists a nonzero vector $\mathbf{a}$ with Hamming weight $wt(\mathbf{a}) \geq n-1$ such that

$$f(\mathbf{x}) \oplus f(\mathbf{x} \oplus \mathbf{a}) \text{ is constant}; \tag{3.31}$$

and

$$\text{for every } \mathbf{b} \neq 0 \text{ or } \mathbf{a}, \text{ the function } f(\mathbf{x}) \oplus f(\mathbf{x} \oplus \mathbf{b}) \text{ is balanced.} \tag{3.32}$$

We need two lemmas [73, Proposition 1 and Lemma 3] for the proof of the theorem. As usual, for vectors $\mathbf{u} = (u_1,\ldots,u_n)$ and $\mathbf{v} = (v_1,\ldots,v_n)$ with entries in $\{0, 1\}$, we write $\mathbf{u} \leq \mathbf{v}$ if $u_i \leq v_i$ for each i, and $\bar{\mathbf{u}}$ is the complement of $\mathbf{u}$.

Lemma 3.35. *Suppose* $0 \le k \le n$. *A Boolean function* f *in* n *variables satisfies* $PC(k)$ *if and only if for every* n*-vector* $\mathbf{u}$ *with* $wt(\mathbf{u}) = k$ *and every* n*-vector* $\mathbf{v}$

$$\sum_{\mathbf{w} \le \bar{\mathbf{u}}} W(\hat{f})(\mathbf{w} \oplus \mathbf{v})^2 = 2^{wt(\bar{\mathbf{u}})+n}.$$

The same equality holds for every $\mathbf{u}$ *with* $wt(\mathbf{u}) \le k$.

Proof. By the definition of the Walsh transform, the left-hand side of the equation in the lemma is

$$\begin{aligned}
\sum_{\mathbf{w} \le \bar{\mathbf{u}}} \left(\sum_{\mathbf{x} \in \mathbb{V}_n} (-1)^{f(\mathbf{x}) \oplus \mathbf{x} \cdot (\mathbf{w} \oplus \mathbf{v})} \right)^2 &= \sum_{\mathbf{w} \le \bar{\mathbf{u}}} \sum_{\mathbf{x}, \mathbf{y} \in \mathbb{V}_n} (-1)^{f(\mathbf{x}) \oplus f(\mathbf{y}) \oplus (\mathbf{x} \oplus \mathbf{y}) \cdot (\mathbf{w} \oplus \mathbf{v})} \\
&= \sum_{\mathbf{x}, \mathbf{y} \in \mathbb{V}_n} (-1)^{f(\mathbf{x}) \oplus f(\mathbf{y}) \oplus (\mathbf{x} \oplus \mathbf{y}) \cdot \mathbf{v}} \sum_{\mathbf{w} \le \bar{\mathbf{u}}} (-1)^{(\mathbf{x} \oplus \mathbf{y}) \cdot \mathbf{w}} \\
&= 2^{wt(\bar{\mathbf{u}})} \sum_{\mathbf{x}, \mathbf{y} \in \mathbb{V}_n, \mathbf{x} \oplus \mathbf{y} \le \mathbf{u}} (-1)^{f(\mathbf{x}) \oplus f(\mathbf{y}) \oplus (\mathbf{x} \oplus \mathbf{y}) \cdot \mathbf{v}} \\
&= 2^{wt(\bar{\mathbf{u}})} \sum_{\mathbf{s} \le \mathbf{u}} (-1)^{\mathbf{s} \cdot \mathbf{v}} \sum_{\mathbf{x} \in \mathbb{V}_n} (-1)^{f(\mathbf{x}) \oplus f(\mathbf{x} \oplus \mathbf{s})} \\
&= 2^{wt(\bar{\mathbf{u}})+n}.
\end{aligned}$$

The final equality follows from Lemma 3.24, which gives

$$\sum_{\mathbf{x} \in \mathbb{V}_n} (-1)^{f(\mathbf{x}) \oplus f(\mathbf{x} \oplus \mathbf{s})} = 0 \quad \text{for nonzero } \mathbf{s} \le \mathbf{u}.$$

The last sentence in the theorem follows by the same proof. ■

Lemma 3.36. *Assume that for a nonnegative integer* m *we have* $2^m = a^2 + b^2 + c^2 + d^2$, *where* a, b, c, d *are nonnegative integers. Then for* m *even, either one of* a, b, c, d *is* $2^{m/2}$ *and the others are* 0, *or each of* a, b, c, d *is* $2^{m/2-1}$; *and for* m *odd, two of* a, b, c, d *are* $2^{(m-1)/2}$ *and the other two are* 0.

Proof. The result follows by induction on $m \ge 3$, using the fact that if a, b, c, d are not all even, then 8 does not divide $a^2 + b^2 + c^2 + d^2$. ■

Proof (*Proof of* Theorem 3.34). We first prove the assertions in the theorem about the functions which satisfy $PC(n-2)$. In the case n even, any bent function satisfies $PC(n)$, and so $PC(n-2)$. In the case n odd, it follows from Lemmas 3.25 and 3.26 that any function of the form (3.28), (3.29) or (3.30) satisfies (3.31) and (3.32) (because g is bent), and therefore must satisfy $PC(n-2)$. We next show that the converse is true in both cases. By Lemma 3.35, if f satisfies $PC(n-2)$ then for every n-vector $\mathbf{u}$ with weight $n-2$ and every n-vector $\mathbf{v}$ we have

$$W(\hat{f})(\mathbf{v})^2 + W(\hat{f})(\mathbf{v} \oplus \mathbf{e}_i)^2 + W(\hat{f})(\mathbf{v} \oplus \mathbf{e}_j)^2 + W(\hat{f})(\mathbf{v} \oplus \mathbf{e}_i \oplus \mathbf{e}_j)^2 = 2^{n+2}, \tag{3.33}$$

where i and j are the positions in $\bar{\mathbf{u}}$ where 1 appears ($\mathbf{e}_k$ is the vector in $\mathbb{V}_n$ whose only nonzero entry is in position k). Thus (3.33) holds for any distinct i and j.

Assume first that n is even; then Lemma 3.36 gives two possibilities for the summands in (3.33). If the first possibility could hold, then we would have $W(\hat{f})(\mathbf{v}) = \pm 2^{n/2+1}$ for some $\mathbf{v} \in \mathbb{V}_n$. Thus by (3.33) $W(\hat{f})(\mathbf{v} \oplus \mathbf{z}) = 0$ must hold for every vector $\mathbf{z}$ of weight 1 or 2. If we let $\mathbf{w} = \mathbf{e}_i \oplus \mathbf{e}_j \oplus \mathbf{e}_k$ be any vector of weight 3, then by applying Lemma 3.36 to (3.33) with $\mathbf{v}$ replaced by $\mathbf{v} \oplus \mathbf{e}_k$ we see that $W(\hat{f})(\mathbf{v} \oplus \mathbf{w}) = \pm 2^{n/2+1}$. We conclude that if $n \geq 4$ and i, j, k, t are any four distinct indices, then both $W(\hat{f})(\mathbf{v} \oplus \mathbf{e}_i \oplus \mathbf{e}_j \oplus \mathbf{e}_k)$ and $W(\hat{f})(\mathbf{v} \oplus \mathbf{e}_i \oplus \mathbf{e}_j \oplus \mathbf{e}_t)$ are equal to $\pm 2^{n/2+1}$. This contradicts (3.33) with $\mathbf{v}$ replaced by $\mathbf{v} \oplus \mathbf{e}_i \oplus \mathbf{e}_k$ and $\{\mathbf{e}_i, \mathbf{e}_j\}$ replaced by $\{\mathbf{e}_k, \mathbf{e}_t\}$, since then the left-hand side of (3.33) would be $\geq 2^{n+3}$. Thus the second possibility in (3.36) must hold, that is $W(\hat{f})(\mathbf{v}) = \pm 2^{n/2}$ for every $\mathbf{v}$. This means that f is bent (Definition 5.1).

Now assume that n is odd and that f satisfies $PC(n-2)$. By Lemma 3.35, for every $\mathbf{v}$ and any distinct i and j, two of the integers $W(\hat{f})(\mathbf{v})$, $W(\hat{f})(\mathbf{v} \oplus \mathbf{e}_i)$, $W(\hat{f})(\mathbf{v} \oplus \mathbf{e}_j)$ and $W(\hat{f})(\mathbf{v} \oplus \mathbf{e}_i \oplus \mathbf{e}_j)$ are equal to $\pm 2^{(n+1)/2}$ and the other two are 0. First, assume that for some $i, 1 \leq i \leq n$, and for some $\mathbf{v}$ we have $W(\hat{f})(\mathbf{v})$ and $W(\hat{f})(\mathbf{v} \oplus \mathbf{e}_i)$ equal to $\pm 2^{(n+1)/2}$. Since we can replace $f(\mathbf{x})$ by $f(\mathbf{x}) \oplus \mathbf{v} \cdot \mathbf{x}$, we may assume with no loss of generality that $\mathbf{v} = 0$. Next, we deduce by induction on $wt(\mathbf{w})$ that for every vector $\mathbf{w} = (w_1, \ldots, w_n)$ with $w_i = 0$, the numbers $W(\hat{f})(\mathbf{w})$ and $W(\hat{f})(\mathbf{w} \oplus \mathbf{e}_i)$ are equal to $\pm 2^{(n+1)/2}$ if $wt(\mathbf{w})$ is even and are equal to 0 otherwise.

Suppose now that $i < n$ and define

$$E_i = \left\{ \mathbf{w} : \sum_{j \neq i} w_j = 0 \right\}$$

(note that the sum is always in $\mathbb{F}_2$). By our assumptions we have $W(\hat{f})(\mathbf{w}) = 0$ for all $\mathbf{w} \notin E_i$ and $W(\hat{f})(\mathbf{w}) = \pm 2^{(n+1)/2}$ for all $\mathbf{w} \in E_i$. Therefore by the inverse Walsh transform (2.2) we have for all $\mathbf{u} \in \mathbb{V}_n$

$$\hat{f}(\mathbf{u}) = 2^{-n} \sum_{\mathbf{w} \in E_i} W(\hat{f})(\mathbf{u})(-1)^{\mathbf{w} \cdot \mathbf{u}}. \tag{3.34}$$

Every element of E_i can be written as $(\mathbf{w}', h_i(\mathbf{w}'))$, where $\mathbf{w}' = (w'_1, \ldots, w'_{n-1}) \in \mathbb{V}_{n-1}$ and $h_i(\mathbf{w}') = \sum_{j \neq i} \mathbf{w}'_j$. It follows from (3.34) that

$$(-1)^{f(\mathbf{u})} = 2^{-n} \sum_{\mathbf{w}' \in \mathbb{V}_{n-1}} W(\hat{f})(\mathbf{w}', h_i(\mathbf{w}'))(-1)^{\mathbf{w}' \cdot \mathbf{u}' \oplus u_n h_i(\mathbf{w}')}, \tag{3.35}$$

where $\mathbf{u}' = (u_1, \ldots, u_{n-1})$. Let $\mathbf{e}'_i$ denote the vector in $\mathbb{V}_{n-1}$ whose only nonzero entry is in position i; then

$$\mathbf{w}' \cdot \mathbf{u}' \oplus u_n h_i(\mathbf{w}') = \mathbf{w}' \cdot (\mathbf{u}' \oplus u_n \overline{\mathbf{e}'_i})$$

so it follows that the right-hand side of (3.35) is $g(\mathbf{u}' \oplus u_n \overline{\mathbf{e}'_i})$ for some function g in n variables. It follows that

$$f(\mathbf{u}) = g(\mathbf{u}' \oplus u_n \overline{\mathbf{e}'_i}) \quad \text{for all } \mathbf{u}; \tag{3.36}$$

thus to complete the proof in the case $i < n$ we need to show that g is bent, that is, f has the form (3.29). Suppose g is not bent, so by Lemma 3.25 there exists a nonzero $\mathbf{b}$ in $\mathbb{V}_{n-1}$ such that $g(\mathbf{u}') \oplus g(\mathbf{u}' \oplus \mathbf{b})$ is not balanced. Then the functions $f(\mathbf{u}) \oplus f(\mathbf{u} \oplus (\mathbf{b}, 0))$ and $f(\mathbf{u}) \oplus f(\mathbf{u} \oplus (\mathbf{b} \oplus \overline{\mathbf{e}'_i}, 1))$ are also not balanced (because by (3.36) each of these functions is unbalanced in $\mathbf{u}'$ with the same unequal count of zeros and ones if we fix $u_n = 0$ or $u_n = 1$). Since f satisfies $PC(n-2)$, by Lemma 3.25 the vectors $(\mathbf{b}, 0)$ and $(\mathbf{b} \oplus \overline{\mathbf{e}'_i}, 1)$ must both have weight $\geq n-1$. This is a contradiction if $n \geq 5$. If $n = 3$, then g is a quadratic nonaffine function in four variables, and every such function is bent.

We do the proof in the case $i = n$ by exchanging n and $n-1$ in (3.29) with $i = n-1$. This gives the functions of the form (3.30).

To complete the proof that any f satisfying $PC(n-2)$ must have one of the forms (3.28), (3.29) or (3.30), the only remaining case is that in which no i and $\mathbf{v}$ exist such that $W(\hat{f})(\mathbf{v})$ and $W(\hat{f})(\mathbf{v} \oplus \mathbf{e}_i)$ are equal to $\pm 2^{(n+1)/2}$. We choose $\mathbf{v}$ such that $W(\hat{f})(\mathbf{v}) = \pm 2^{(n+1)/2}$. As before, we can take $\mathbf{v} = 0$ without loss of generality, and so by our assumptions $W(\hat{f})(\mathbf{e}_i) = 0$ for every i. We deduce by induction on $wt(\mathbf{w})$ that for every $\mathbf{w}$, $W(\hat{f})(\mathbf{w}) = \pm 2^{(n+1)/2}$ if $wt(\mathbf{w})$ is even and $W(\hat{f})(\mathbf{w}) = 0$ if $wt(\mathbf{w})$ is odd. Define $E = \{\mathbf{w} : wt(\mathbf{w}) \text{ is even}\}$. With $\mathbf{u}' = (u_1, \ldots, u_{n-1})$ as before, for every $\mathbf{u}$ in $\mathbb{V}_n$ we have (3.34) with E_i replaced by E, whence

$$\hat{f}(\mathbf{u}) = 2^{-n} \sum_{\mathbf{w}' \in \mathbb{V}_{n-1}} W(\hat{f})(\mathbf{w}', h(\mathbf{w}'))(-1)^{\mathbf{w}' \cdot \mathbf{u}' \oplus u_n h(\mathbf{w}')},$$

where $h(\mathbf{w}') = \sum_{i=1}^{n-1} \mathbf{w}'_i$. Therefore

$$(-1)^{f(\mathbf{u})} = 2^{-n} \sum_{\mathbf{w}' \in \mathbb{V}_{n-1}} W(\hat{f})(\mathbf{w}', h(\mathbf{w}'))(-1)^{\mathbf{w}' \cdot (\mathbf{u}' \oplus u_n (1, \ldots, 1))}, \tag{3.37}$$

and this implies that $f(\mathbf{u}) = g(\mathbf{u}' \oplus u_n(1, \ldots, 1))$, where $g(\mathbf{u}') = f(\mathbf{u}', 0)$ must be bent (since it is a function in an even number $n-1$ of variables and satisfies $PC(n-2)$). Thus we have a function f of form (3.28) in this case.

To complete the proof of Theorem 3.34, we need to show that for n odd the only functions which satisfy $PC(n-1)$ are those of form (3.28); this was already proved in [370, Theorem 2, p. 144]. ■

We remark that from (3.31) and (3.32) we see that Theorem 3.34 implies that any function satisfying $PC(n-2)$ has exactly one linear structure $\mathbf{a}$ (see Section 5.4), and

Table 3.1 Values of the Walsh transform for $f(\mathbf{x}) = x_1x_2 \oplus x_1x_3 \oplus x_2x_3 \oplus x_1 \oplus 1$

w	$W(\hat{f})(\mathbf{w})$
000	−4
001	0
010	0
011	4
100	0
101	−4
110	−4
111	0

$wt(\mathbf{a})$ is $n-1$ or n. Furthermore, all of the functions of forms (3.28), (3.29) or (3.30) are partially bent in the sense of [64] (see also Section 5.16).

Theorem 3.34 shows that the problem of counting the functions which satisfy $PC(k)$ for large k is connected with the problem of counting bent functions, about which not much is known (see Section 5.14).

To illustrate Theorem 3.34, we give Table 3.1 which contains the Walsh transform values for the function

$$f(x_1, x_2, x_3) = x_1x_2 \oplus x_1x_3 \oplus x_2x_3 \oplus x_1 \oplus 1$$

of Example 3.1.

We already saw that $f(\mathbf{x})$ satisfies $PC(2)$ and Theorem 3.34 confirms this. We also see that $f(\mathbf{x}) = (x_1 \oplus x_2)(x_1 \oplus x_3) \oplus 1 = g(x_1 \oplus x_2, x_1 \oplus x_3) \oplus 1$, where $g(y_1, y_2) = y_1y_2$ is bent, in accordance with (3.28) in Theorem 3.34 (a renumbering of the variables in (3.28) is needed here).

CHAPTER

Correlation immune and resilient Boolean functions

4

4.1 INTRODUCTION

To myself I seem to have been only like a boy playing on the seashore, and diverting myself in now and then finding a smoother pebble or a prettier shell than ordinary, whilst the great ocean of truth lay all undiscovered before me.

Isaac Newton (1642–1727)

A Boolean function $f(\mathbf{x})$ in n variables is said to be *correlation immune of order* k, $1 \leq k \leq n$, if for any fixed subset of k variables the probability that, given the value of $f(\mathbf{x})$, the k variables have any fixed set of values, is always 2^{-k}, no matter what the choice of the fixed set of k values is; in other words, $f(\mathbf{x})$ is correlation immune of order k if its values are statistically independent of any subset of k input variables. If the chosen subset of k variables is $\{x(i_1),\ldots,x(i_k)\}$, then the definition of correlation immunity of order k is equivalent to the information theory condition that the information obtained about the values of $x(i_1),\ldots,x(i_k)$ given $f(\mathbf{x})$ is zero, that is

$$I(x(i_1),\ldots,x(i_k)|f(\mathbf{x})) = 0 \tag{4.1}$$

in the usual notation. We recall that basic information theory shows that the mutual information function $I(A \mid B)$ is symmetric, so (4.1) is equivalent to

$$I(f(\mathbf{x}) \mid x(i_1),\ldots,x(i_k)) = 0.$$

Siegenthaler [422] first defined correlation immunity, and he asserts [422, p. 777] that it was the intuitive content of (4.1) which led him to his definition. To give an example, the following function with $n = 3$ is easily seen to be correlation immune of order 1, but not of order 2.

Example 4.1 (*A 3-variable function, correlation immune of order* 1).

Input	000	001	010	011	100	101	110	111
Output	1	1	1	0	0	1	1	1

4.2 BASIC PROPERTIES OF CORRELATION IMMUNITY

It is useful to collect various conditions equivalent to correlation immunity (of order 1) in the following lemma.

Lemma 4.2. *A function $f(\mathbf{x})$ in n variables is correlation immune (order 1) if and only if any of the following conditions hold. (Equivalent conditions for correlation immunity of order k, $1 \le k \le n$, hold, but are omitted for brevity.)*

(a) *If $\Omega_f = \{\mathbf{x} \in \mathbb{V}_n : f(\mathbf{x}) = 1\}$, then for each $1 \le i \le n$, we have $|\{\mathbf{x} \in \Omega_f : x_i = 1\}| = |\Omega_f|/2$.*
(b) *For each $1 \le i \le n$, $f(\mathbf{x}) \oplus x_i$ is a balanced function.*
(c) *For each $1 \le i \le n$, $\Pr(x_i = 1 | f(\mathbf{x}) = 1) = 1/2 = \Pr(x_i = 0 | f(\mathbf{x}) = 1)$.*
(d) *Let f_{0i} and f_{1i} denote the functions in $n-1$ variables obtained from $f(\mathbf{x})$ by setting $x_i = 0$ or 1, respectively. Then for each $i = 1, 2, \ldots, n$, the functions f_{0i} and f_{1i} have the same Hamming weight.*
(e) *All of the Walsh transforms*

$$W(\hat{f})(\mathbf{w}) = \sum_{\mathbf{x} \in \mathbb{V}_n} (-1)^{f(\mathbf{x}) \oplus \mathbf{x} \cdot \mathbf{w}}, \quad wt(\mathbf{w}) = 1,$$

are equal to zero.
(f) *For each $i = 1, 2, \ldots, n$, $\Pr(f(\mathbf{x}) = 1 \mid x_i = 1) = \Pr(f(\mathbf{x}) = 1 \mid x_i = 0) = wt(f)/2^n$.*

Proof. All of the equivalent forms follow easily from the definition of correlation immunity. ■

The extension of Lemma 4.2(e) to higher orders of correlation immunity is important enough to warrant separate treatment. This is done in

Lemma 4.3. *A function $f(\mathbf{x})$ in n variables is correlation immune of order k, $1 \le k \le n$, if and only if all of the Walsh transforms*

$$W(\hat{f})(\mathbf{w}) = \sum_{\mathbf{x} \in \mathbb{V}_n} (-1)^{f(\mathbf{x}) \oplus \mathbf{x} \cdot \mathbf{w}}, \quad 1 \le wt(\mathbf{w}) \le k,$$

are equal to zero.

Proof. The proof is based on the observation that the Walsh transform $W(\hat{f})(\mathbf{w})$ is the crosscorrelation between $f(\mathbf{x})$ and the linear function $\ell_{\mathbf{w}}(\mathbf{x}) = \mathbf{w} \cdot \mathbf{x}$. We let the k-vector $\mathbf{y}$ be defined by

$$\mathbf{y} = (x(i_1), x(i_2), \ldots, x(i_k)),$$

where $x(i_1), x(i_2), \ldots, x(i_k)$ are the variables in $\ell_{\mathbf{w}}(\mathbf{x})$. Then we look at the Walsh transform in k variables of the conditional probability $\Pr(\mathbf{y} \mid \mathbf{z})$, where $\mathbf{z}$ is a possible value of $f(\mathbf{x})$. We see by the definition of the expectation that

$$\sum_{\mathrm{y}} \Pr(\mathrm{y}|\mathrm{z})(-1)^{\mathbf{w}\cdot\mathbf{x}} = E[(-1)^{\mathbf{w}\cdot\mathbf{x}}|f(\mathbf{x})=\mathrm{z}] = E[(-1)^{\mathbf{w}\cdot\mathbf{x}}] = \sum_{\mathrm{y}} \Pr(\mathrm{y})(-1)^{\mathbf{w}\cdot\mathbf{x}};$$

the two summations are equal by our correlation immunity hypothesis. Thus Pr(y | z) and Pr(y) are identical since their Walsh transforms are identical (any function can be recovered by inverting its Walsh transform — see Chapter 2, Definition 2.5). This means that the crosscorrelation between $f(\mathbf{x})$ and $\ell_{\mathbf{w}}(\mathbf{x})$ is zero, which gives the lemma. ■

The above short proof of Lemma 4.3 is due to Brynielsson [50]. This work is apparently unpublished, but the proof is reported by Rueppel [389] in Simmons [427, p. 94]. The original proof of Lemma 4.3 was by Xiao and Massey [467]; they refer to Lemma 4.3 as a 'spectral characterization' of correlation immune functions, because the lemma gives a restriction on the Walsh spectrum (see Chapter 2).

Another proof of Lemma 4.3, based on combinatorics and linear algebra rather than probability theory, has been given by Sarkar [394].

We see from the proof of Lemma 4.3 that the functions $f(\mathbf{x})$ and $\ell_{\mathbf{w}}(\mathbf{x}) = \mathbf{w}\cdot\mathbf{x}$ are statistically independent if and only if the Walsh transform $W(\hat{f})(\mathbf{w}) = 0$. It follows from Definition 2.7 of the correlation value $c(f(\mathbf{x}),\ell_{\mathbf{w}}(\mathbf{x}))$ that

$$\begin{aligned} c(f(\mathbf{x}),\ell_{\mathbf{w}}(\mathbf{x})) &= 1 - 2\Pr(f(\mathbf{x}) \neq c(f(\mathbf{x}),\ell_{\mathbf{w}}(\mathbf{x}))) \\ &= \Pr(f(\mathbf{x}) = c(f(\mathbf{x}),\ell_{\mathbf{w}}(\mathbf{x}))) - \Pr(f(\mathbf{x}) \neq c(f(\mathbf{x}),\ell_{\mathbf{w}}(\mathbf{x}))) \\ &= 2^{-n}\sum_{\mathbf{x}\in\mathbb{V}_n}(-1)^{f(\mathbf{x})\oplus\mathbf{w}\cdot\mathbf{x}} = 2^{-n}W(\hat{f})(\mathbf{w}). \end{aligned} \tag{4.2}$$

Thus Lemma 4.3 says that achieving correlation immunity for $f(\mathbf{x})$ is the same as getting zero correlation of $f(\mathbf{x})$ with certain linear functions $\ell_{\mathbf{w}}(\mathbf{x})$.

It is impossible to guarantee that $f(\mathbf{x})$ will not have a nonzero correlation with any linear function, that is, we cannot achieve $c(f(\mathbf{x}),\ell_{\mathbf{w}}(\mathbf{x}))=0$ for every vector $\mathbf{w}$, because of the following lemma, which was apparently first proved by Meier and Staffelbach [318, p. 559].

Lemma 4.4. *For any Boolean function $f(\mathbf{x})$, the total square correlation of $f(\mathbf{x})$ with the set of all linear functions is equal to one, that is,*

$$\sum_{\mathbf{w}\in\mathbb{V}_n} c(f(\mathbf{x}),\mathbf{w}\cdot\mathbf{x})^2 = 1.$$

Proof. By (4.2) we have

$$\sum_{\mathbf{w}\in\mathbb{V}_n} c(f(\mathbf{x}),\mathbf{w}\cdot\mathbf{x})^2 = 2^{-2n}\sum_{\mathbf{w}\in\mathbb{V}_n} W(\hat{f})(\mathbf{w})^2,$$

and now the lemma follows from Parseval's equation (2.16). ■

Lemma 4.4 gives an essential limitation on the amount of correlation immunity that a Boolean function can have. If we manage to secure zero correlation of $f(\mathbf{x})$ with various linear functions, then necessarily we will have nonzero correlations with some other linear functions. This is connected with the tradeoff between the order of correlation immunity and the degree of $f(\mathbf{x})$, discussed in Section 4.6. For example, if $f(\mathbf{x})$ in n variables has maximum order of correlation immunity $n-1$, then $f(\mathbf{x})$ must be $x_1 \oplus \cdots \oplus x_n$ or $x_1 \oplus \cdots \oplus x_n \oplus 1$, so $f(\mathbf{x})$ has the maximum possible correlation to the sum of all the variables.

Because of Lemma 4.4 and (4.2), it is natural to seek those Boolean functions $f(\mathbf{x})$ such that the largest possible value of $|W(\hat{f})(\mathbf{w})|$ is as small as possible. Meier and Staffelbach [318] called these functions perfect nonlinear (see Section 5.4 for a detailed discussion of the motivation for this terminology). They also observed that these functions are exactly the same as the already known bent functions [385], to which Chapter 5 is devoted.

4.3 LFSRs AND CORRELATION IMMUNITY

Cryptographers care about correlation immunity because its absence in Boolean functions which are used in a cryptosystem can allow effective attacks on the system. Siegenthaler [423] gave a method for attacking a stream cipher whose keystream is generated by using two or more linear feedback shift registers (LFSRs for short; see Section 2.11 and the book of Golomb [211] for a readable account of the basic theory of LFSRs) whose outputs are combined by a nonlinear Boolean function. The output of the nonlinear function is a bitstream which serves as the keystream for the stream cipher. This type of stream cipher is called a nonlinear combination generator. This topic is discussed in more detail in Section 6.3.

We illustrate the method of [423] by explaining a simplified correlation attack on a stream cipher in which the output sequences $\{x_i(n) : n = 1,2,\ldots\}$, $i = 1,2,3$, from three LFSRs are combined by using the nonlinear function $f(x_1,x_2,x_3)$ given in Example 4.5.

Example 4.5 (*The function* $f(\mathbf{x}) = x_1x_2 \oplus x_1x_3 \oplus x_2x_3$).

Input	000	001	010	011	100	101	110	111
Output	0	0	0	1	0	1	1	1

The function $f(\mathbf{x})$ is not correlation immune (using Lemma 4.2(c)) since

$$\Pr(x_i = 1 \mid f(\mathbf{x}) = 1) = 3/4 \quad \text{for each } i = 1,2,3. \tag{4.3}$$

This major failure of correlation immunity allows the known plaintext correlation attack which we now describe. Of course, we also have

$$\Pr(x_i = 0 | f(x) = 0) = 3/4 \quad \text{for each } i = 1,2,3 \tag{4.4}$$

and we use both (4.3) and (4.4) in our attack.

The keystream for the stream cipher is the sequence $z_i = f(x_1(i), x_2(i), x_3(i))$, $i = 1, 2, \ldots$. For simplicity, we assume that the attacker knows the length $r(i)$ and the position of the feedback connections (or 'taps') for each of the three LFSRs. Thus the secret key for the keystream is just the three initial inputs ($r(1) + r(2) + r(3) = R$, say, bits). We also assume that the periodic output of each LFSR has maximum period $2^{r(i)} - 1$; if this is not true, then the attack is made easier. The goal is to use a sizeable quantity of known plaintext to determine the R input bits with a computation of size much less than 2^R (the size of the keyspace made up of all possible inputs to the three LFSRs).

Since we have a large amount of known plaintext, we can derive the corresponding amount of keystream. We xor a sufficiently long block of keystream (say 500 bits) with the $2^{r(1)} - 1$ blocks of the same length obtained from the possible phase shifts of the period of the first LFSR. Because of (4.3) and (4.4), when the phase shift block corresponds to the actual initial input to the LFSR, we expect the xor to contain about 75% ones, in contrast to the 50% we would expect if the keystream bits were uncorrelated with the LFSR bits. Thus we recover the initial input of the first LFSR with high probability. We can confirm our choice of the initial input by xor with other sufficiently long keystream blocks. Now we repeat this process with the other two LFSRs and we recover all of the R secret bits which define the keystream. The computational effort for this attack is clearly $O\left(\sum_{i=1}^{3} 2^{r(i)}\right)$, which is a large improvement on the effort $O(2^R)$ needed to search the keyspace.

Plainly this attack can be extended to the case in which we do not know the positions of the feedback connections for the LFSRs. We simply guess the positions of the connections and test our guesses with the above attack. Now the computational effort for the attack becomes $\sum_{i=1}^{3} t_i 2^{r(i)}$, where t_i is the number of LFSRs of length $r(i)$ which have the maximum period. It is well known that the number of LFSRs of length r with the maximum period is $\phi(2^r - 1)/r$ (see, for example, [211, Chapter III, part 5]), where ϕ is Euler's function.

The attack can also be extended to the case where only a large amount of ciphertext is available, and to the case where the inputs $x_i(n)$ are generated by independent and uniformly distributed random variables, instead of by LFSRs. These generalizations are given in [423]. Note that the main reason for the usefulness of correlation attacks in this setting is the fact that the attack on a keystream generated by applying a nonlinear combining function to multiple inputs can be broken down into subattacks, one on each of the inputs. This is an example of a 'divide and conquer' attack. The correlation attack may succeed even if one or more of the inputs is immune to correlation attack; this situation arises in [423] in the discussion of an attack on a generator proposed by Geffe [183]. The Geffe generator combines three LFSRs exactly as above, but the nonlinear function in Example 4.5 is replaced by $f(\mathbf{x}) = x_1x_2 \oplus x_1x_3 \oplus x_2$. Now Equations (4.3) and (4.4) only hold for $i = 2$ and 3, but these failures of correlation immunity are still enough to allow the generator to be successfully attacked.

Siegenthaler [424] used these ideas to describe a correlation attack on stream ciphers whose keystream is produced by using some of the stages of an LFSR of length L as inputs to a nonlinear function $f(x_1, \ldots, x_n)$. Thus the keystream would be the sequence

$z_i = f(x_1(i), \ldots, x_n(i)), i = 1,2,\ldots$, where $x_j(i)$ is the contents of stage $s(j)$ of the shift register after i shifts from the initial state and $1 \leq s(1) \leq s(2) \leq \cdots \leq s(n) \leq L$. This setup is called a *nonlinear filter generator* (see Section 6.4).

The abbreviation 'ML' in the title of [424] stands for 'maximum length', meaning that the period of the LFSR of length L is as large as possible, namely $2^L - 1$. We use the standard terminology *m-sequence* for such a sequence. As in [423], the work in [424] requires an exhaustive search over all $2^L - 1$ phases of the LFSR in order to find the largest correlation. This means that the correlation attack will not be feasible, even for relatively modest sizes of the parameter L, because too much computation will be required.

This disadvantage was overcome by Meier and Staffelbach [317], who gave two new correlation attacks where exhaustive search of the phases of the LFSR is not required, provided that the number of feedback connections is small, say less than 10. In 1989, Meier and Staffelbach estimated that the attack of [424] would be applicable only for L less than about 50, whereas the attacks of [317] would work for L up to about 1000 (given a small number of feedback connections). Meier and Staffelbach [316] gave a preliminary account of the results of [317] in 1988. Independently, also in 1988, a correlation attack based on ideas similar to those in [317] was given by Zeng and Huang [475]. An improved version of this attack method was given by Zeng et al. [477]. All attacks similar to these have come to be known by the name *fast correlation attack*.

In order to illustrate the basic ideas in a fast correlation attack, we give an account of the attack devised by Chepyzhov and Smeets [110]. This attack is described as a 'minor improvement' on [317] in [109], but we think the authors there are unduly modest. The attack of [110] shows the two basic features contained in all fast correlation attacks:

1. an iterative error correction algorithm;
2. a method for finding low weight parity check polynomials.

The problem analyzed in [110] is illustrated in Figure 4.1. The box labeled LFSR in the figure is a linear feedback shift register with a primitive feedback polynomial $g(x)$ and output sequence $\{a_i\}$. The box labeled BSC in the figure is a binary symmetric channel which leaves a bit from the sequence $\{a_i\}$ unchanged with probability $1 - p$, and adds $1 \pmod 2$ to that bit with probability p. The resulting keystream sequence is $\{z_i\}$. We assume $p < 0.5$, so there is some correlation between the bits a_i and z_i.

There is no loss of generality here, since the plaintext input to any stream cipher can be regarded as the output of a BSC which produces a nonrandom bitstring with $p \neq 0.5$, and then by symmetry we may take $p < 0.5$.

In Figure 4.1 the sequence $\{a_i\}$ is the output of the LFSR, which is sent through a BSC which adds noise to it. The resulting keystream sequence $\{z_i\}$ is observed by the cryptanalyst, whose goal is to reconstruct the initial state of the LFSR from a sufficiently long observed string of bits from $\{z_i\}$, say N bits. If we define $r = \deg g(x)$ then we must have $r < N$ and we want N as small as possible. We assume N is fixed and will determine later how small we can choose N to be.

We define a parity check polynomial (as in [317]) for the LFSR to be any polynomial $h(x)$ with degree $\geq r$ such that $h(0) = 1$ and for any sequence $\{a_i\}$ (whose formal power series $\sum_{i=0}^{\infty} a_i x^i$ we denote by $a(x)$) generated by the LFSR we have that $h(x)a(x)$ is

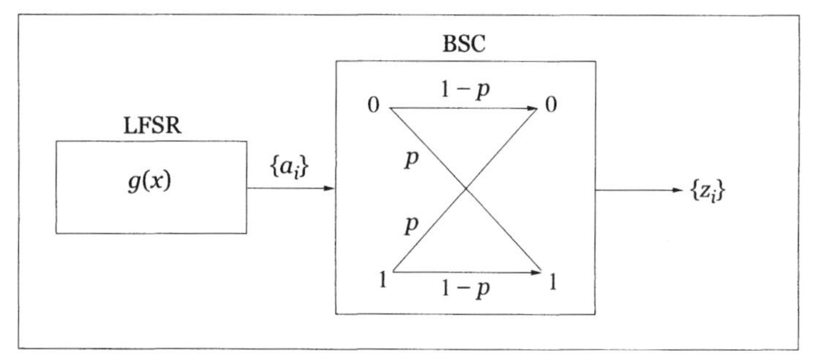

FIGURE 4.1

The correlation attack model for recovery of the initial state of an LFSR.

a polynomial of degree $<\deg h(x)$. For brevity we will simply refer to a parity check polynomial $h(x)$ as a check for the LFSR. From LFSR theory, we know that if the polynomial $h(x)$ is a check, then $g(x)$ divides $h(x)$.

To simplify the exposition of the attack, we assume that we have obtained a set $\{h^{(i)}(x) : 1 \leq i \leq M\}$ of M checks, and we define the coefficients of the $h^{(i)}(x)$ by

$$h^{(i)}(x) = \sum_{k=0}^{\deg h^{(i)}} h_k^{(i)} x^k.$$

This set is called a *set of M independent checks of length L with at most t taps* if the following three conditions hold:

$$\text{for all } i = 1, 2, \ldots, M, \quad h_0^{(i)} = 1 \text{ and } \deg h^{(i)} < L, \tag{4.5}$$

$$\text{for all } h^{(i)}(x), |\{h_k^{(i)} : h_k^{(i)} \neq 0, k \geq 1\}| \leq t, \tag{4.6}$$

$$\text{for all } u, v \text{ with } u \neq v, \{k > 0 : h_k^{(u)} = 1\} \cap \{k > 0 : h_k^{(v)} = 1\} = \emptyset. \tag{4.7}$$

In other words, we have that the Hamming weight of the polynomials in the set of checks is not more than $t+1$ and two different checks have different nonzero coefficients except for their constant term. We give an example for $g(x) = 1 + x$, $L = 8$, $t = 3$.

Example 4.6 (*Three checks for* $g(x) = 1 + x$ *with length* 8 *and* ≤ 3 *taps*).

$$h^{(1)}(x) = 1 + x + x^5 + x^7$$
$$h^{(2)}(x) = 1 + x^2 + x^3 + x^4$$
$$h^{(3)}(x) = 1 + x^6.$$

Now we describe the iterative error correction algorithm which uses the checks. Given a bit a_i with $i > L$ from the LFSR output, for any check $h(x) = \sum_{k=0}^{L-1} h_k x^k$ we have

$$a_i = \sum_{k \geq 1, h_k = 1} a_{i-k} = a_{i(1)} + a_{i(2)} + \cdots + a_{i(t)},$$

where the $i(j)$ denote the values of the indices $i - k$ for which $h_k = 1$. If we have a set $\{h^{(j)}(x) : 1 \leq j \leq M\}$ of M independent checks, then we have for this bit a_i

$$\begin{array}{l} a_i + a_{i(1,1)} + a_{i(1,2)} + \cdots + a_{i(1,t)} = 0 \\ a_i + a_{i(2,1)} + a_{i(2,2)} + \cdots + a_{i(2,t)} = 0 \\ \cdots \quad \cdots \quad \cdots \quad \cdots \quad \cdots \\ a_i + a_{i(M,1)} + a_{i(M,2)} + \cdots + a_{i(M,t)} = 0, \end{array} \tag{4.8}$$

where for fixed j the subscripts $i(j,k)$ denote the values of the indices $i-k$ when $h_k^{(j)} = 1$.

By (4.7), all of the indices $i(j,k)$ are different. Clearly the Equations (4.8) hold for any bit a_i with $i > L$. If we replace a_i by z_i and $a_{i(j,k)}$ by $z_{i(j,k)}$ then some of the Equations (4.8) may be false because a_m is not necessarily equal to the corresponding z_m.

Using the check Equations (4.8) and the ideas of Gallager [181] on low-density parity check codes, it is possible to compute a 'corrected' value $z_{(i,1)}$ for each z_i which will satisfy $\Pr(z_{(i,1)} = a_i) > 1 - p$.

This gives us a new sequence $\{z_{(i,1)} : 0 \leq i \leq N-1\}$ in place of the originally observed keystream bits z_i. Repeating the procedure on this new sequence we obtain another N-bit set $\{z_{(i,2)}\}$ and as we iterate the procedure over and over we obtain new N-bit sets such that the number of disagreements between $\{z_{(i,s)}\}$ and $\{z_i\}$ tends to zero as s increases. To prove this it is clearly enough to show that the first iteration must lead to a smaller number of expected disagreements. So we must analyze this first iteration.

We think of the 'noisy' output bits z_i as random variables, which we denote by Z_i. We also define the random variable

$$Z_i^* = \sum_{j=1}^{t} h_j Z_{i-j}.$$

When no errors occur, then $Z_i = Z_i^* = a_i$. Therefore we also need the random variable

$$Y_i = Z_i + Z_i^*,$$

which satisfies $Y_i = 0$ when $Z_i = a_i$ for all i. We let $q = 1 - p$ as is customary, so

$$\Pr(Z_i = a_i) = q > 0.5$$

by our assumption $p < 0.5$. Thus we obtain

$$\begin{aligned}
\Pr(Z_i^* = a_i) &= \Pr(\text{number of bit changes is even}) \\
&= \sum_{i=0, i \text{ even}}^{t} \binom{t}{i} p^i q^{t-i} = 1 - \sum_{i=0, i \text{ odd}}^{t} \binom{t}{i} p^i q^{t-i} \\
&= \frac{1}{2}\left(\sum_{i=0}^{t} \binom{t}{i} p^i q^{t-i} + \sum_{i=0}^{t} \binom{t}{i} (-p)^i q^{t-i} \right) \\
&= \frac{1 + (2\epsilon)^t}{2} > 0.5,
\end{aligned}$$

where

$$\epsilon = \frac{1 - 2p}{2}. \tag{4.9}$$

Thus if we define Q by $Q = \Pr(Z_i^* = a_i)$, we have

$$Q = \Pr(Z_i^* = a_i) = \frac{1 + (2\epsilon)^t}{2}. \tag{4.10}$$

We define D (a random variable) to be the number of nonzeros in $\{Y_i : 0 \leq i \leq N - 1\}$. In the first error correction step, we determine the sequence $\{z_{(i,1)}\}$ defined by

$$\begin{aligned}
z_{(i,1)} &= z_i \quad \text{when } D < T \\
z_{(i,1)} &= z_i + 1 \quad \text{when } D \geq T,
\end{aligned} \tag{4.11}$$

where T is some integer in the range $[0, M]$ to be chosen later. We now have

$$\begin{aligned}
\Pr(z_{(i,1)} = a_i, D < T) &= \Pr(z_{(i,1)} = z_i = a_i, D < T) \\
&= q \sum_{d=0}^{T-1} \binom{M}{d} (1 - Q)^d Q^{M-d}
\end{aligned}$$

and similarly

$$\Pr\left(z_{(i,1)} = a_i, D \geq T\right) = p \sum_{d=T}^{M} \binom{M}{d} Q^d (1 - Q)^{M-d}.$$

Adding these two probabilities gives

$$\Pr(z_{(i,1)} = a_i) = q + \sum_{d=T}^{M} \binom{M}{d} (Q(1 - Q))^{M-d} \left(pQ^{2d-M} - q(1 - Q)^{2d-M} \right). \tag{4.12}$$

Let $Q(1) = \Pr(z_{(i,1)} = a_i)$; we get on average more correct symbols after the first error correction step if

$$Q(1) = \Pr(z_{(i,1)} = a_i) \geq \Pr(z_i = a_i) = q.$$

From (4.12) it follows that if

$$pQ^{2T-M} - q(1-Q)^{2T-M} > 0,$$

then for all $i \geq T$

$$pQ^{2i-M} - q(1-Q)^{2i-M} > 0.$$

Thus we have proved the following lemma.

Lemma 4.7. *Let* $Q = \Pr(Z_i^* = a_i)$ *and* $Q(1) = \Pr(z_{(i,1)} = a_i)$, *where* $z_{(i,1)}$ *is as in* (4.11) *and* T *in* (4.11) *is defined by*

$$T = \max\{d \in [0, M] : (Q/(1-Q))^{M-2d} > q/(1-q)\}.$$

Then

$$Q(1) > Q \quad \textit{if and only if } (Q/(1-Q))^M > q/(1-q). \tag{4.13}$$

Now we can find the minimum necessary number of parity check polynomials.

Theorem 4.8. *For the convergence of the iterative error correction algorithm to the correct sequence* $\{a_i\}$, *we need at least*

$$M = \lfloor \log(q/(1-q))/\log(Q/(1-Q)) \rfloor + 1$$

independent checks, where $\lfloor\ \rfloor$ *is the greatest integer function. If* $q = 0.5 + \epsilon$ *(so* ϵ *is given by* (4.9)*) and* $Q = (1 + (2\epsilon)^t)/2$, *then we need* $O((2\epsilon)^{1-t})$ *checks.*

Proof. The least allowable value of M follows from (4.13). The big-O estimate for this value of M follows at once from the given values of q and Q by using the approximation $1 + x$ for $\log x$. ■

When the conditions for the convergence of the iterative error correction algorithm are satisfied, we know from [181] that the required number of iterations is usually quite small. Since for each of the N observed bits we need to evaluate M check equations like (4.8), the iterative error correction procedure has complexity approximately $O(MN)$. In particular, the smaller the number of taps t, the fewer checks we need.

We are left with the problem of finding the M low weight parity check polynomials which are needed for the error correction algorithm. This turns out to be a problem which can best be analyzed with the help of coding theory. To avoid technicalities of coding theory, we refer the reader to the exposition in [110, pp. 181–184].

Much further work has been done on fast correlation attacks since their introduction in 1988. We do not give any details of this ongoing research here, but simply list some relevant papers: [173,330–332,10,115,192,361,205,240,241,109,61,204,242,328,329, 200]. A good survey of the literature as of 1993 is given by Golić [189].

4.4 COUNTING CORRELATION IMMUNE FUNCTIONS

It is of interest to count the number of Boolean functions in n variables that are correlation immune of order k, $1 \le k \le n$; we let $CI(n,k)$ denote this number. In contrast to the situation for $SAC(k)$ functions (see Chapter 3), here we obtain an asymptotic formula as $n \to \infty$ for every fixed value of k. The formula was proved by Denisov [148] (but with a mistake not corrected until his paper [149] years later; the mistake was omitting the factor A_k) and is given in:

Theorem 4.9. *As $n \to \infty$, an asymptotic formula for $CI(n,k)$ is*

$$CI(n,k) \asymp \frac{\exp_2(2^n - k)}{A_k e^{(((n-k)\binom{n}{k}-n)/2)\log 2+\sum_{i=1}^{k}\log(\pi/2)^{1/2}\binom{n}{i})}},$$

where A_k is defined by $A_1 = 1, A_k = \left(1+\sum_{i=2}^{k}(i-1)^2\binom{n}{i}\right)^{1/2}$ for $k \ge 2$ and where k is any fixed integer, $1 \le k \le n$.

Proof. The proof is too complicated to give here, though the corrected proof in [149] is significantly simpler than the one in [148]. The key idea is the use of characteristic functions of random vectors as defined in probability theory. ■

Unfortunately the results in [148] were not widely known when published and this led to various estimates for $CI(n,k)$ in the later literature which are much weaker than Theorem 4.9 when k is fixed and n is large, for example in Yang and Guo [468, Section 2], Schneider [402, Theorem 20] (we remark that [402, Theorem 2] was already proved in [180] and that a more extensive version of this paper by Schneider appears in [403]) and Park et al. [355]. The bounds in these papers are of some interest in the case where n is not large, but in that case better bounds have been obtained by Carlet and Klapper [92] (the result will be mentioned in Section 5.14) and by Carlet and Gouget [88].

The second Denisov paper [149] also contains a result like Theorem 4.9 for the case of balanced correlation immune functions. We state this next, but again the proof is too complicated to give here. We note that in the case where n is not large, the best bounds on $BCI(n,k)$ are given in [92] and [88] (see Section 5.14).

Theorem 4.10. *Let $BCI(n,k)$ denote the number of Boolean functions in n variables which are balanced and correlation immune of order k, $1 \le k \le n$. As $n \to \infty$ an asymptotic formula for $BCI(n,k)$ is*

$$BCI(n,k) \asymp \frac{\exp_2(2^n)}{e^{(\binom{n}{k}\frac{n-k}{2})\log 2+\sum_{i=0}^{k}\log(\pi/2)^{1/2}\binom{n}{i})}},$$

where k is any fixed integer.

We remark that the proofs in [149] give Theorems 4.9 and 4.10 with $k^2 = o(n)$ instead of the hypothesis that k is fixed.

In theory, it is possible to give an exact formula for $CI(n,1)$ if we can count certain kinds of orthogonal arrays (see the definition in Section 4.7), because of the following theorem.

Theorem 4.11. *Let $A(2j,n)$ be the number of orthogonal arrays $OA(2j,n,2,1)$. Then*

$$CI(n,1) = \sum_{j=0}^{2^n-1} \frac{A(2j,n)}{(2j)!}.$$

Proof. This follows from Lemma 4.13 below and the interpretation of correlation immunity of order 1 in terms of orthogonal arrays, as explained in Section 4.7. ■

A complicated explicit formula for $A(2j,n)$ is given in [479]; from this, various values of $CI(n,1)$ can be calculated for small n, but it does not seem possible to deduce a reasonably simple explicit formula for $CI(n,1)$.

4.5 RESILIENT FUNCTIONS

In many papers in the literature, a function which is balanced and correlation immune of order k is said to be a *k-resilient function*. We will occasionally use this terminology, which was first introduced by Chor et al. [111] in connection with the *bit extraction problem* (first named and defined by Vazirani [454, p. 376]). Since this problem is relevant to some cryptography problems, we discuss it here.

The bit extraction problem can be thought of as a three-move game between a *user* and an *adversary*. Three integer parameters m, n and t are involved in the game, which is defined as follows: first the user picks a function $f : \mathbb{V}_n \to \mathbb{V}_m$. Next, the adversary picks t locations in the n-bit input string for f and fixes the bit values at these locations. The user knows neither the locations nor the values of the fixed bits chosen by the adversary. The remaining $n-t$ input bits are determined by independent tosses of a fair coin. Finally, the user applies the function f to the entire string. The goal of the user is to cause the output of the function to be uniformly distributed in $\mathbb{V}_m$, while the goal of the adversary is to prevent this. The question is which of the two players has a winning strategy. We say that the user extracts m (random) bits if the user has a winning strategy in the game.

Plainly, the user has a winning strategy in the following two extreme cases:

(a) $m = 1$ and $t \le n-1$ (the user defines f as the xor of all n input bits)

(b) $t = 1$ and $m \le n-1$ (the user defines $f(x_1,\ldots,x_n) = (y_1,\ldots,y_m)$ by putting

$$y_i = x_i \oplus x_{i+1} \text{ or } y_i = x_i \oplus x_n, \text{ for each } i = 1,2,\ldots,m).$$

In both (a) and (b) we have $m \leq n-t$. On the other hand, the adversary has a winning strategy if $m > n-t$, since the $n-t$ random input bits cannot produce more than $n-t$ random output bits.

Now we generalize the concept of k-resilient functions to functions $f : \mathbb{V}_n \to \mathbb{V}_m$. We say such an f is an *(n,m,t)-resilient function* if every possible output m-tuple of f is equally likely to occur when the values of t arbitrary input bits are fixed by an adversary and the remaining $n-t$ bits are chosen independently at random. Thus an $(n,1,t)$-resilient function is simply a t-resilient Boolean function. Clearly, in the bit extraction problem as defined above, the user has a winning strategy if he can find an (n,m,t)-resilient function.

Thus a solution to the bit extraction problem enables two parties Ann and Bob to agree on a secret random string of length m (which could then, for example, be used as a session key in a cryptographic protocol) if they start with a random string of length n, which has had t bits compromised by being fixed by an adversary. Ann and Bob know the number but not the location of the compromised bits. From this cryptographic point of view, the important question to be answered is, given n and t in the bit extraction problem, what is the largest m for which an (n,m,t)-resilient function exists? This question is analyzed in [111]. This question is also a special case of the privacy amplification problem considered by Bennett et al. [24]. That paper and the preliminary conference version [23] introduce a definition equivalent to that of (n,m,t)-resilient functions and independently prove some of the results of [111].

An interesting interpretation of the bit extraction problem in terms of colorings of the vertices of the n-dimensional cube is given by Friedman [180].

4.6 TRADEOFF BETWEEN CORRELATION IMMUNITY AND DEGREE

Here we look at another illustration of the general principle that a Boolean function cannot simultaneously have too many cryptographically desirable properties. In this instance we examine the tradeoff between high order correlation immunity and high degree for a Boolean function. The first result in this direction goes back to Siegenthaler [422, Theorem 1], who proved Theorem 4.12 below. This result was extended by Xiao and Massey [467], who proved Theorem 4.14 below. We give the much simpler proofs of both theorems which were found by Sarkar [394]. A different simple proof for the second assertion in Theorem 4.12 was given by Zhang and Zheng [480, p. 1746].

Theorem 4.12. *If f is a Boolean function in n variables, which is correlation immune of order k, then the degree of f is at most $n-k$. If also f is balanced and $k < n-1$, then the degree of f is at most $n-k-1$.*

Proof. A (generalized) truth table for $f(x_1,\ldots,x_n)$ is an array with 2^n rows and $n+1$ columns, where each of the first n columns has values for one of the variables x_i and the first n entries of the 2^n rows are the coordinates of all the n-vectors in the lexicographic order $0,\ldots,0;0,\ldots,0,1;\ldots$ and finally $1,\ldots,1$. The last column gives the output values $f(x_1,\ldots,x_n)$. Since f is correlation immune of order k, if we choose any k variables and make them the leftmost ones in the truth table, then, by the generalization to order k of Lemma 4.2(d), the last column of the truth table is the concatenation of 2^k strings, each of length 2^{n-k} and of equal Hamming weight. We now suppose that the degree of f is $n-i$ for some $i<k$ and deduce a contradiction. We shall identify a Boolean function with the output column of any truth table for the function.

Since f is correlation immune of order k, it is also correlation immune of orders i and $i+1$. By our assumption the algebraic normal form of f has at least one term T of degree $n-i$ and no terms of greater degree. Let $y_1,\ldots,y_i$ be the variables not appearing in T and let y be any other variable among the x_i. We arrange the truth table for f so that the variables $y_1,\ldots,y_i,y$ are the leftmost ones. This gives a division of the output column into 2^{i+1} strings $s(0),s(1),\ldots,s(2^{i+1}-1)$, all of whose weights are equal. Define the strings $g(j)=s(2j)s(2j+1)$ for $0\leq j\leq 2^i-1$. Then

$$wt(g(j))=wt(s(2j))+wt(s(2j+1))=2wt(s(2j))$$

and so $wt(g(j))$ must be even. Thus the string $g(0)$ of even weight corresponds to the function of $n-i$ variables obtained from f by setting the variables $y_1,\ldots,y_i$ equal to 0. The term T in f does not contain any of the variables $y_1,\ldots,y_i$ and so must be in the algebraic normal form of $g(0)$. Thus $g(0)$ represents a function of $n-i$ variables with degree $n-i$ and so this function must have odd weight, which is a contradiction.

Now suppose f is balanced and has degree $n-k$ for $k<n-1$. Let T be a term of degree $n-k$ and let $y_1,\ldots,y_k$ be the variables not in T. When these variables are made the leftmost ones in the truth table for f, then the output column can be divided into 2^k strings $s(0),s(1),\ldots,s(2^k-1)$ of equal weight. Each of these 2^k functions has $n-k$ variables and contains the term T, so each function has degree $n-k$ and so has odd weight, say w. Therefore $wt(f)=2^k w$, with w odd. Since f is balanced, this gives $w=2^{n-k-1}$, which is even for $k<n-1$. This is a contradiction. ■

Lemma 4.13. *If f is a Boolean function in n variables which is correlation immune of order k, then 2^k divides $wt(f)$.*

Proof. This follows from the remark concerning Lemma 4.2 (*d*) in the proof of Theorem 4.12. ■

Theorem 4.14. *If f is a Boolean function in n variables which is correlation immune of order k, then the algebraic normal form of f either has no terms of degree $n-k$ or has all possible terms of degree $n-k$.*

Proof. We shall identify a Boolean function with the output column of any truth table for the function. We show that if f has one term of degree $n-k$, then it has all possible terms of degree $n-k$. For this it suffices to show that if f has a term $y_1\cdots y_{n-k}$ and y is any other variable, then f also has the term $y_1\cdots y_{n-k-1}y$. So we suppose without loss of generality that f has $x_1\cdots x_{n-k}$ and we prove that then f also has $x_1\cdots x_{n-k-1}x_{n-k+1}$. Arrange the truth table for f so that $x_n,x_{n-1},\ldots,x_1$ is the order of the variables from left to right. Now consider the output column to be divided into 2^{k+1} strings $s(0),s(1),\ldots,s(2^{k+1}-1)$. Each of the $s(i)$ is a function of $n-k-1$ variables.

Let $g(j) = s(2j)s(2j+1)$ for $0 \leq j \leq 2^k - 1$. Each $g(j)$ is a function of the $n-k$ rightmost variables $x_{n-k}, \ldots, x_1$ and is obtained from f by setting the leftmost variables $x_n, \ldots, x_{n-k+1}$ equal to one of the 2^k possible choices. Thus each $g(j)$ has the term $x_{n-k} \cdots x_1$ and so $wt(g(j))$ is odd. Because f is correlation immune of order k, all of the weights $wt(g(j))$ are the same, by the generalization to order k of Lemma 4.2(d). Let w be this common weight; then

$$\begin{aligned} w = wt(g(j)) &= wt(s(2j)) + wt(s(2j+1)) \\ &= wt(s(2j+2)) + wt(s(2j+3)) = wt(g(j+1)). \end{aligned} \tag{4.14}$$

Now consider the truth table for f with variables in the order $x_n, x_{n-1}, \ldots, x_{n-k+2}, x_{n-k}, x_{n-k+1}, x_{n-k-1}, \ldots, x_1$ from left to right. Now the output column for f is the concatenation of strings $s(0)s(2)s(1)s(3) \cdots s(4i)s(4i+2)s(4i+1)s(4i+3) \cdots s(2^k - 4)s(2^k-2)s(2^k-3)s(2^k-1)$. Define $h(2j) = s(4j)s(4j+2)$ and $h(2j+1) = s(4j+1)s(4j+3)$ for $0 \leq j \leq 2^k - 1$. Each $h(j)$ is a function of the $n-k$ rightmost variables $x_{n-k+1}, x_{n-k-1}, \ldots, x_1$. We show that the weight of each $h(i)$ is odd and hence all of them have the term $x_{n-k+1}x_{n-k-1} \cdots x_1$. This ensures that f has the same term, which is our goal.

To prove $wt(h(i))$ is odd, first note that since f is correlation immune of order k, all of the values $wt(h(j))$ are equal. Thus $wt(h(j)) = wt(h(j+1))$, which implies

$$wt(s(2j)) + wt(s(2j+2)) = wt(s(2j+1)) + wt(s(2j+3)). \tag{4.15}$$

By (4.14) and (4.15) we get

$$wt(s(2j)) = wt(s(2j+3)) \quad \text{and} \quad wt(s(2j+1)) = wt(s(2j+2)).$$

Now since $wt(s(2j)) + wt(s(2j+1)) = w$ is odd, we have that exactly one of $wt(s(2j))$ and $wt(s(2j+2))$ is odd and exactly one of $wt(s(2j+1))$ and $wt(s(2j+3))$ is odd. Hence, $wt(h(2j))$ and $wt(h(2j+1))$ are both odd. ■

4.7 CONNECTIONS WITH ORTHOGONAL ARRAYS

The notion of orthogonal array goes back to a 1947 paper by Rao [376] dealing with the design of experiments. An m by n array (or matrix) with entries from a set of q elements is called an *orthogonal array* of size m with n constraints, q levels, strength k and index r if any set of k columns of A contains all q^k possible row vectors exactly r times. Thus $m = rq^k$. We denote such an array by $OA(m,n,q,k)$. There is an extensive literature on constructions and applications of orthogonal arrays (see [223]).

The first statement of a connection between resilient functions and orthogonal arrays appears in Chor et al. [111, p. 402]. The connection between correlation immunity and orthogonal arrays is simpler, so we discuss that first. The first exposition of the latter connection was given by Camion et al. [53, Theorem 3.1]; their result is Theorem 4.15.

We define the *support table* of a Boolean function $f(\mathbf{x})$ in n variables to be the array with $wt(f)$ rows and n columns made up of all the vectors $\mathbf{x}$ such that $f(\mathbf{x}) = 1$, that is all vectors in the inverse image $f^{-1}(1)$. The support table is not unique unless we fix an ordering of its rows.

Theorem 4.15. *A Boolean function f in n variables is correlation immune of order k if and only if its support table is an $OA(wt(f), n, 2, k)$.*

Proof. Our proof follows an argument of Stinson [444]. Assume first that f is correlation immune of order k. Let A_1 be the $wt(f)$ by n support table of f and let A_0 be the $2^n - wt(f)$ by n array made up of the vectors in the inverse image $f^{-1}(0)$. Let $\{i(1), \ldots, i(k)\}$ be any set of k column indices and let $(z_1, \ldots, z_k)$ be any binary k-vector. Fix j arbitrarily and let $\lambda(b)$ denote the number of rows in A_b for which z_j occurs in column $i(j)$. Since every possible n-vector is a row in either A_0 or A_1, $\lambda(0) + \lambda(1) = 2^{n-k}$. Therefore

$$\Pr(f(\mathbf{x}) = 1 \mid x_{i(j)} = z_j, 1 \leq j \leq k) = \frac{\lambda(1)}{2^{n-k}} = \frac{wt(f)}{2^n},$$

where the second equality follows because f is correlation immune of order k (see Lemma 4.2(f)). Thus $\lambda(1) = wt(f)/2^k$, which proves that the support table A_1 is an $OA(wt(f), n, 2, k)$. The same argument incidentally shows that A_0 is an $OA(|f^{-1}(0)|, n, 2, k)$.

Now assume we have an $OA(wt(f), n, 2, k) = A$ which is the support table of a Boolean function $f(\mathbf{x})$ in n variables. This means $f(\mathbf{x}) = 1$ if and only if $\mathbf{x}$ is a row of A. It follows easily that the function f is correlation immune of order k. ■

The following example is from [53, p. 89]. Column i corresponds to the variable x_i and we see that any two columns contain all four possible 2-vectors (as rows) exactly twice. Thus $f(\mathbf{x})$ is correlation immune of order 2 and the support table is an $OA(8, 4, 2, 2)$.

Example 4.16 (*Support table of $f(x_1, \ldots, x_4) = x_1 \oplus x_3 \oplus x_4$*).

1 0 0 0
1 1 0 0
0 0 1 0
0 1 1 0
0 0 0 1
0 1 0 1
1 0 1 1
1 1 1 1

4.8 CONSTRUCTING CORRELATION IMMUNE FUNCTIONS

There is a very simple method for constructing correlation immune functions of order 1: simply define the 'first half' $\{f(v_0),\ldots,f(v_{2^{n-1}-1})\}$ of the function f arbitrarily, and then define the 'second half' of the function by taking the bits of the first half in reverse order. Now by Lemma 4.2(a) or (d), f is correlation immune. This is a 'folklore' construction, but it is mentioned by Mitchell [334, p. 163] in the context of counting correlation immune functions. It is easy to generate many balanced correlation immune functions in this way, but it is not easy to control other desirable cryptographic properties such as high nonlinearity or large degree. Thus the more specialized constructions which we discuss below are more useful for cryptographic applications.

In the original paper [422] on correlation immunity, Siegenthaler already gave a recursive construction for correlation immune functions of order k:

Theorem 4.17. *Let* $\mathbf{x} = (x_1,\ldots,x_n)$ *and suppose that* $f_1(\mathbf{x})$ *and* $f_2(\mathbf{x})$ *are correlation immune functions of order k such that* $\Pr(f_1(\mathbf{x})=1)=\Pr(f_2(\mathbf{x})=1)=p$. *Then the function* f *of* $n+1$ *variables defined by*

$$f(\mathbf{x},x_{n+1}) = x_{n+1}f_1(\mathbf{x}) + (x_{n+1}+1)f_2(\mathbf{x}) \tag{4.16}$$

is also correlation immune of order k and satisfies $\Pr(f(\mathbf{x})=1)=p$.

Proof. Let $\mathbf{y}=(x_{i(1)},\ldots,x_{i(k)})$ be made up of an arbitrary choice of k of the variables x_i, and let $\mathbf{y}_0=(y_1,\ldots,y_k)$ be any fixed binary k-vector. Then, since f_1 and f_2 do not depend on x_{n+1} we have for either fixed choice of the bit b and $i=1$ or 2

$$\Pr(f_i=1|\mathbf{y}=\mathbf{y}_0,x_{n+1}=b)=\Pr(f_i=1|\mathbf{y}=\mathbf{y}_0)=\Pr(f_i=1), \tag{4.17}$$

where the second equality follows from our hypothesis that f_i is correlation immune of order k (using the generalization of Lemma 4.2 (f) to order k).

Now (4.16) and (4.17) imply

$$\Pr(f=1|\mathbf{y}=\mathbf{y}_0,x_{n+1}=1)=\Pr(f_1=1)$$

and

$$\Pr(f=1|\mathbf{y}=\mathbf{y}_0,x_{n+1}=0)=\Pr(f_2=1),$$

but the two right-hand-side probabilities are equal to p by our hypotheses, so we obtain

$$\Pr(f=1|\mathbf{y}=\mathbf{y}_0,x_{n+1}=b)=\Pr(f=1)=p.$$

This clearly implies that the value of f is independent of the choice of any subset of k of the $n+1$ input variables, so f is correlation immune of order at least k. ■

From a cryptographic viewpoint, Theorem 4.17 is most interesting when $p=1/2$, so f_1 and f_2 are k-resilient functions. In this case, Camion et al. [53, Theorem 3.2] give a more precise form of Theorem 4.17:

Theorem 4.18. *Let* $\mathbf{x} = (x_1, \ldots, x_n)$ *and suppose that* $f_1(\mathbf{x}), f_2(\mathbf{x})$ *and* $f(\mathbf{x}, x_{n+1})$ *are related by Equation* (4.16). *Then for* $k < n-1$, f *is* $(k+1)$*-resilient if and only if the following two conditions hold:*

(i) f_1 *and* f_2 *are* k*-resilient functions*
(ii) *for all* $\mathbf{v}$ *in* $\mathbb{V}_n$ *with* $wt(\mathbf{v}) = k+1$ *we have the Walsh transform equation*

$$W(f_1)(\mathbf{v}) + W(f_2)(\mathbf{v}) = 0. \tag{4.18}$$

Also, if the degrees of f, f_1 *and* f_2 *are equal (so the degree of* $f_1 + f_2$ *is less than the degree of* f*), then* f *has its maximum degree* $n+1-(k+2)$ *if and only if* f_1 *and* f_2 *have their maximum degree* $n-(k+1)$.

Proof. Let $\mathbf{w} = (\mathbf{v}, d)$ be a vector in $\mathbb{V}_{n+1}$; then

$$\begin{aligned} W(f)(\mathbf{w}) &= \sum_{(\mathbf{x}, x_{n+1}) \in \mathbb{V}_{n+1}} f(\mathbf{w})(-1)^{\mathbf{v}\cdot\mathbf{x} \oplus d x_{n+1}} \\ &= \sum_{\substack{\mathbf{x} \in \mathbb{V}_n \\ x_{n+1}=1}} f_1(\mathbf{x})(-1)^{\mathbf{v}\cdot\mathbf{x} \oplus d} + \sum_{\substack{\mathbf{x} \in \mathbb{V}_n \\ x_{n+1}=0}} f_2(\mathbf{x})(-1)^{\mathbf{v}\cdot\mathbf{x}}, \end{aligned}$$

that is

$$W(f)(\mathbf{w}) = (-1)^d W(f_1)(\mathbf{v}) + W(f_2)(\mathbf{v}). \tag{4.19}$$

Suppose first that f satisfies (i) and (ii). Then from (4.19) and (i) we have

$$W(f)(0) = W(f_1)(0) + W(f_2)(0) = 2^n,$$

so f is balanced. If $\mathbf{w} = (\mathbf{v}, d)$ has $0 < wt(\mathbf{v}) < k+1$, then $W(f)(\mathbf{w}) = 0$ from (4.19) and (i); if $\mathbf{w}$ has $wt(\mathbf{v}) = k+1$ and $d = 0$, then $W(f)(\mathbf{w}) = 0$ from (4.19) and (ii). This implies f is $(k+1)$-resilient.

Now suppose that f is $(k+1)$-resilient. Then for all $\mathbf{w} = (\mathbf{v}, d)$ with $1 \le wt(\mathbf{w}) \le k+1$, (4.19) gives

$$0 = (-1)^d W(f_1)(\mathbf{v}) + W(f_2)(\mathbf{v}). \tag{4.20}$$

For $\mathbf{w} = (0, 1)$, (4.20) gives $W(f_1)(0) = W(f_2)(0)$. Since f is balanced, for $\mathbf{w} = 0$ (4.19) gives $W(f)(0) = wt(f) = 2^n = W(f_1)(0) + W(f_2)(0)$. Therefore f_1 and f_2 are balanced. If $0 < wt(\mathbf{v}) < k+1$, then (4.20) gives $W(f_1)(\mathbf{v}) = W(f_2)(\mathbf{v})$ when $d = 1$ and gives $W(f_1)(\mathbf{v}) + W(f_2)(\mathbf{v}) = 0$ when $d = 0$. Therefore $W(f_1)(\mathbf{v}) = W(f_2)(\mathbf{v}) = 0$, so (i) is true.

Further, if $wt(\mathbf{v}) = k+1$ and $d = 0$, then (4.18) follows from (4.20) and so (ii) is true. The last statement in the theorem follows at once from Theorem 4.12 and (4.16). ∎

Example 4.19. *Functions* f_1, f_2, f *of equal degree in Theorem* 4.18 *with* $n = 4$, $k = 1$*; each function has the largest possible degree consistent with its order of correlation immunity*

$f_1 = x_1x_2 \oplus x_3 \oplus x_4$	1-resilient
$f_2 = x_1x_2 \oplus x_2 \oplus x_3 \oplus x_4$	1-resilient
$f = x_5 f_1 \oplus (x_5 \oplus 1) f_2 = x_1x_2 \oplus x_2x_5 \oplus x_2 \oplus x_3 \oplus x_4$	2-resilient

Camion et al. [53, Proposition 4.2] also gave a nonrecursive method for constructing correlation immune functions of order at least k:

Theorem 4.20. *Given m and n with $1 \leq m < n$, define $\mathbf{x} = (x_1, \ldots, x_n)$ and*

$$r = n - m, \quad \mathbf{u} = (x_1, \ldots, x_m) \quad \text{and} \quad \mathbf{v} = (x_{m+1}, \ldots, x_n).$$

Let $g(\mathbf{u})$ be an arbitrary Boolean function in m variables and let $\phi(\mathbf{u})$ be any function $\phi : \mathbb{V}_m \rightarrow \mathbb{V}_r$ with

$$w = \min\{wt(\phi(\mathbf{u})) : \mathbf{u} \in \mathbb{V}_m\} \geq 1. \tag{4.21}$$

Define a Boolean function $f(\mathbf{x})$ by

$$f(\mathbf{x}) = f(\mathbf{u}, \mathbf{v}) = \mathbf{v} \cdot \phi(\mathbf{u}) + g(\mathbf{u}).$$

Then $f(\mathbf{x})$ is balanced and correlation immune of order k with $k \geq w - 1$.

Proof. We have

$$\sum_{\mathbf{x} \in \mathbb{V}_n} (-1)^{f(\mathbf{x})} = \sum_{\mathbf{u} \in \mathbb{V}_m} (-1)^{g(\mathbf{u})} \sum_{\mathbf{v} \in \mathbb{V}_r} (-1)^{\mathbf{v} \cdot \phi(\mathbf{u})} = 0,$$

since the sums over $\mathbf{v}$ are always 0 because $\phi(\mathbf{u}) \neq 0$ by (4.21). Thus $f(\mathbf{x})$ is balanced.

For any $k \leq w - 1$ and any choice of $\mathbf{b}$ in $\mathbb{V}_m$, $\mathbf{a}$ in $\mathbb{V}_r$ with $1 \leq wt(\mathbf{b}, \mathbf{a}) \leq k$, we have

$$\begin{aligned} W(f)(\mathbf{b}, \mathbf{a}) &= \sum_{\mathbf{u} \in \mathbb{V}_m, \mathbf{v} \in \mathbb{V}_r} (-1)^{f(\mathbf{u},\mathbf{v}) \oplus (\mathbf{b},\mathbf{a}) \cdot (\mathbf{u},\mathbf{v})} \\ &= \sum_{\mathbf{u} \in \mathbb{V}_m, \mathbf{v} \in \mathbb{V}_r} (-1)^{\phi(\mathbf{u}) \cdot \mathbf{v} \oplus g(\mathbf{u}) \oplus \mathbf{b} \cdot \mathbf{u} \oplus \mathbf{a} \cdot \mathbf{v}} \\ &= \sum_{\mathbf{u} \in \mathbb{V}_m} (-1)^{\mathbf{b} \cdot \mathbf{u} \oplus g(\mathbf{u})} \sum_{\mathbf{v} \in \mathbb{V}_r} (-1)^{(\phi(\mathbf{u}) \oplus \mathbf{a}) \cdot \mathbf{v}}. \end{aligned}$$

Since $wt(\mathbf{a}) \leq k$ and $wt(\phi(\mathbf{u})) \geq w \geq k + 1$, we get $\phi(\mathbf{u}) \oplus \mathbf{a} \neq 0$, so the sums over $\mathbf{v}$ are 0. Thus $W(f)(\mathbf{b}, \mathbf{a}) = 0$, so by Lemma 4.3, f is correlation immune of order k. ■

One virtue of the functions in Theorem 4.20 (they are called Maiorana–McFarland or MM functions; see Sections 3.7 and 5.8) is that it is possible to exercise some control over some properties, for example the nonlinearity and propagation characteristics, of the functions given by the construction. This is discussed in the papers of Carlet [73, 75,80], Chee et al. [108] (note that the journal version [107] is almost identical with this paper) and Seberry et al. [408]. The paper of Cusick [128] gives another method for constructing functions which are balanced and correlation immune of order k; here the degree and nonlinearity of the functions can be controlled. A more complete analysis of this method is given by Carlet [80, Proposition 2].

4.9 TRADEOFF BETWEEN CORRELATION IMMUNITY AND NONLINEARITY

The first paper to give a detailed analysis of the correlation immunity for a large set of Boolean functions is Seberry et al. [408]. In that paper a different construction for the functions in [53] (see Theorem 4.20 above) is given, and that new construction facilitates the analysis of properties such as nonlinearity. The first paper which explicitly shows the tradeoff between correlation immunity and nonlinearity for arbitrary Boolean functions is Chee et al. [108]. Their result is proved in our next two next lemmas, taken from [108, pp. 391–392].

Lemma 4.21. *Define* $\eta(f) = |\{\mathbf{w} \in \mathbb{V}_n : W(\hat{f})(\mathbf{w}) \neq 0\}|$ *for a Boolean function in n variables f. Then*

$$\mathcal{N}_f \leq 2^{n-1} - 2^{n-1}\eta(f)^{-1/2}.$$

Proof. By Parseval's equation (2.16) we have

$$\eta(f) \max_{\mathbf{w}\in\mathbb{V}_n} |W(\hat{f})(\mathbf{w})|^2 \geq \sum_{\mathbf{w}\in\mathbb{V}_n} W(\hat{f})(\mathbf{w})^2 = 2^{2n},$$

so $\max|W(\hat{f})(\mathbf{w})| \geq 2^n\eta(f)^{-1/2}$. Now the lemma follows from Theorem 2.17. ■

Lemma 4.22. *If f is any Boolean function in n variables which is correlation immune of order k and* $\mu(n,k) = 2^n - \sum_{i=1}^{k}\binom{n}{i}$, *then*

$$\mathcal{N}_f \leq 2^{n-1} - 2^{n-1}\mu(n,k)^{-1/2}.$$

Proof. Since f is correlation immune of order k, Lemma 4.3 implies that $\eta(\hat{f})$ in Lemma 4.21 satisfies

$$\begin{aligned} \eta(f) &= 2^n - |\{\mathbf{w} \in \mathbb{V}_n : W(\hat{f})(\mathbf{w}) = 0\}| \\ &\leq 2^n - |\{\mathbf{w} \in \mathbb{V}_n : 1 \leq wt(\mathbf{w}) \leq k\}| \\ &= \mu(n,k). \end{aligned}$$

So Lemma 4.22 follows from Lemma 4.21. ■

It is important to note that Lemma 4.22 is usually a weak estimate. A much stronger result for balanced functions is given in Theorem 4.26.

The result of Lemma 4.21, with the same proof, is given independently by Zheng and Zhang [485, Lemma 3]. They also explicitly characterize the functions f which give equality in the inequality of Lemma 4.21. They call these functions plateaued functions and they are defined as follows. We use the vectors $\alpha_0 = (0,\ldots,0), \alpha_1 = (0,\ldots,0,1),\ldots,\alpha_{2^n-1} = (1,\ldots,1)$ from Definition 2.1 and the Sylvester–Hadamard matrices H_n from (2.17). If we let the rows of H_n in order be $\ell_0,\ell_1,\ldots,\ell_{2^n-1}$, then by

Lemma 2.26 we know that for each i, ℓ_i is the sequence (as defined in Section 2.1) of the linear function $\alpha_i \cdot (x_1,\ldots,x_n) = \alpha_i \cdot \mathbf{x}$, that is ℓ_i written as a vector is

$$\ell_i = ((-1)^{\alpha_i \cdot \alpha_0}, (-1)^{\alpha_i \cdot \alpha_1}, \ldots, (-1)^{\alpha_i \cdot \alpha_{2^n-1}}).$$

We let

$$\zeta = ((-1)^{f(\alpha_0)}, (-1)^{f(\alpha_1)}, \ldots, (-1)^{f(\alpha_{2^n-1})})$$

denote the sequence of the function $f(\mathbf{x})$. We know from the definition of the Walsh transform $W(\hat{f})$ (see Section 2.9) that

$$W(\hat{f})(\alpha_i) = \zeta \cdot \ell_i, \quad 0 \le i \le 2^n - 1, \tag{4.22}$$

so

$$|\{\mathbf{w} : W(\hat{f})(\mathbf{w}) = 0\}| = |\{i : \zeta \cdot \ell_i = 0, 0 \le i \le 2^n - 1\}|. \tag{4.23}$$

Now we can give the definition from [485, Definition 8] (they state the definition in terms of the dot products $\zeta \cdot \ell_i$, which is equivalent by (4.22) and (4.23)):

Definition 4.23. *Let f be a Boolean function in n variables with sequence ζ. If there exists an even integer r such that $|\{\mathbf{w} : W(\hat{f})(\mathbf{w}) \neq 0\}| = 2^r$ and each integer $W(\hat{f})(\mathbf{w})^2$ has the value 0 or 2^{2n-r}, then f is called a* plateaued function of order r *on $\mathbb{V}_n$. We also simply call f a plateaued function if reference to the order r is not needed.*

Lemma 4.24. *Let f be a Boolean function in n variables. Equality holds in the inequality of Lemma 4.21 if and only if the function f is a plateaued function.*

Proof. Define

$$M_f = \max_{\mathbf{w} \in \mathbb{V}_n} |W(\hat{f})(\mathbf{w})|. \tag{4.24}$$

First, assume that f is a plateaued function of order r. Then $M_f = 2^{n-r/2}$ and Theorem 2.17 shows that equality holds in Lemma 4.21.

Next assume that equality holds in Lemma 4.21. Then by Theorem 2.17

$$2^n = |\{\mathbf{w} : W(\hat{f})(\mathbf{w}) \neq 0\}|^{1/2} M_f = \eta(f)^{1/2} M_f.$$

Now both $\eta(f)^{1/2}$ and M_f must be integers and in fact powers of 2. Therefore $\eta(f) = 2^r$ for some even integer $r \le n$ and $M_f = 2^{n-r/2}$. It follows from Parseval's equation (2.16) that the only nonzero value of $W(\hat{f})(\mathbf{w})$ is 2^{2n-r}, so f is a plateaued function of order r. ■

Corollary 4.25. *Let f be a Boolean function in n variables. Define M_f by* (4.24) *and define $\eta(f)$ as in Lemma* 4.21. *Then $\eta(f)^{1/2}M_f \geq 2^n$ with equality if and only if f is a plateaued function.*

Proof. This follows from the proof of Lemma 4.24. ∎

In the special case of balanced functions, our next theorem gives a form of the tradeoff between correlation immunity and nonlinearity which was proved independently by Sarkar and Maitra [395, Theorem 2], Tarannikov [447, Theorem 3.1] and Zheng and Zhang [482, Theorems 5 and 6]. We give the proof from [447]. The proof of [395] is more complicated because their result is stronger in some cases. The proofs of [482] are also more complicated, because they prove much more than the inequality in Theorem 4.26. They also discuss when equality can occur, and they show that under certain conditions the inequality still holds even if the function is not balanced.

Theorem 4.26. *If $f(\mathbf{x})$ is a balanced function in n variables which is correlation immune of order $k \leq n-2$, then*

$$\mathcal{N}_f \leq 2^{n-1} - 2^{k+1}.$$

Proof. If $k = n-2$, then by Theorem 4.12 we have that f is affine, so $\mathcal{N}_f = 0$, as desired. Let $f(\mathbf{x}|x_{i(1)} = c_1, \ldots, x_{i(t)} = c_t)$ denote the function in $n-t$ variables obtained from $f(\mathbf{x})$ by setting $x_{i(1)}, \ldots, x_{i(t)}$ equal to $c_1, \ldots, c_t$, respectively; we will call this function a *subfunction* (in $n-t$ variables) of $f(\mathbf{x})$.

If $k \leq n-3$, we may assume without loss of generality that $f(\mathbf{x})$ is correlation immune of order k but not of order $k+1$. Then by the generalization of Lemma 4.2(d) f has a subfunction $f(\mathbf{x}|x_{i(1)} = a_1, \ldots, x_{i(k+1)} = a_{k+1}) = g$, say, in $n-k-1$ variables such that $wt(g) = w \neq 2^{n-k-2}$. We may assume $w < 2^{n-k-2}$ since

$$wt(f) = 2^{n-1} = \sum_{(c_1,\ldots,c_{k+1})} wt(f(\mathbf{x}|x_{i(1)} = c_1, \ldots, x_{i(k+1)} = c_{k+1})),$$

where the sum is over all of the 2^{k+1} possible choices of $c_1, \ldots, c_{k+1}$. If this sum has a term greater than 2^{n-k-2}, then it also has a term less than 2^{n-k-2}.

Consider a subfunction $f(\mathbf{x}|x_{i(1)} = b_1, \ldots, x_{i(k+1)} = b_{k+1}) = h$, say, where the given fixed vector $\mathbf{a} = (a_1, \ldots, a_{k+1})$ and the vector $\mathbf{b} = (b_1, \ldots, b_{k+1})$ differ in only the jth coordinate. Now since $f(\mathbf{x})$ is correlation immune of order k, $wt(g) + wt(h) = wt(f(\mathbf{x}|x_{i(1)} = a_1, \ldots, x_{i(j-1)} = a_{j-1}, x_{i(j+1)} = a_{j+1}, \ldots, x_{i(k+1)} = a_{k+1})) = 2^{n-k-1}$, by the generalization of Lemma 4.2(d). This gives

$$wt(h) = 2^{n-k-1} - w. \tag{4.25}$$

Now consider a subfunction $f(\mathbf{x}|x_{i(1)} = b_1, \ldots, x_{i(k+1)} = b_{k+1}) = h_1$, say, where $\mathbf{b}$ differs from the given $\mathbf{a}$ in exactly the two coordinates $j(1)$ and $j(2)$, say $x_{j(1)} = c, x_{j(2)} = d$ in $\mathbf{a}$, so $x_{j(1)} = c \oplus 1$ and $x_{j(2)} = d \oplus 1$ in $\mathbf{b}$. Let h_2 and h_3 denote the subfunctions with the same fixed values as h_1, except that $x_{j(1)} = c$ in h_2 and $x_{j(2)} = d$ in h_3. By the argument above which gives (4.25) we have

$$wt(h_2) = wt(h_3) = 2^{n-k-1} - w. \tag{4.26}$$

Since $f(\mathbf{x})$ is correlation immune of order k, we have

$$wt(h_1) + wt(h_3) = 2^{n-k-1}. \tag{4.27}$$

Combining (4.26) and (4.27) gives $wt(h_1) = w$. Proceeding by induction, we have for any vector $\mathbf{b}$,

$$wt(f(\mathbf{x}|x_{i(1)} = b_1, \ldots, x_{i(k+1)} = b_{k+1})) = \begin{cases} h & \text{if } d(\mathbf{a},\mathbf{b}) \text{ even} \\ 2^{n-k-1} - h & \text{if } d(\mathbf{a},\mathbf{b}) \text{ odd} \end{cases}$$

Now define the affine function ℓ in n variables by

$$\ell = \sum_{j=1}^{k+1} x_{i(j)} \oplus A,$$

where $A \equiv wt(\mathbf{a}) \pmod 2$ is 0 or 1. Then

$$\begin{aligned} d(f,\ell) &= \sum_{(b_1,\ldots,b_{k+1})} d\left(f(\mathbf{x}|x_{i(1)} = b_1, \ldots, x_{i(k+1)} = b_{k+1}), \sum_{j=1}^{k+1} b_{i(j)} \oplus A\right) \\ &= \sum_{\mathbf{b}, d(\mathbf{a},\mathbf{b})\ even} wt(f(\mathbf{x}|x_{i(1)} = b_1, \ldots, x_{i(k+1)} = b_{k+1})) \\ &\quad + \sum_{\mathbf{b}, d(\mathbf{a},\mathbf{b})\ odd} 2^{n-k-1} - wt(f(\mathbf{x}|x_{i(1)} = b_1, \ldots, x_{i(k+1)} = b_{k+1})) \\ &= 2^k w + 2^k w = 2^{k+1} w, \end{aligned}$$

where the second equality is true since $\sum_{j=1}^{k+1} b_{i(j)} \oplus A$ is 0 when $d(\mathbf{a},\mathbf{b})$ is even and is 1 when $d(\mathbf{a},\mathbf{b})$ is odd. Hence

$$\mathcal{N}_f \le d(f,\ell) = 2^{k+1} w \le 2^{k+1}(2^{n-k-2} - 1) = 2^{n-1} - 2^{k+1}.$$

■

Theorem 4.27. *Let $f(\mathbf{x})$ be a balanced function in n variables which is correlation immune of order $k \le n-2$. Then equality is possible in Theorem 4.26 only if $f(\mathbf{x})$ has its maximum possible degree $n-k-1$. If* $\deg f < n-k-1$, *then $\mathcal{N}_f \le 2^{n-1} - 2^{k+2}$.*

Proof. Let $g = f(\mathbf{x}|x_{i(1)} = a_1, \ldots, x_{i(k+1)} = a_{k+1})$ be the subfunction in the proof of Theorem 4.26 which has $wt(g) = w < 2^{n-k-2}$. We know that $\deg g \le \deg f \le n-k-1$ by Theorem 4.12. If in fact $\deg f < n-k-1$, then g must have even weight because

g is a function of $n-k-1$ variables. Therefore $w \leq 2^{n-k-2}-2$ and by the proof of Theorem 4.26 $\mathcal{N}_f \leq 2^{k+1}w \leq 2^{n-1}-2^{k+2}$. ■

We remark that, because of Theorem 2.27 (which says $\mathcal{N}_f \leq 2^{n-1}-2^{n/2-1}$ for any Boolean function in n variables), Theorem 4.26 does not give any useful information unless $k \geq n/2-2$. The case $k = n/2-2$, n even, is discussed in detail in Chapter 5. Thus in considering the question whether equality can hold in Theorem 4.26 we can confine ourselves to the case $k > n/2-2$. A complete answer to this question is not yet known, but partial results are obtained in [356,395,447,449,165,482,483]. Note that [165] was written after [449] and depends on the latter paper, though it was published earlier.

Theorem 4.27 has been refined by Carlet [76] so that the upper bound also depends on the degree (the result is better only if $d \leq n/2$):

Theorem 4.28. *If $f(\mathbf{x})$ is a balanced function of degree d in n variables which is correlation immune of order $k \leq n-2$, then*

$$\mathcal{N}_f \leq 2^{n-1} - 2^{k+1+\lfloor (n-k-2)/d \rfloor}. \tag{4.28}$$

Proof. The somewhat complicated proof is in [76, Sections 3 and 4]. A different proof is given by Carlet and Sarkar [95]. Both proofs actually give a result which is a bit stronger than (4.28). ■

The inequality (4.28) can be further refined [105] if $f(\mathbf{x})$ is known to satisfy some propagation criteria (see Section 3.5) in addition to the hypotheses of Theorem 4.28. Other work on the connections between propagation criteria, correlation immunity and nonlinearity is given in [58] and [484] (there is an amusing slip in the recording of reference 3 in this paper, where 'Projet Codes' is mistaken for a person).

There is some work [357] on nonlinearity for generalized resilient functions with range $\mathbb{V}_m$ for $m > 1$, but in accordance with our general policy to consider only Boolean functions in this book, we do not discuss this.

CHAPTER

Bent Boolean functions

5

5.1 INTRODUCTION

Profound study of nature is the most fertile source of mathematical discoveries.
Joseph Fourier (1768–1830)

Shannon [418] introduced the concepts of confusion and diffusion as a fundamental technique to achieve security in cryptographic systems. Confusion is reflected in nonlinearity of certain cryptographic primitives; most linear systems are easily breakable. In this context it is important to have criteria which reflect nonlinearity. There are a variety of such criteria. The best known criterion is the so-called perfect nonlinearity as introduced by Meier and Staffelbach in [318]. It turns out that this concept is equivalent to the bent property discovered by Rothaus in the mid-1960s but published some 10 years later in 1976 in [385]. Bent functions can also be defined in terms of difference sets; this perspective is explained in McFarland [311] and Dillon [151–153].

It is desirable that a nonlinearity criterion remains invariant under a large group of transformations. For many applications, this symmetry group should contain the group of affine transformations. The perfect nonlinear or bent functions are indeed invariant under this group, as we shall prove.

Depending on the applications, some kinds of functions are considered weak. The linear or affine functions are considered cryptographically weak, and the criteria we shall introduce can be expressed in terms of a distance to these weak functions.

We also give various characterizations of the bent property, and we investigate the connection between this property and the algebraic normal form. We use bent functions to construct balanced Boolean functions with high nonlinearity, which satisfy the SAC, as well.

Bent functions have practical applications in spread spectrum communications (see Olsen et al. [352] and Simon et al. [426, pp. 342–349]), in cryptography (see Meier and Staffelbach [318] and Nyberg [345,347]) and in coding theory (see MacWilliams and Sloane [287, pp. 426–432 and 456–465]). The bent property is also closely related to the SAC property, investigated in Chapter 3, which is a diffusion property. Few construction methods for bent functions are known. The oldest methods are due to Rothaus, Maiorana, McFarland, and Dillon; these are described by Dobbertin [157], who also gives a new method. Other new methods have been described by Adams and Tavares [2] and by Carlet [65–67].

5.2 DEFINITIONS AND BACKGROUND

Definition 5.1. *A Boolean function f in n variables is called* bent *if and only if the Walsh transform coefficients of $\hat{f}$ are all $\pm 2^{n/2}$, that is, $W(\hat{f})^2$ is constant.*

Remark 5.2. *We immediately notice that bent functions exist only for even dimensions, that is $n = 2k$.*

Example 5.3 (*Examples of bent functions*).

1. $f(\mathbf{x}) = x_1x_2$ *on* $\mathbb{V}_2$; $W(\hat{f})(\mathbf{u}) = \pm 2$.
2. $f(\mathbf{x}) = 1 \oplus x_1x_2 \oplus x_1x_3 \oplus x_1x_4 \oplus x_2x_3 \oplus x_2x_4 \oplus x_3x_4$ *on* $\mathbb{V}_4$; $W(\hat{f})(\mathbf{u}) = \pm 4$.

Definition 5.4. *For each function $f : \mathbb{V}_n \to \mathbb{R}$ we associate the $2^n \times 2^n$ matrix (we shall use this matrix extensively in Chapter 8)*

$$M_f = (f(\mathbf{u} \oplus \mathbf{v}))_{\mathbf{u},\mathbf{v}}. \tag{5.1}$$

For a bent function f, we define a Boolean function $\mathcal{F}$ such that

$$\frac{W(\hat{f})(\mathbf{u})}{2^{n/2}} = (-1)^{\mathcal{F}(\mathbf{u})} = \hat{\mathcal{F}}(\mathbf{u}), \tag{5.2}$$

and we call it the dual *of f (Dillon calls it the* 'Fourier transform' *of f and Carlet calls it the* 'dual' *of f).*

McFarland (see [151,310]) found the next result.

Theorem 5.5. *Let the Sylvester-Hadamard matrix H_n be defined as in Section 2.9, and α_i, $0 \le i \le 2^n - 1$, as in Definition 2.1. If $f : \mathbb{V}_n \to \mathbb{R}$, then*

$$H_n M_f H_n^{-1} = \mathrm{diag}\left(W(f)(\alpha_0), W(f)(\alpha_1), \ldots, W(f)(\alpha_{2^n-1})\right). \tag{5.3}$$

Proof. We recall that $H_n^{-1} = 2^{-n}H_n^t$. The $(\mathbf{u},\mathbf{v})$ entry in the matrix $H_n M_f H_n^{-1}$ is

$$\begin{aligned}
\frac{1}{2^n}\sum_{(\mathbf{s},\mathbf{t})\in\mathbb{V}_n^2} H_n(\mathbf{u},\mathbf{s})f(\mathbf{s}\oplus\mathbf{t})H_n(\mathbf{t},\mathbf{v}) &= \frac{1}{2^n}\sum_{\mathbf{w}\in\mathbb{V}_n} f(\mathbf{w})\sum_{\mathbf{s}\in\mathbb{V}_n} H_n(\mathbf{u},\mathbf{s})H_n(\mathbf{s}\oplus\mathbf{w},\mathbf{v})\\
&= \frac{1}{2^n}\sum_{\mathbf{w}\in\mathbb{V}_n} f(\mathbf{w})H_n(\mathbf{w},\mathbf{v})\sum_{\mathbf{s}\in\mathbb{V}_n} H_n(\mathbf{u},\mathbf{s})H_n(\mathbf{v},\mathbf{s})\\
&= \begin{cases} W(f)(\mathbf{u}) & \text{if } \mathbf{u}=\mathbf{v}\\ 0 & \text{otherwise,}\end{cases}
\end{aligned} \tag{5.4}$$

using the results of Section 2.9. ■

Definition 5.6. *Let G be an Abelian group of order v and D a subset of G of order k. D is a (v,k,λ,n)*-difference set *in G if for every nonzero element g in G the equation $g = d_i - d_j$ (or $d_i d_j^{-1}$ in multiplicative notation) has exactly λ solutions $(d_i, d_j) \in D^2$. We define $n = k - \lambda$.*

Remark 5.7. *Since each of the $v-1$ nontrivial elements of G occurs λ times among the $k(k-1)$ nontrivial elements of D, one must have*

$$\lambda(v-1) = k(k-1). \tag{5.5}$$

The *incidence matrix* associated with the subset D is the $v \times v$ matrix Υ_D, defined by

$$\Upsilon_D(x,y) = \begin{cases} 1 & \text{if } x-y \in D \\ 0 & \text{otherwise.} \end{cases} \tag{5.6}$$

As a consequence of the definition for the incidence matrix we have:

Lemma 5.8. *D is a (v,k,λ,n)-difference set if and only if*

$$\Upsilon_D^2 = nI_v + \lambda J,$$

where J is the $v \times v$ matrix with all entries 1.

The complement of a difference set is also a difference set.

Lemma 5.9. *If D is a (v,k,λ,n)-difference set in G, then its complement $\bar{D} = G - D$ is a $(v, v-k, v-2k+\lambda, n)$-difference set in G.*

Proof. It follows from Lemma 5.8. ■

There are special difference sets which exist only in Abelian groups of square order.

Theorem 5.10. *D is a (v,k,λ,n)-difference set with $v = 4n$ if and only if $J - 2\Upsilon_D$ is a Hadamard matrix. (We call D a* Hadamard difference set.*)*

The following result was noted by Kesava Menon [252].

Theorem 5.11. *A Hadamard difference set has the parameters*

$$\begin{aligned}(v,k,\lambda,n) &= (4N^2, 2N^2 - N, N^2 - N, N^2) \quad \textit{or} \\ &\quad\ (4N^2, 2N^2 + N, N^2 + N, N^2).\end{aligned} \tag{5.7}$$

Proof. From the equations $n = k - \lambda, k(k-1) = \lambda(v-1)$ and $v = 4n$ we get

$$\begin{aligned} 0 &= k(k-1) - \lambda(v-1) = k^2 - k - (k-n)(4n-1) \\ &= k^2 - 4nk + n(4n-1) = (k-2n)^2 - n, \end{aligned}$$

which proves the theorem. ■

Two difference sets D_1 and D_2 in the Abelian group G are called *equivalent* if there exists an automorphism α of G such that

$$D_1^{\alpha} = D_2 + g, \tag{5.8}$$

for some g in G. If (5.8) holds for $D_1 = D_2 = D$, then the group automorphism α is called a *multiplier* of D. A multiplier of the form

$$g \mapsto g^t, \quad t \in \mathbf{Z}$$

is called a *numerical multiplier*.

H.B. Mann and R.L. McFarland (see [151]) showed that every multiplier of a difference set must fix at least one translate of that difference set. We denote by $M(D)$ the subgroup of multipliers of D, in the group of automorphisms of G. It is easy to see that if $D_1^{\alpha} = D_2 + g$, then $M(D_1) = \alpha M(D_2)\alpha^{-1}$, that is D_1 and D_2 are isomorphic.

The bent function on $\mathbb{V}_4$ from Example 5.3 is of particular interest, because it corresponds to a difference set by $D = f^{-1}(1)$ (as we shall see), which happens to be the only (16, 6, 2, 4)-difference set in $\mathbb{V}_4$ up to an equivalence. Thus, by noting that

$$f = 1 \oplus x_1x_2 \oplus x_1x_3 \oplus x_1x_4 \oplus x_2x_3 \oplus x_2x_4 \oplus x_3x_4 = 1110100010000001,$$

we get that

$$D = \{(0,0,0,0),(0,0,0,1),(0,0,1,0),(0,1,0,0),(1,0,0,0),(1,1,1,1)\},$$

that is, $D = \{\alpha_0, \alpha_1, \alpha_2, \alpha_4, \alpha_8, \alpha_{15}\}$, using the lexicographical order notation of $\mathbb{V}_4$ (see Definition 2.1). For the above difference set, McFarland proved that $M(D)$ has order 720.

The interested reader can find more on difference sets in [140,151,244,466] and the references therein.

5.3 CHARACTERIZATIONS OF THE BENT PROPERTY

The following theorem contains many equivalent definitions of the bent property.

Theorem 5.12. *Let $f : \mathbb{V}_n \to \mathbb{F}_2$ be a Boolean function. The following statements are equivalent:*

(i) *f is bent.*
(ii) *Let η_f be the $(1,-1)$-sequence of f of Definition 2.1. For any affine function ℓ, $\eta_f \cdot \eta_\ell = \pm 2^{n/2}$.*
(iii) *The dual $\mathcal{F}$ as in (5.2) is bent.*
(iv) *The matrix $M_{\hat{f}} = \left(\hat{f}(\mathbf{u} \oplus \mathbf{v})\right)_{\mathbf{u},\mathbf{v}}$ is a Hadamard matrix.*
(v) *The nonlinearity of f is $\mathcal{N}_f = 2^{n-1} - 2^{n/2-1}$.*

(vi) *$f(\mathbf{x}) \oplus \alpha \cdot \mathbf{x}$ has $2^{n-1} \pm 2^{\frac{n}{2}-1}$ zeros for all $\alpha \in \mathbb{V}_n$.*

(vii) *The directional derivative $f_{\mathbf{v}}(\mathbf{x}) = f(\mathbf{x} \oplus \mathbf{v}) \oplus f(\mathbf{x})$ is balanced for all nonzero $\mathbf{v}$ in $\mathbb{V}_n$.*

(viii) *$f^{-1}(1)$ (or $f^{-1}(0)$) is a Hadamard difference set in $\mathbb{V}_n$.*

(ix) *$\chi\left(f^{-1}(1)\right) = \pm 2^{n/2-1}$ for all nonprincipal characters χ of $\mathbb{V}_n$.*

Proof. (i) $\Leftrightarrow$ (ii) This equivalence is just a restatement of the definition.

(i) $\Leftrightarrow$ (iii) The coefficients of the Walsh transform of $\mathcal{F}$ are $\pm 2^{n/2}$.

(i) $\Leftrightarrow$ (iv) Assume f is bent. From Theorem 5.5 we know that

$$H_n M_{\hat{f}} H_n^{-1} = \text{diag}(W(\hat{f})(\alpha_0), \ldots, W(\hat{f})(\alpha_{2^n-1})), \tag{5.9}$$

which implies, by applying the transposition operation,

$$\left(H_n^{-1}\right)^t M_{\hat{f}}^t H_n^t = H_n M_{\hat{f}}^t H_n^{-1} = \text{diag}(W(\hat{f})(\alpha_0), \ldots, W(\hat{f})(\alpha_{2^n-1})). \tag{5.10}$$

Multiplying the Equations (5.9) and (5.10) we get

$$H_n (M_{\hat{f}})(M_{\hat{f}}^t) H_n^{-1} = 2^n I_{2^n},$$

which renders

$$(M_{\hat{f}})(M_{\hat{f}}^t) = 2^n I_{2^n},$$

that is, the matrix $M_{\hat{f}}$ is Hadamard. The converse is obvious.

(i) $\Leftrightarrow$ (v) Recall the computations we have done at the end of Section 2.9. Let $H_n = (h_{ij})_{ij}$ and ℓ_i be the ith row of H_n. Write $\ell_{i+2^n} = -\ell_i$, $i = 1, 2, \ldots, 2^n$, and so, $\ell_1, \ell_2, \ldots, \ell_{2^{n+1}}$ represent all affine functions on $\mathbb{V}_n$.

Let ϕ_i be the affine function corresponding to ℓ_i. We already showed in Theorem 2.27 that $\mathcal{N}_f \leq 2^{n-1} - 2^{\frac{n}{2}-1}$ for any function on $\mathbb{V}_n$. Since f is bent, it follows that

$$\eta_f \cdot \ell_i = \pm 2^{n/2}. \tag{5.11}$$

Using Equation (5.11) and the relation (2.20), $d(f, g) = 2^{n-1} - \frac{1}{2}\eta_f \cdot \eta_g$, we deduce

$$d(f, \phi_i) = 2^{n-1} \pm 2^{n/2-1},$$

which implies $\mathcal{N}_f = 2^{n-1} - 2^{n/2-1}$.

Conversely, we suppose that f attains the upper bound for nonlinearity. Then $(\eta_f \cdot \ell_i)^2 = 2^n, i = 1, 2, \ldots, 2^{n+1}$. If this were not true, then there would exist $1 \leq i_1, i_2 \leq 2^n$ such that $(\eta_f \cdot \ell_{i_1})^2 > 2^n$ and $(\eta_f \cdot \ell_{i_2})^2 < 2^n$. Thus $\eta_f \cdot \ell_{i_1} > 2^{n/2}$ or $\eta_f \cdot \ell_{i_1+2^n} > 2^{n/2}$, so we may suppose without loss of generality that $\eta_f \cdot \ell_{i_1} > 2^{n/2}$. Furthermore, we infer that

$d(f,\ell_{i_1}) < 2^{n-1} - 2^{n/2-1}$ and hence $\mathcal{N}_f < 2^{n-1} - 2^{n/2-1}$. This is a contradiction which shows that we must have

$$\eta_f \cdot \ell_i = \pm 2^{n/2}, \quad i = 1, 2, \ldots, 2^{n+1},$$

which obviously implies that f is bent.

(i) $\Leftrightarrow$ (vi) Let $N_{\mathbf{v}}$ be the number of zeros of $g(\mathbf{x}) = f(\mathbf{x}) \oplus \mathbf{v}\cdot\mathbf{x}$. Then

$$\begin{aligned} W(\hat{f})(\mathbf{v}) &= \sum_{\mathbf{x}\in\mathbb{V}_n} (-1)^{f(\mathbf{x})\oplus\mathbf{v}\cdot\mathbf{x}} \\ &= N_{\mathbf{v}} - (2^n - N_{\mathbf{v}}) = 2N_{\mathbf{v}} - 2^n, \end{aligned}$$

or

$$N_{\mathbf{v}} = 2^{n-1} + 2^{-1} W(\hat{f})(\mathbf{v}).$$

Since $W(\hat{f})(\mathbf{v}) = \pm 2^{n/2}$, then $N_{\mathbf{v}} = 2^{n-1} \pm 2^{n/2-1}$. For the converse we reverse the argument.

(iv) $\Leftrightarrow$ (vii) Since $M_{\hat{f}} = \left(\hat{f}(\mathbf{u}\oplus\mathbf{v})\right)_{\mathbf{u},\mathbf{v}}$ is a Hadamard matrix, then

$$\sum_{\mathbf{w}\in\mathbb{V}_n} (-1)^{f(\mathbf{u}\oplus\mathbf{w})\oplus f(\mathbf{w}\oplus\mathbf{v})} = 0 \quad \text{for any } \mathbf{u} \neq \mathbf{v}. \tag{5.12}$$

When $\mathbf{w}$ runs through $\mathbb{V}_n$, $\mathbf{w}\oplus\mathbf{v}$ runs through $\mathbb{V}_n$ as well. Thus the above equation can be written as

$$\sum_{\mathbf{w}\in\mathbb{V}_n} (-1)^{f(\mathbf{u}\oplus\mathbf{v}\oplus\mathbf{w})\oplus f(\mathbf{w})} = 0 \quad \text{for any } \mathbf{u} \neq \mathbf{v}, \tag{5.13}$$

that is, $f_{\mathbf{u}\oplus\mathbf{v}}$ is balanced. The converse is straightforward.

(i) $\Leftrightarrow$ (viii) This follows from Theorem 5.10, since the $(\mathbf{u},\mathbf{v})$ entries of the matrix $J - 2\Upsilon_D$ are

$$\begin{cases} -1 & \text{if } \mathbf{u}-\mathbf{v}\in D \\ 1 & \text{otherwise} \end{cases} = \begin{cases} -1 & \text{if } \mathbf{u}\oplus\mathbf{v}\in f^{-1}(1) \\ 1 & \text{otherwise} \end{cases} = \begin{cases} -1 & \text{if } f(\mathbf{u}\oplus\mathbf{v}) = 1 \\ 1 & \text{otherwise} \end{cases} = (-1)^{f(\mathbf{u}\oplus\mathbf{v})},$$

where $D = f^{-1}(1)$.

(i) $\Leftrightarrow$ (ix) We note that if $\chi(\mathbf{x}) = (-1)^{\mathbf{v}\cdot\mathbf{x}}$, $\mathbf{v}\neq 0$, is a nonprincipal character of $\mathbb{V}_n$, then for any $f : \mathbb{V}_n \to \mathbb{F}_2$

$$W(\hat{f})(\mathbf{v}) = \sum_{\mathbf{u}\in\mathbb{V}_n} (-1)^{f(\mathbf{u})\oplus\mathbf{u}\cdot\mathbf{v}} = \chi(f^{-1}(0)) - \chi(f^{-1}(1)) = -2\chi(f^{-1}(1)).$$

■

Carlet and Guillot [89,90] found an alternate characterization of the bent property.

Theorem 5.13. *Let f be a Boolean function on $\mathbb{V}_n$, $n = 2k$. Then f is bent if and only if there exist k-dimensional subspaces $E_1,\dots,E_r$ of $\mathbb{V}_n$ (r is not constrained) and integers $m_1,\dots,m_r$, such that for any element $\mathbf{x} \in \mathbb{V}_n$ the following holds:*

$$\sum_{i=1}^{r} m_i \phi_{E_i}(\mathbf{x}) \equiv 2^{k-1}\delta_0(\mathbf{x}) + f(\mathbf{x}) \pmod{2^k},$$

where δ_0 is the Dirac symbol on $\mathbb{V}_n$ ($\delta_0(\mathbf{x}) = 1$ if $\mathbf{x} = 0$, and 0 otherwise), and ϕ_{E_i} is the characteristic function of E_i in $\mathbb{V}_n$.

The congruence occurring in Theorem 5.13 can be replaced by equality if one allows the spaces E_i to have dimension k or $k+1$.

Theorem 5.14. *Let f be a Boolean function on $\mathbb{V}_n$, $n = 2k$. Then f is bent if and only if there exist subspaces $E_1,\dots,E_r$ of $\mathbb{V}_n$ of dimension k or $k+1$, and integers $m_1,\dots,m_r$ such that for any element $\mathbf{x} \in \mathbb{V}_n$ the following holds:*

$$\sum_{i=1}^{r} m_i \phi_{E_i}(\mathbf{x}) = \pm 2^{k-1}\delta_0(\mathbf{x}) + f(\mathbf{x}).$$

5.4 MEIER AND STAFFELBACH'S APPROACH

Let us recall the notion of linear structure: we say that a Boolean function f has *linear structures* if there exists a nonzero $\mathbf{a}$ such that the expression $f(\mathbf{x} \oplus \mathbf{a}) \oplus f(\mathbf{x})$ is constant, as a function of $\mathbf{x}$.

Meier and Staffelbach [318] investigated a class of functions whose definition is motivated by considering linear structures.

Definition 5.15. *A Boolean function $f : \mathbb{V}_n \to \mathbb{F}_2$ is called* perfect nonlinear with respect to linear structures *(or briefly* perfect nonlinear*) if for every nonzero vector $\mathbf{a} \in \mathbb{V}_n$ the values $f(\mathbf{x} \oplus \mathbf{a})$ and $f(\mathbf{x})$ are equal for exactly half of the arguments $\mathbf{x} \in \mathbb{V}_n$.*

Let $LS(n)$ denote the subset of Boolean functions having linear structures. We first note that $LS(n)$ properly contains the set of all affine functions.

In [318] Meier and Staffelbach showed.

Theorem 5.16. *The class of perfect nonlinear functions is the class of functions f with $\sigma(f) = 2^{n-2}$, where*

$$\sigma(f) = \min_{S \in LS(n)} d(f, S).$$

In the same paper, the authors also showed that the class of bent and perfect nonlinear functions coincide, which is the result of Theorem 5.12(vii) above.

5.5 DEGREE OF A BENT FUNCTION

For a Boolean function we have $\hat{f}(\mathbf{u}) = (-1)^{f(\mathbf{u})} = 1 - 2f(\mathbf{u})$, where the second equality holds if we consider the functions as real functions. From Theorem 2.10 we know that for an arbitrary subspace S of $\mathbb{V}_n$ we have, using $\hat{f}$ instead of f,

$$\sum_{\mathbf{u}\in S} W(\hat{f})(\mathbf{u}) = 2^{\dim S} \sum_{\mathbf{u}\in S^{\perp}} \hat{f}(\mathbf{u}). \tag{5.14}$$

Taking S to be the set of all vectors $\mathbf{u}$ 'included' in $\mathbf{v}$, that is, $\mathbf{u} \leq \mathbf{v}$, the above equation transforms into (see Corollary 2.11)

$$\sum_{\mathbf{u}\leq\mathbf{v}} W(\hat{f})(\mathbf{u}) = 2^{wt(\mathbf{v})} \sum_{\mathbf{u}\leq\bar{\mathbf{v}}} \hat{f}(\mathbf{u}).$$

Furthermore, if f is bent, by using Equation (5.2) that defines the dual $\mathcal{F}$,

$$W(\hat{f})(\mathbf{u}) = 2^{n/2}\hat{\mathcal{F}}(\mathbf{u}) = 2^{n/2}(1 - 2\mathcal{F}(\mathbf{u})),$$

we obtain

$$2^{wt(\mathbf{v})} - 2\sum_{\mathbf{u}\leq\mathbf{v}} f(\mathbf{u}) = 2^{-wt(\mathbf{v})} \sum_{\mathbf{u}\leq\bar{\mathbf{v}}} \left(2^{n/2} - 2^{n/2+1}\mathcal{F}(\mathbf{u})\right). \tag{5.15}$$

Rewriting (5.15) we arrive at the next lemma.

Lemma 5.17. *If f is a bent Boolean function on $\mathbb{V}_n$, then regarding f as a real-valued function we have*

$$\sum_{\mathbf{u}\leq\mathbf{v}} f(\mathbf{u}) = 2^{wt(\mathbf{v})-1} - 2^{\frac{n}{2}-1} + 2^{wt(\mathbf{v})-\frac{n}{2}} \sum_{\mathbf{u}\leq\bar{\mathbf{v}}} \mathcal{F}(\mathbf{u}). \tag{5.16}$$

Now we are able to prove a very important result concerning the degree of a bent function, namely:

Theorem 5.18. *For $n = 2$, the degree of a bent function on $\mathbb{V}_2$ is 2. For $n > 2$, the degree of a bent function is at most $n/2$.*

Proof. The first part is obvious. Now, let f be a bent function and $n > 2$,

$$f(\mathbf{x}) = \sum_{\mathbf{v}\in\mathbb{V}_n} g(\mathbf{v}) x_1^{v_1} \cdots x_n^{v_n}, \tag{5.17}$$

where the coefficients are given by

$$g(\mathbf{v}) = \sum_{\mathbf{u}\leq\mathbf{v}} f(\mathbf{u})$$

(see [287, Theorem 1, p. 372]). Thus, the monomial $x_1^{v_1} \cdots x_n^{v_n}$ is present in $f(\mathbf{x})$ if and only if $g(\mathbf{v})$ is odd. But if $wt(\mathbf{v}) > n/2$ and $n > 2$, then the last sum in (5.16) is also even and $g(\mathbf{v})$ is zero in $\mathbb{F}_2$. Therefore f has degree at most $n/2$. ■

Corollary 5.19. *If f is bent of degree $n/2$, then its dual $\mathcal{F}$ is also bent of degree $n/2$.*

A very interesting approach which further connects the degree of a bent function to the degree of its dual is due to X.-D. Hou [228].

Lemma 5.20. *Let f be a bent function on $\mathbb{V}_{2k}$. Then*

$$\deg \mathcal{F} \leq \frac{(\deg f - 2)k + \deg f}{\deg f - 1}.$$

As an easy consequence of this lemma, we can deduce the following double inequality

$$\frac{2k - \deg f}{k + 1 - \deg f} \leq \deg \mathcal{F} \leq \frac{(\deg f - 2)k + \deg f}{\deg f - 1}.$$

Hou used Lemma 5.20 to deduce the following result.

Theorem 5.21. *Let π be a permutation of $\mathbb{V}_k$. Then*

$$\deg \pi^{-1} \leq \frac{(\deg \pi - 1)k + 1}{\deg \pi},$$

where $\deg \pi = \max_i \deg \pi_i$, *for* $\pi = (\pi_1, \ldots, \pi_k) : \mathbb{V}_k \to \mathbb{V}_k$.

5.6 NEW FROM OLD BENT FUNCTIONS

There are many ways to construct bent functions on $\mathbb{V}_{m+n}$ starting from bent functions on $\mathbb{V}_n$ and $\mathbb{V}_m$. Most known constructions are given by Yarlagadda and Hershey [471]: Rothaus's construction, the eigenvectors of the Sylvester-Hadamard matrices [470], constructions based on Kronecker algebra, concatenation, dyadic shifts, linear transformations of variables. In this section, constructions will be given based on concatenation and linear (affine) transformation and dyadic shifts.

Theorem 5.22. *Let f and g be Boolean functions on $\mathbb{V}_m$ and $\mathbb{V}_n$, respectively. Then the Boolean function $h : \mathbb{V}_{m+n} \to \mathbb{F}_2$ defined by $h(\mathbf{x}, \mathbf{y}) = f(\mathbf{x}) \oplus g(\mathbf{y})$ is bent if and only if f and g are bent.*

Proof. The idea is to use the fact that $M_{\hat{h}} = M_{\hat{f}} \otimes M_{\hat{g}}$, where $\otimes$ is the Kronecker product. We write $\mathbf{z} \in \mathbb{V}_{m+n}, \mathbf{z} = (\mathbf{x}, \mathbf{y})$, where $\mathbf{x} \in \mathbb{V}_m, \mathbf{y} \in \mathbb{V}_n$. Then

$$\begin{aligned} W(\hat{h})(\mathbf{z}) &= \sum_{\mathbf{t}\in\mathbb{V}_{m+n}} (-1)^{\mathbf{z}\cdot\mathbf{t}\oplus h(\mathbf{t})} \\ &= \sum_{\mathbf{r}\in\mathbb{V}_m}\sum_{\mathbf{s}\in\mathbb{V}_n} (-1)^{\mathbf{x}\cdot\mathbf{r}\oplus\mathbf{y}\cdot\mathbf{s}\oplus f(\mathbf{r})\oplus g(\mathbf{s})} \\ &= W(\hat{f})(\mathbf{x})W(\hat{g})(\mathbf{y}). \end{aligned} \tag{5.18}$$

If f and g are bent, then $W(\hat{f})(\mathbf{x}) = \pm 2^{m/2}$ and $W(\hat{g})(\mathbf{y}) = \pm 2^{m/2}$. Thus $W(\hat{h})(\mathbf{z}) = \pm 2^{(m+n)/2}$, and so h is bent.

Conversely, we assume that h is bent. We must prove that f, g are bent, so we suppose for instance that f is not bent. It follows that there exists $\mathbf{u} \in \mathbb{V}_m$ such that $|W(\hat{f})(\mathbf{u})| > 2^{m/2}$. Thus for any $\mathbf{v} \in \mathbb{V}_n$, $|W(\hat{g})(\mathbf{v})| < 2^{n/2}$, since $2^{(m+n)/2} = |W(\hat{f})(\mathbf{u})||W(\hat{g})(\mathbf{v})|$. Using Parseval's equation (Corollary 2.19) we get a contradiction. ∎

Corollary 5.23. *The function*

$$f(\mathbf{x}) = x_1x_2 \oplus x_3x_4 \oplus \cdots \oplus x_{2k-1}x_{2k}, \quad k \geq 1, \tag{5.19}$$

is bent.

Theorem 5.22 is the result that prompted Dillon [151] to point out that functions that occur in that way are 'rather uninteresting', since they can be decomposed into functions defined on lower dimensional vector spaces. Define a Boolean function $f : \mathbb{V}_n \to \mathbb{F}_2$ to be *decomposable* [151], [287, Chapter 14, p. 429] (Rothaus [385] calls such a function irreducible) if it is linearly equivalent to a sum of functions in disjoint sets of variables, that is, there exists a nonsingular $n \times n$ matrix T, $n = m + k$, $1 \leq m < n$, and two functions $g : \mathbb{V}_m \to \mathbb{F}_2$, $h : \mathbb{V}_k \to \mathbb{F}_2$ such that $f(\mathbf{x}T) = g(x_1, \ldots, x_m) \oplus h(x_{m+1}, \ldots, x_n)$.

Lemma 5.24. *For $n \geq 6$, every bent function of degree $k = n/2$ on $\mathbb{V}_n$ is indecomposable.*

Proof. We give the simple argument of Rothaus [385]. If the bent function f of degree k is linearly equivalent to the sum of two functions g and h in disjoint variables, then, since both g, h must be bent by Theorem 5.22, their degrees are less than k, which means that their sum cannot have degree k. That is a contradiction and the lemma is proved. ∎

The above lemma is not true for $n = 4$ as we can see by taking the bent function $x_1x_2 \oplus x_3x_4$ on $\mathbb{V}_4$.

Using Dickson's theorem (cf. [287, Chapter 15, Theorem 4, p. 438]) one can show that any quadratic bent function is an affine transformation of the above 'principal quadratic' function. We define a matrix to be *symplectic* if it is symmetric and has zeros on the principal diagonal. We state Dickson's theorem without proof.

Theorem 5.25. *If B is a symplectic $n \times n$ matrix of rank $2h$, then there exists an invertible binary matrix R such that RBR^t has zeros everywhere except on the two diagonals immediately above and below the main diagonal, and there has* $1010\cdots100\cdots0$ *with h ones.*

Every second order Boolean function can by an affine transformation of variables be reduced to

$$\sum_{i=1}^{h} x_{2i-1}x_{2i} \oplus \ell(x_{2h+1},\ldots,x_n), \tag{5.20}$$

where ℓ is an affine function.

Concatenation under certain conditions produces also some bent functions of higher dimension (see Preneel et al. [371, p. 168]).

Proposition 5.26. *The concatenation f in $n+2$ variables of 4 bent functions g_i $(1 \le i \le 4)$ in n variables is bent if and only if*

$$W(\hat{g}_1)(\mathbf{w})W(\hat{g}_2)(\mathbf{w})W(\hat{g}_3)(\mathbf{w})W(\hat{g}_4)(\mathbf{w}) = -2^{2n}, \quad \textit{for all } \mathbf{w} \in \mathbb{V}_n.$$

Proposition 5.27. *Let f be the concatenation $f = g_1||g_2||g_3||g_4$, with $g_i : \mathbb{V}_n \to \mathbb{F}_2$. If f, g_1, g_2, g_3 are bent, then g_4 is bent. If $g_1 = g_2$, then $g_3 = \bar{g}_4$; If $g_1 = g_2 = g_3$, then $g_4 = \bar{g}_1$.*

From a given bent function we can produce 23 new other ones using the next result.

Proposition 5.28. *If the concatenation of* 4 *arbitrary vectors is bent, then the concatenation of all* 4! *permutations of these vectors is bent.*

The class of bent functions is stable under addition of affine functions. This also follows from Theorem 5.12, but we give a direct proof here.

Theorem 5.29. *If f is bent, then $f \oplus \ell$ is bent for any affine function ℓ.*

Proof. It suffices to consider the case of $\ell(\mathbf{x}) = \ell_{\mathbf{v}}(\mathbf{x}) = \mathbf{x}\cdot\mathbf{v}$, since adding 1 to a function will simply complement the truth table. We note that if $\ell_{\mathbf{v}}(\mathbf{x}) = \mathbf{x}\cdot\mathbf{v}$ and $g(\mathbf{x}) = f(\mathbf{x}) \oplus \mathbf{x}\cdot\mathbf{v}$, then

$$\begin{aligned} W(\hat{g})(\mathbf{u}) &= \sum_{\mathbf{x}\in\mathbb{V}_n} (-1)^{g(\mathbf{x})\oplus\mathbf{u}\cdot\mathbf{x}} \\ &= \sum_{\mathbf{x}\in\mathbb{V}_n} (-1)^{f(\mathbf{x})\oplus(\mathbf{v}\oplus\mathbf{u})\cdot\mathbf{x}} \\ &= W(\hat{f})(\mathbf{u}\oplus\mathbf{v}). \end{aligned} \tag{5.21}$$

Thus if f is bent, then g is bent for any $\mathbf{v} \in \mathbb{V}_n$. ∎

A natural question now is whether by performing an affine transformation on the variables of a bent function we produce also a bent function. The answer is positive (we shall prove it next) if the transformation is nonsingular, thus answering a requirement for good cryptographic functions.

Let A be an $n \times n$ invertible matrix over $\mathbb{F}_2$, and B a vector in $\mathbb{V}_n$. Write $g(\mathbf{x}) = f(A\mathbf{x} \oplus B)$. Then

$$\begin{aligned} W(\hat{g})(\mathbf{u}) &= \sum_{\mathbf{x}\in\mathbb{V}_n}(-1)^{g(\mathbf{x})\oplus\mathbf{u}\cdot\mathbf{x}} = \sum_{\mathbf{x}\in\mathbb{V}_n}(-1)^{f(A\mathbf{x}\oplus B)\oplus\mathbf{u}\cdot\mathbf{x}} \\ &= \sum_{\substack{\mathbf{w}\in\mathbb{V}_n,\\ \mathbf{x}=A^{-1}\mathbf{w}\oplus A^{-1}B}} (-1)^{f(\mathbf{w})\oplus\mathbf{u}A^{-1}\cdot\mathbf{w}\oplus\mathbf{u}\cdot A^{-1}B} \\ &= (-1)^{\mathbf{u}\cdot A^{-1}B}\, W(\hat{f})(\mathbf{u}A^{-1}). \end{aligned} \tag{5.22}$$

Hence we have show:

Theorem 5.30. *Let A be an $n \times n$ invertible matrix over $\mathbb{F}_2$, and v a vector in $\mathbb{V}_n$. If f is bent, then $g(\mathbf{x}) = f(A\mathbf{x} \oplus v)$ is also bent.*

5.7 ROTHAUS'S CONSTRUCTION

In his small but beautiful paper [385], Rothaus identified two large general classes of bent functions on $\mathbb{V}_n$, $n = 2k$. These are given as follows:

(F1) Let $x_1, y_1, \ldots, x_k, y_k$ be independent variables. Also, let $P(\mathbf{x})$ be an arbitrary polynomial, where $\mathbf{x} = (x_1, \ldots, x_k)$. Then

$$f_1(x_1, y_1, \ldots, x_k, y_k) = \sum_{i=1}^{k} x_i y_i \oplus P(\mathbf{x}) \tag{5.23}$$

is a bent function. This generalizes Corollary 5.23.

(F2) Let $x_1, x_2, \ldots, x_k, x_{k+1}, y_{k+1}$ be independent variables and $P(\mathbf{x})$ be an arbitrary polynomial, where $\mathbf{x} = (x_1, \ldots, x_k)$. Now, let $A(\mathbf{x}), B(\mathbf{x}), C(\mathbf{x})$ be bent functions such that $A(\mathbf{x}) \oplus B(\mathbf{x}) \oplus C(\mathbf{x})$ is also bent. Then the polynomial

$$\begin{aligned} f_2(\mathbf{x}, x_{k+1}, y_{k+1}) &= A(\mathbf{x})B(\mathbf{x}) \oplus B(\mathbf{x})C(\mathbf{x}) \oplus C(\mathbf{x})A(\mathbf{x}) \\ &\quad \oplus [A(\mathbf{x}) \oplus B(\mathbf{x})]y_{k+1} \oplus [A(\mathbf{x}) \oplus C(\mathbf{x})]x_{k+1} \oplus x_{k+1}y_{k+1} \end{aligned} \tag{5.24}$$

is a bent function.

For a function f belonging to the first family of bent functions (F1), the dual can be calculated by

$$\mathcal{F}_1(x_1, y_1, \ldots, x_k, y_k) = \sum_{i=1}^{k} x_i y_i \oplus P(\mathbf{y}), \quad \text{where } y = (y_1, \ldots, y_k).$$

Using the first family we deduce:

Theorem 5.31. *There exist bent functions on $\mathbb{V}_n$ of every degree $2 \le d \le n/2 = k$. Moreover, the functions*

$$\begin{aligned} f_2(\mathbf{x},\mathbf{y}) &= \mathbf{x}\cdot\mathbf{y} \\ f_3(\mathbf{x},\mathbf{y}) &= \mathbf{x}\cdot\mathbf{y} \oplus x_1x_2x_3 \\ &\cdots\cdots\cdots\cdots \\ f_k(\mathbf{x},\mathbf{y}) &= \mathbf{x}\cdot\mathbf{y} \oplus x_1x_2\cdots x_k \end{aligned} \tag{5.25}$$

are pairwise inequivalent bent functions on $\mathbb{V}_n$ (recall Definition 2.15*).*

5.8 MAIORANA AND MCFARLAND'S CONSTRUCTION

Maiorana (see [151]) generalized the Rothaus construction by finding one more class of bent functions.

(F3) The function

$$f(\mathbf{x},\mathbf{y}) = \pi(\mathbf{x})\cdot\mathbf{y} \oplus g(\mathbf{x}), \quad \mathbf{x},\mathbf{y} \in V_{n/2}$$

is bent, where g is arbitrary and π is a permutation on $\mathbb{V}_n$.

We can calculate the dual for a bent function in this class by

$$\mathscr{F}(\mathbf{x},\mathbf{y}) = \mathbf{x}\cdot\pi^{-1}(\mathbf{y}) \oplus g\left(\pi^{-1}(\mathbf{y})\right).$$

A very useful characterization for the permutations of $\mathbb{V}_n$ was given by Dillon in [151].

Lemma 5.32. *The function $\pi(\mathbf{x}) = (P_1(\mathbf{x}),\ldots,P_n(\mathbf{x}))$ is a permutation of $\mathbb{V}_n$ if and only if every nonzero linear combination of the P_i s is a balanced function on $\mathbb{V}_n$.*

Remark 5.33. *Dillon points out in [151] that Maiorana and McFarland arrived independently at the same construction, so it is only fitting to call this the class of Maiorana–McFarland (MM) type.*

The previous construction was generalized by Carlet [80,82] (see the proposition below) and in one direction by Stănică [434]; the latter result is proved in Section 5.12.

Proposition 5.34. *Let $n - r + s$ be even, $r \le s$, and ϕ a mapping from $\mathbb{V}_s$ to $\mathbb{V}_r$ such that, for every $\mathbf{a} \in \mathbb{V}_r$, the set $\phi^{-1}(\mathbf{a})$ is an $(n-2r)$-dimensional affine subspace of $\mathbb{V}_s$. Let g be any Boolean functions on $\mathbb{V}_s$ whose restriction on $\phi^{-1}(\mathbf{a})$ is bent for every $\mathbf{a} \in \mathbb{V}_r$, if $n > 2r$ (no restriction is imposed if $n = 2r$). Then the function $f_{\phi,g} = \mathbf{x}\cdot\phi(\mathbf{y}) \oplus g(\mathbf{y})$ is bent on $\mathbb{V}_n$.*

McFarland in [311] discovered a very general class of difference sets that produce new bent functions.

Theorem 5.35. *Let E be a vector space of dimension $s+1$ over the finite field $\mathbb{F}_q$, let $H_1,\ldots,H_r, r=\frac{q^{s+1}-1}{q-1}$ be the hyperplanes in E, and let $e_1,e_2,\ldots,e_r$ be any r elements of E. Let K be an arbitrary group of order $r+1$ and let $k_1,\ldots,k_r$ be any r distinct elements of K. Let $C_i=H_i+(e_i,k_i)$ denote the coset of H_i in the direct product $G=E\times K$ which contains the element (e_i,k_i). Then $D=C_1\cup C_2\cup\cdots\cup C_r$ is a difference set in G with parameters*

$$(v,k,\lambda,n)=\left(q^{s+1}\left[\frac{q^{s+1}-1}{q-1}+1\right],q^s\left[\frac{q^{s+1}-1}{q-1}\right],q^s\left[\frac{q^s-1}{q-1}\right],q^{2s}\right). \quad (5.26)$$

Specializing the above theorem we arrive at

(F4) Let $H_1,\ldots,H_{2^k-1}$ be k-dimensional subspaces of $\mathbb{V}_n, n=2k$, such that

$$H_i\cap H_j=\{0\},\quad 1\le i<j\le 2^k-1,$$

and let $H_i^*=H_i-\{0\}$. Then $D=\cup H_i^*$ is a $(4^k,2\cdot 4^{k-1}-2^{k-1},4^{k-1}-2^{k-1},4^{k-1})$-difference set in $\mathbb{V}_{2k}$ and the characteristic function of D is a bent function on $\mathbb{V}_{2k}$.

Again the dual can be computed easily.

Proposition 5.36. *The dual of the characteristic function of $\cup H_i$ is the characteristic function of $\cup H_i^{\perp}$.*

By applying a linear transformation if necessary, we may assume that

$$H_1=\{(\mathbf{x},0),\mathbf{x}\in\mathbb{V}_k\},\quad H_2=\{(0,\mathbf{y}),\mathbf{y}\in\mathbb{V}_k\},$$

so the characteristic function of $D=H_1^*\cup H_2^*$ contains $x_1\cdots x_k$ and $x_{k+1}\cdots x_{2k}$. Thus we have proved:

Proposition 5.37. *Every bent function in* (F4) *has degree k.*

5.9 DILLON'S CONSTRUCTION

In his very nice survey paper [152], Dillon constructed a class of bent functions which he proved are not equivalent to any bent function in the Maiorana–McFarland family. He introduced the *Partial Spreads class* $\mathcal{PS}$ as the set of all sums modulo 2 of the characteristic functions (which take the value 1 on the subspace and 0 elsewhere) of $2^{n/2-1}$ (the $\mathcal{PS}^-$ class) or $2^{n/2-1}+1$ (the $\mathcal{PS}^+$ class) subspaces of dimension $n/2$ which intersect in 0 only (and so their direct sum is $\mathbb{V}_n$). All the functions in the $\mathcal{PS}^-$ class have maximum algebraic degree, namely $n/2$. Explicit characterization of these classes is not easy. A subclass labeled $\mathcal{PS}_{ap}$ of $\mathcal{PS}^-$ can be described on $\mathbb{V}_n$ which is identified with $\mathbb{F}_{2^{n/2}}\times\mathbb{F}_{2^{n/2}}$, a two-dimensional vector space over $\mathbb{F}_{2^{n/2}}$: a function f belongs to

$\mathcal{PS}_{ap}$ if it is of the form

$$f(x,y) = g\left(xy^{2^{n/2}-2}\right),$$

where g is a balanced function on $V_{n/2}$ with $g(0) = 0$. Moreover, Dillon [152] proved the following theorem which gives explicitly some of the functions that belong to the subclass $\mathcal{PS}_{ap}$.

Theorem 5.38. *Let $F = \mathbb{F}_2$, $K = \mathbb{F}_{2^{2k}}$ and $K^* = L^*H$, where L^* (resp. H) is the group of $(2^k+1)th$ (respectively $(2^k-1)st$) powers in K. Let α be any element of K such that exactly 2^{k-1} elements of the coset αH have trace 1 with respect to the extension K/F. Then the function*

$$\mathrm{Tr}_{K/F}(\alpha x^{2^k-1}) \tag{5.27}$$

is the characteristic function of a difference set in K. Furthermore, for $m > 3$ these difference sets are not equivalent to any in the Maiorana-McFarland family.

Dillon gives an example on $\mathbb{F}_{2^8}$ of a bent function, constructed using the idea in the above theorem and he extended that example to $\mathbb{V}_{2k} = \mathbb{V}_8 \oplus \mathbb{V}_{2k-8}$, by applying a procedure similar to our Theorem 5.22. In the paper [153], Dillon states a conjecture (see Theorem 5.39 below), which is proved in the paper of Lachaud and Wolfmann [266], although they do not state this explicitly. Dillon informed us [154] that the conjecture was also settled in the affirmative by N. Katz in a private communication.

Theorem 5.39. *For all $k > 1$, there exists a nonzero element α in $\mathbb{F}_{2^k}$ such that*

$$\sum_{z \in \mathbb{F}^*_{2^k}} (-1)^{\mathrm{Tr}_{\mathbb{F}_{2^k}/\mathbb{F}_2}(\alpha z + z^{-1})} = -1. \tag{5.28}$$

Since the conjecture is true, the function

$$\mathrm{Tr}_{K/F}(\alpha x^{2^k-1})$$

on $K = \mathbb{F}_{2^{2k}}$ is the characteristic function of a difference set in K which (for $k > 3$) is not equivalent to any difference set in the Maiorana-McFarland class.

We order the elements of $\mathbb{F}_{2^n}$ by a power of a primitive element β, that is

$$b_1 = 0 \text{ (that is, } \beta^{-\infty}), \quad b_2 = 1 = \beta^0, \ldots, b_{2^n} = \beta^{2^n-2}.$$

On $\mathbb{F}_{2^8}$ we consider the Dillon bent function (see [153])

$$f(x) = \mathrm{Tr}(x^{15}) = x^{15} + x^{30} + x^{60} + x^{120} + x^{135} + x^{195} + x^{225} + x^{240}, \tag{5.29}$$

with values in $\mathbb{F}_2$. Using the order determined by a solution of the primitive polynomial $x^8 + x^6 + x^5 + x^4 + 1$ and a computer program in Mathematica, we write f in the bit form as

$$\begin{aligned} f = &|0110100011000101 1|0110100011000101 1|0110100011000101 1|\\ &0110100011000101 1|0110100011000101 1|0110100011000101 1|\\ &0110100011000101 1|0110100011000101 1|0110100011000101 1|\\ &0110100011000101 1|0110100011000101 1|0110100011000101 1|\\ &0110100011000101 1|0110100011000101 1|0110100011000101 1|0. \end{aligned}$$

We inserted the vertical lines to emphasize the repeated strings. Regarded as a function on $\mathbb{V}_n$, the above string has low nonlinearity. Indeed, our calculation revealed the nonlinearity of the above string to be 88, which is even lower than the nonlinearity of some balanced Boolean functions (see Section 5.15).

We call a function *m-periodic* (multiplicative periodic) of *m-period* T if $f(x)=f(Tx)$ for any x. We analyze now all the functions in Dillon's class. Let

$$f_\alpha(x)=\mathrm{Tr}_{K/F}(\alpha x^{2^k-1})=\sum_{i=0}^{n-1}\alpha^{2^i}x^{(2^k-1)2^i}$$

(the powers are taken modulo $2^{2k}-1$) be the bent function of Theorem 5.38.

In the following theorem we show that any function in Dillon's class displays this periodicity property if one looks at it in the n-dimensional space $\mathbb{F}_2^n$ as opposed to $\mathbb{F}_{2^n}$. We take β to be a primitive element of $\mathbb{F}_{2^{2k}}$ over $\mathbb{F}_2$.

Theorem 5.40. *Let $f_\alpha:\mathbb{F}_{2^{2k}}\to\mathbb{F}_2$ be a function in Dillon's class (given in Theorem 5.38). Then f_α is affine equivalent to an m-periodic function of period β^{2^k+1}.*

Proof. Let $F=\mathbb{F}_2$, $K=\mathbb{F}_{2^{2k}}$. Since any element of K^* (in particular α) is a product of a (2^k+1)-power and a (2^k-1)-power, we can write $\alpha=a_1^{2^k+1}a_2^{2^k-1}$, and so, the function f_α becomes

$$f_\alpha(x)=\mathrm{Tr}_{K/F}(\alpha x^{2^k-1})=\mathrm{Tr}_{K/F}(a_1^{2^k+1}a_2^{2^k-1}x^{2^k-1})=\mathrm{Tr}_{K/F}(a_1^{2^k+1}(a_2x)^{2^k-1}).$$

Thus any function in Dillon's class is equivalent to a function of the form

$$g(x)=\mathrm{Tr}_{K/F}(a^{2^k+1}x^{2^k-1}).$$

Any element of K^* is a power of β, so $a=\beta^{i_0}, 0\le i_0<2^k-1$. Hence, g can be written as

$$g(\beta^j)=\mathrm{Tr}_{K/F}(\beta^{i_0(2^k+1)+j(2^k-1)}),\quad 0\le j\le 2^n. \tag{5.30}$$

We must show that the m-period of the above function is equal to β^{2^k+1}. For this we must prove that

$$\mathrm{Tr}_{K/F}(\beta^{i_0(2^k+1)+j(2^k-1)})=\mathrm{Tr}_{K/F}(\beta^{i_0(2^k+1)+(j+(2^k+1))(2^k-1)}),$$

which is implied by $i_0(2^k+1)+j(2^k-1) \equiv i_0(2^k+1)+(j+(2^k+1))(2^k-1) \pmod{2^n-1}$, and this is certainly true. ■

We do not know if one can practically use this observation, so at this stage it is of theoretical interest only.

5.10 DOBBERTIN'S CONSTRUCTION

Dobbertin [157] found a new class of bent functions that strictly includes both $\mathcal{PS}_{ap}$ and the MM class. Let σ be a balanced Boolean function on $L := \mathbb{F}_{2^k}$, and $\phi, \psi : L \to L$ be such that ϕ is one to one and ψ is arbitrary. Define the function

$$f_{\sigma,\phi,\psi}(x, \phi(y)) = \begin{cases} \sigma\left(\dfrac{x+\psi(y)}{y}\right) & \text{if } y \neq 0 \\ 0 & \text{otherwise.} \end{cases} \tag{5.31}$$

If σ is affine, Dobbertin showed that the above Equation (5.31) gives precisely the Maiorana–McFarland functions. Dillon's construction gives precisely $f_{\sigma,1_L,0}$, where 1_L is the identity on L. In this case, one can show that $f_{\sigma,\phi,\psi}$ and $f_{\sigma,1_L,0}$ are affinely equivalent. As examples of bent functions which are of neither type, Dobbertin gave the following. Let $\phi(x) = x^d$, $\psi(x) = x^{d'}$ (or $\psi = 0$) for $d, d' < 2^n - 1$, and a nontrivial subspace U of L and $y_0 \in L \setminus U$ be given such that:

- ϕ is bijective, that is $\gcd(d, 2^n - 1) = 1$;
- ϕ and ψ are not affine, that is, d, d' are not powers of 2;
- ϕ and ψ are affine on $y_0 + U$.

Define $\sigma : L \to \mathbb{F}_2$ as a nonaffine Boolean function by

$$\sigma(x) = \tau\rho(x) + \mathrm{Tr}(xy_0),$$

where $\rho : L \to U$ is an onto linear mapping chosen such that the map above is a one to one correspondence between all Boolean functions σ with the support supp $W(f) \subseteq y_0 + U$ and all Boolean functions τ on U. The function $f = f_{\sigma,\phi,\psi}$ is neither of Dillon nor of Maiorana–McFarland type. Explicitly, f will have the form

$$f(x, y^d) = \begin{cases} \tau\rho\left(\dfrac{x}{y} + y^{d'-1}\right) + \mathrm{Tr}\left(\left(\dfrac{x}{y} + y^{d'-1}\right) y_0\right) & \text{if } y \neq 0 \\ 0 & \text{otherwise.} \end{cases}$$

5.11 CARLET'S CONSTRUCTION

Carlet [65,67] introduced two new classes of bent functions by modifying the Maiorana–McFarland class. The first class, $\mathscr{D}_0$, is the set of all Boolean functions on $\mathbb{V}_n$ ($n=2k$), of the form

$$f(\mathbf{x},\mathbf{y})=\mathbf{x}\cdot\pi(\mathbf{y})\oplus\delta_0(\mathbf{x}),$$

where δ_0 is the Dirac function (see Theorem 5.13), and π is a permutation on $\mathbb{V}_{n/2}$. The dual of each such function is $\mathbf{y}\cdot\pi^{-1}(\mathbf{x})\oplus\delta_0(\mathbf{y})$. These functions are bent and, interestingly enough, this class strictly includes the MM class, as well as Dillon's class of bent functions. A superset of $\mathscr{D}_0$, namely $\mathscr{D}$, contains all functions of the form $f(\mathbf{x},\mathbf{y})=\mathbf{x}\cdot\pi(\mathbf{y})\oplus 1_{E_1}(\mathbf{x})1_{E_2}(\mathbf{y})$, where π is a permutation on $\mathbb{V}_{n/2}$, E_1,E_2 are two linear subspaces of $\mathbb{V}_{n/2}$ such that $\pi(E_2)=E_1^{\perp}$, of characteristic functions $1_{E_1},1_{E_2}$.

The second class of bent functions, say $\mathscr{C}$, is the set of all functions

$$f(\mathbf{x},\mathbf{y})=\mathbf{x}\cdot\pi(\mathbf{y})\oplus 1_U(\mathbf{x}),$$

where U is any linear subspace of $\mathbb{V}_{n/2}$ such that, for any $\mathbf{a}\in\mathbb{V}_{n/2}$, the set $\pi^{-1}(\mathbf{a}\oplus U^{\perp})$ is a flat (affine subspace). If one takes another Boolean function g on $\mathbb{V}_{n/2}$ whose restriction to every flat $\pi^{-1}(\mathbf{a}\oplus U^{\perp})$ is affine, then $f(\mathbf{x},\mathbf{y})=\mathbf{x}\cdot\pi(\mathbf{y})\oplus 1_U(\mathbf{x})\oplus g(\mathbf{y})$ is bent, as well [60].

5.12 EXTENDED MAIORANA–MCFARLAND CLASS

We know that if f is a bent function, so is $f\oplus\ell$, where ℓ is an affine transformation on $\mathbb{V}_n$; also, if we compose f with any nonsingular affine transformation on $\mathbb{V}_n$, the resulting function is still bent (see Theorems 5.29 and 5.30). We call the set of all functions of MM (Maiorana-McFarland) type subjected to these two operations the *extended MM class*. Canteaut et al. [60, Lemma 33] state the following easy result, which gives a simple, necessary and sufficient condition for a bent function to belong to the extended MM class.

Lemma 5.41. *Let $f:\mathbb{V}_n\to\mathbb{F}_2$ be a bent function. The following properties are equivalent:*

(i) *f is in the extended MM class.*
(ii) *There exists a subspace U of dimension $n/2$ such that the function f is affine on every coset (flat) of U.*

In what follows we always assume $n\geq 8$ and we use the term *block* to mean a string of 4 consecutive bits v_i in f given by its truth table

$$f=(v_0,v_2,\ldots,v_{2^n-1}). \tag{5.32}$$

We shall use capital letters to denote blocks or strings made up by concatenation of consecutive blocks. If X and X' are such strings of blocks with the same length, then the notation (in this section only)

$$(X|X') \tag{5.33}$$

will mean that X and X' occupy the same positions in the first and second half of f (where X need not necessarily start with the first position in f). Below we will often use the eight special blocks contained in the set T defined by

$$T := \begin{Bmatrix} A=0,0,1,1; \bar{A}=1,1,0,0; B=0,1,0,1; \bar{B}=1,0,1,0; \\ C=0,1,1,0; \bar{C}=1,0,0,1; D=0,0,0,0; \bar{D}=1,1,1,1. \end{Bmatrix} \tag{5.34}$$

It is convenient to have the following lemma (part of the 'folklore' of the theory of Boolean functions) that characterizes the affine functions [434].

Lemma 5.42 (*Folklore lemma*). *The truth table of an affine function $f=(v_0,\ldots,v_{2^n-1})$ is a linear string of length 2^n made up of 4-bit blocks $I_1,\ldots,I_{2^{n-2}}$ given as follows:*

1. *The first block I_1 is one of $A,B,C,D,\bar{A},\bar{B},\bar{C}$ or $\bar{D}$.*
2. *The second block I_2 is I_1 or $\bar{I}_1$.*
3. *The next two blocks I_3, I_4 are I_1, I_2 or $\bar{I}_1$, $\bar{I}_2$.*
4. *The next four blocks $I_5,\ldots,I_8$ are $I_1,\ldots,I_4$ or $\bar{I}_1,\ldots,\bar{I}_4$.*

...

$n-1$. *The last 2^{n-3} blocks $I_{2^{n-3}+1},\ldots,I_{2^{n-2}}$ are $I_1,\ldots,I_{2^{n-3}}$ or $\bar{I}_1,\ldots,\bar{I}_{2^{n-3}}$.*

Proof. Any affine function f has the form

$$f(\mathbf{x})=a_1x_1\oplus\cdots\oplus a_nx_n\oplus a_0,\quad a_i\in\mathbb{F}_2\ (0\le i\le n).$$

By our lexicographical convention for the values v_i, the values of f for which $x_1=0$ are $v_0,\ldots,v_{2^{n-1}-1}$, the values of f for which $x_1=x_2=0$ are $v_0,\ldots,v_{2^{n-2}-1}$, and so on. Thus, by choosing $I_1 = v_0v_1v_2v_3$ from the eight possibilities we determine the coefficients a_{n-1}, a_n and a_0. Now the next block I_2 must be I_1 or $\bar{I}_2$, depending on whether a_{n-2} is 0 or 1, respectively. The next two blocks I_3,I_4 must be I_1,I_2 or $\bar{I}_1,\bar{I}_2$, depending on whether a_{n-3} is 0 or 1, respectively. Continuing in this way, we prove the lemma. ■

In view of Lemma 5.42, we can say that any affine function is *based on* one of the blocks A,B,C,D.

In order to describe our construction of bent functions, we need to extend the notation (5.33). If H,I,J,K are strings of the same length made up of concatenations of consecutive blocks, then the notation

$$(H\ldots J|I\ldots K) \tag{5.35}$$

means that the pairs H,I and J,K each occupy the same positions in the first and second halves of the function (with 2^n bits) given by (5.35), but H and J need not be adjacent strings. We shall use the term *segment* for a portion of a function such as the portion defined in (5.35).

A method for explicitly constructing a class of bent functions is given in the following theorem [434]. As usual, $H \oplus I$ denotes the xor of the strings H and I.

Theorem 5.43. *For $n = 2k \geq 8$ and $1 \leq i \leq 2^{2k-1}$, let M_i, N_i, P_i, Q_i be arbitrary affine functions made up of 2^{k-4} blocks. Define f to be the concatenation of the 2^{k-1} segments S_i given by*

$$S_i = (M_i N_i \dots P_i Q_i | P_i Q_i \dots \bar{M}_i \bar{N}_i). \tag{5.36}$$

In each segment exactly two of the blocks are based on each of the letters A,B,C,D. The strings $M_i N_i \dots P_i Q_i$ may be placed arbitrarily in the segments S_i of the form (4), *subject to the conditions:*

(i) *Concatenation of the 2^{k-1} segments gives 2^{2k-2} blocks, each of which occupies a different position in the function f.*
(ii) *For each $i = 1,2,\dots,2^{k-2}$, we have*

$$S_{i+2^{k-2}} = (N_i \bar{M}_i \dots Q_i \bar{P}_i | Q_i \bar{P}_i \dots \bar{N}_i M_i).$$

(iii) *For any pair i,j with $1 \leq i < j \leq 2^{k-2}$, all of the strings $M_i \oplus M_j, N_i \oplus N_j, P_i \oplus P_j, Q_i \oplus Q_j$ are balanced.*

Then f is a bent function.

An example of a function which satisfies the conditions of Theorem 5.43 for $n = 8$ is

$$DAAC\ ABCD\ BCDB\ CDBA\ A\bar{D}C\bar{A}\ B\bar{A}D\bar{C}\ C\bar{B}B\bar{D}\ D\bar{C}A\bar{B}$$
$$AC\bar{D}\bar{A}\ CD\bar{A}\bar{B}\ DB\bar{B}\bar{C}\ BA\bar{C}\bar{D}\ C\bar{A}\bar{A}D\ D\bar{C}\bar{B}A\ B\bar{D}\bar{C}B\ A\bar{B}\bar{D}C.$$

In order to give an idea of the combinatorial arguments that go into the proof of Theorem 5.43, we will briefly describe the proof of the following theorem, which is essentially a very special case of Theorem 5.43.

Theorem 5.44. *Let T_1 denote a subset of four elements of T, such that one element is chosen from each of the four pairs $A,\bar{A}; B,\bar{B}; C,\bar{C}; D,\bar{D}$. Suppose $n = 2k \geq 8$ and for each $i = 1,2,\dots 2^{k-3}$ and for each M in T_1, let $H_i^{(M)}$ denote an affine function based on the block corresponding to M and made up of 2^{k-3} blocks. Define f to be the concatenation of the 2^{k-1} segments U_{4i+j-4} $(1 \leq i \leq 2^{k-3}, 1 \leq j \leq 4)$ given by*

$$U_{4i+j-4} = (H_i^{(M)} \dots H_i^{(M)} | H_i^{(M)} \dots \bar{H}_i^{(M)}), \tag{5.37}$$

where $j = 1,2,3,4$ corresponds to $H_i^{(M)}$ based on A,B,C,D, respectively. The strings $H_i^{(M)}$ can be placed arbitrarily in the segments U_{4i+j-4} of form (5.37), *subject to the conditions:*

(i) *Concatenation of the* 2^{k-1} *segments gives* 2^{2k-2} *blocks, each of which occupies a different position in the function* f.

(ii) *For* $i=1$, *the corresponding segments* $U_1,\ldots,U_4$ *contain* 2^{k-3} *pairs of form* $(M|M)$ *and* 2^{k-3} *pairs, all of form* $(M|\bar{M})$ *or all of form* $(\bar{M}|M)$. *For* $i>1$, *each of the segments has* 2^{k-4} *pairs of each of the forms* $(M|M)$, $(\bar{M}|M)$, $(\bar{M}|\bar{M})$, $(M|\bar{M})$.

(iii) *For each* M *in* T_1 *and for any pair* i,j *with* $1\le i<j\le 2^{k-3}$, *the strings* $H_i^{(M)}\oplus H_j^{(M)}$ *are balanced.*

Then f *is a bent function.*

An example of a function which satisfies the conditions of Theorem 5.44 for $n=8$ is

$$\begin{array}{c} DDDD\ BBBB\ AAAA\ CCCC\ D\bar{D}D\bar{D}\ B\bar{B}B\bar{B}\ A\bar{A}A\bar{A}\ C\bar{C}C\bar{C} \\ DD\bar{D}\bar{D}\ BB\bar{B}\bar{B}\ AA\bar{A}\bar{A}\ CC\bar{C}\bar{C}\ D\bar{D}\bar{D}D\ B\bar{B}\bar{B}B\ \bar{A}AA\bar{A}\ \bar{C}CC\bar{C}. \end{array} \tag{5.38}$$

Now we turn to the proof of Theorem 5.44. If S is a given string and g is a given Boolean function, we shall sometimes use the notation

S^g = the string of bits in g which occupy the same positions as the bits in the string S.

If an affine function ℓ has its second half the same as its first half, we say that ℓ is *double*. If some bits of an affine function ℓ agree with the corresponding bits in a function f, we say that ℓ *cancels* those bits in f.

In order to prove Theorem 5.44, we need only prove that the function f defined there has nonlinearity at least $2^{2k-1}-2^{k-1}$. The idea of our proof is to count the bits which occur in f but do not occur in the corresponding positions in a given affine function ℓ, that is, we count the bits of f which are *not canceled* by ℓ.

Our proof relies on the following two combinatorial lemmas. In both lemmas, the function f is assumed to satisfy the conditions of Theorem 5.44.

Lemma 5.45. *Given any two segments* U_1, U_2 *using the same block* M *in* f *and any affine function* ℓ *made up of* M *and* $\bar{M}$, *then*

$$C_\ell(U_1U_2)\le 2^{k-1}+2^{k-3}, \tag{5.39}$$

where $C_\ell(T)$ *is the number of blocks in* T *that are canceled by the affine function* ℓ.

Proof. Without loss of generality we may assume that U_1, U_2 and ℓ are made up of A and $\bar{A}$. We first consider the case where $U_1^\ell = U_2^\ell$. We write $U_1=(H\ldots H|H\ldots \bar{H})$.

We let α and β denote the number of blocks in U_1 which are canceled by the corresponding block in ℓ and which have, respectively, the same block or the complementary block A or $\bar{A}$ in the corresponding position in the segment U_2. We let γ and δ denote the number of blocks in U_1 which are not canceled by the corresponding blocks in ℓ

and which have, respectively, the same block or the complementary block A or $\bar{A}$ in the corresponding position in the segment U_2. Thus we have

$$\alpha+\gamma=\beta+\delta=2^{k-2} \tag{5.40}$$

by condition (iii) of Theorem 5.44. Since the number of blocks canceled by ℓ in U_1 and U_2 is $\alpha+\beta$ and $\alpha+\delta$, respectively, the inequality (5.39) is equivalent to

$$2\alpha+\beta+\delta \leq 2^{k-1}+2^{k-3}. \tag{5.41}$$

Using the Equations (5.40) we obtain that (5.41) is equivalent to $2\alpha \leq 2^{k-2}+2^{k-3}$. So, it suffices to prove that $\alpha \leq 2^{k-3}+2^{k-4}$. In what follows, in order to avoid annoying repetitions, any canceled block in the segment U_1 that will be involved in the proof of this case will mean a canceled block which has the same block in the corresponding position in the segment U_2, unless otherwise stated. We write $\alpha=\alpha'+\alpha''$, where α' is the number of blocks from $(H|H)$ and α'' is the number of blocks from $(H|\bar{H})$ canceled by the corresponding block in ℓ.

When ℓ is double, then $\alpha' \leq 2^{k-3}$: if ℓ cancels s blocks in the first half of $(H|H)$ which have the same block in the corresponding position in U_2 (here $s \leq 2^{k-4}$ by condition (iii) of Theorem 5.44), then ℓ does not cancel the rest of the $2^{k-4}-s$ blocks because ℓ has a different block corresponding to their positions. Since this does not change for the second half of $(H|H)$, ℓ cancels there s blocks as well. This is true because in the second half we only need to consider the blocks corresponding to the blocks canceled in the first half: all other canceled blocks have different blocks in the corresponding positions in U_2. Since $s \leq 2^{k-4}$, we get that $\alpha' \leq 2^{k-3}$. When ℓ is not double we obtain $\alpha'=2^{k-4}$: if ℓ cancels s blocks in the first half of $(H|H)$ then ℓ cancels $2^{k-4}-s$ blocks in the second half because ℓ has the corresponding second half equal to the complement of the corresponding first half. Now we analyze the pair of strings $(H|\bar{H})$. When ℓ is double, then $\alpha'' \leq 2^{k-4}$: if ℓ cancels s blocks in the first half of $(H|\bar{H})$, which have the same block in the corresponding position in U_2, $s \leq 2^{k-4}$, then ℓ does not cancel the rest of the $2^{k-4}-s$ blocks because ℓ has different blocks in the positions corresponding to their positions. When ℓ is not double then we obtain $\alpha'' \leq 2^{k-3}$. Anyway, $\alpha \leq 2^{k-3}+2^{k-4}$ and the proof of the lemma is complete in this case. The case where $U_1^\ell=\bar{U}_2^\ell$ is very similar.

To complete the proof of the lemma, we must consider the cases where $U_1^\ell \neq U_2^\ell$ and $U_1^\ell \neq \bar{U}_2^\ell$. These cases depend on counting arguments somewhat similar to the above. We omit the details. ■

Lemma 5.46. *If in a segment U (which is of the form* (5.37) *in Theorem* 5.44*) an affine function ℓ cancels more than half of its blocks, then exactly $2^{k-2}+2^{k-3}$ blocks in U are canceled.*

Proof. By induction we show that if h is an affine function on $\mathbb{V}_m$ and an affine function ℓ cancels more than half of its blocks, then ℓ cancels the whole function h. For $m=3,4$ the assertion is trivially satisfied. Suppose that $m>4$ and ℓ cancels in $h=(H|H)$ (or $h'=(H|\bar{H})$) defined on $\mathbb{V}_m$ more than half of its blocks. Then there is a half of h (or h')

in which ℓ cancels more than half of its blocks. But each pair of strings h or h' is an affine function on $\mathbb{V}_{m-1}$ and according to the first step of induction, all of its blocks are canceled. Since ℓ cancels more than half of the blocks of one half of h (or h'), then ℓ must cancel at least one block in the other half. But this implies that ℓ cancels that half too because ℓ must have one of the forms $(H^\ell|H^\ell)$ or $(H^\ell|\bar{H}^\ell)$.

Now we finish the proof of our lemma. Suppose that ℓ cancels more than half of the blocks of $U = (H \ldots H|H \ldots \bar{H})$. This further implies that ℓ cancels more than half of the blocks of $(H|H)$ or $(H|\bar{H})$. From what we have proved before, since the two strings are affine functions, we get that ℓ cancels one of the pairs of strings $(H|H)$ or $(H|\bar{H})$ completely. Since ℓ could have only one of the forms

$$(H^\ell \ldots H^\ell|H^\ell \ldots H^\ell), \quad (H^\ell \ldots H^\ell|\bar{H}^\ell \ldots \bar{H}^\ell),$$
$$(H^\ell \ldots \bar{H}^\ell|H^\ell \ldots \bar{H}^\ell), \quad (H^\ell \ldots \bar{H}^\ell|\bar{H}^\ell \ldots H^\ell),$$

then ℓ must cancel one of the other quarters of f as well. This completes the proof of our lemma. ∎

Proof (*Proof of* Theorem 5.44). Let f be a function constructed in accordance with the conditions in the theorem and let ℓ be an arbitrary affine function, which we may assume is made up of A and $\bar{A}$ (see Folklore Lemma 5.42). There are exactly $2^{2k-2} - 2^{2k-4}$ blocks other than A or $\bar{A}$ in f and each of these blocks has Hamming distance 2 from A or $\bar{A}$, so the contribution of these blocks to $d(f,\ell)$ is

$$2^{2k-1} - 2^{2k-3}. \tag{5.42}$$

The remaining 2^{2k-4} blocks in f lie in the 2^{k-3} segments made up of A and $\bar{A}$. By Lemmas 5.45 and 5.46, no more than one segment can have more than half of its blocks canceled by ℓ, so at least $2^{k-3} - 1$ segments contribute at least 2^k to $d(f,\ell)$, for a total contribution which is $\geq$

$$2^k(2^{k-3} - 1) = 2^{2k-3} - 2^k. \tag{5.43}$$

If the one remaining segment has more than half of its blocks canceled by ℓ, then by Lemma 5.46 exactly $2^{k-2} + 2^{k-3}$ blocks are canceled, so the contribution to $d(f,\ell)$ by this segment is

$$4(2^{k-1} - (2^{k-2} + 2^{k-3})) = 2^k - 2^{k-1}. \tag{5.44}$$

Adding the contributions (5.42), (5.43) and (5.44) gives $d(f,\ell) \geq 2^{2k-1} - 2^{k-1}$, so f is bent. ∎

It turns out that the bent functions constructed in Theorem 5.44 are not new. It is clear that a function f which satisfies the conditions of Theorem 5.44 is linear in the variables $x_1, x_{k+2}, x_{k+3}, \ldots, x_{2k}$ and so it must be in the extended MM class by Theorem 5.44 and Lemma 5.41. For instance, the function (5.38) can be shown to be

$$f(x_1,\ldots,x_8) = x_1x_5 \oplus x_2x_6 \oplus x_3x_7 \oplus x_4x_8 \oplus x_1x_2x_3,$$

which is obviously in the MM class.

Dobbertin [157] points out that it can be quite difficult to decide whether a given bent function is in the extended MM class, but in Theorem 5.43 we have the advantage of a very explicit representation. It may be that even our Theorem 5.43 does not produce functions outside the MM class, although we are unable at this point to show this.

5.13 NORMAL AND NONNORMAL BENT FUNCTIONS

Dobbertin in [157] introduced the concept of normal Boolean function. We say that a Boolean function f in $n = 2k$ variables is *normal* if there exists a flat U (affine subspace of $\mathbb{V}_n$) of dimension k, on which f is constant. As a consequence, the restriction of f to each proper coset of U is balanced [157]. If there is such a flat U such that the restriction of f to U is affine but nonconstant, we call f a *weakly normal* Boolean function. Certainly, the concept of (weak) normality can be extended to odd dimensions n, as well, by taking the flat U of dimension $\lceil n/2 \rceil$. The concept of normality was introduced by Dobbertin in relation to constructions of bent functions so he had no interest in defining this concept for flats of other than dimension k. In fact, one can show that a bent function cannot be constant on a flat of dimension higher than k.

Since normality turned out to be an important property, it was extended to *(weak) m-normality*, which refers to a Boolean function on $\mathbb{V}_n$ that is (affine but nonconstant) constant on some flat U of dimension m.

The fact that there are nonnormal Boolean functions is easy to see and we reproduce here a counting argument of Blackburn in a private communication to Dobbertin [157]. Given a flat U of dimension k in $\mathbb{V}_{2k}$, there are exactly

$$b_k = 2^{2^{2k}-2^k+1}$$

Boolean functions that are constant on U. It is well-known that there are

$$s_k = \prod_{i=0}^{k-1} \frac{2^{2k}-2^i}{2^k-2^i} \asymp 2^{k^2}$$

subspaces of $\mathbb{V}_{2k}$ of dimension k, and so the number of flats of dimension k is $a_k = 2^k s_k$. An upper bound for the number of normal Boolean functions is

$$u_k = a_k b_k = 2^{2^{2k}-2^k+k+1} s_k \asymp 2^{2^{2k}-2^k+k^2+k+1} < 2^{2^{2k}} \quad \text{if } k \geq 5,$$

and so there are nonnormal Boolean functions on $\mathbb{V}_{2k}$.

The existence of nonnormal bent functions for $n \geq 10$ and weakly nonnormal for $n \geq 14$ on $\mathbb{V}_n$ was proven by Canteaut et al. [60] using an algorithm of Daum et al. [137], thus disproving a conjecture of Dobbertin [157]. These results are based on the work of Dillon and Dobbertin [155] on Kasami functions $f_{\alpha,j}: \mathbb{F}_{2^n} \to \mathbb{F}_2$, defined by $f_{\alpha,j}(x) = \mathrm{Tr}(\alpha x^d)$, where $d = 2^{2j} - 2^j + 1$, $\gcd(n, j) = 1$. They proved that $f_{\alpha,j}$ is bent if 6 does not divide n, $1 < j < n/2$. It was known that all bent functions for $n \leq 6$ are normal and these Kasami functions showed promise for being nonnormal; this was confirmed computationally by Canteaut et al. [60].

Normality behaves well on direct sums. In [86], Carlet et al. characterized when the direct sum of bent functions is normal.

Definition 5.47. *Let $\mathbb{U} \subseteq \mathbb{V}$ be two finite vector spaces over $\mathbb{F}_2$ and $\beta: \mathbb{U} \to \mathbb{F}_2$, $f: \mathbb{V} \to \mathbb{F}_2$ be bent functions. We say that f is a* normal extension *of β, and write $\beta \preceq f$, if $\mathbb{V}$ can be decomposed as $\mathbb{V} = \mathbb{U} \oplus \mathbb{W}_1 \oplus \mathbb{W}_2$ such that*

(i) $\beta(u) = f(u \oplus w_1)$, *for all $u \in \mathbb{U}, w_1 \in \mathbb{W}_2$;*
(ii) $\dim(\mathbb{W}_1) = \dim(\mathbb{W}_2)$.

One can show quite easily that relation $\preceq$ is reflexive, transitive, and antisymmetric.

Theorem 5.48. *If f_i are two bent functions on $\mathbb{V}_{n_i}$, $i = 1, 2$, then the direct sum $f_1 \oplus f_2$ is normal if and only if isomorphic or complementary-isomorphic bent functions β_i exist such that f_i is a normal extension of β_i $(i = 1, 2)$. Moreover, if $\beta \preceq f$, then f is normal if and only if β is normal.*

Proof. See Carlet et al. [86] for a proof. ■

Corollary 5.49. *The direct sum of normal bent functions is always normal. The direct sum of a normal and nonnormal bent function is always nonnormal.*

The above corollary enables us to construct an infinite family of nonnormal bent functions from a nonnormal one (the existence of such being guaranteed by Daum et al. [137] and Canteaut et al. [60]). All the constructions of bent functions until then provided only normal bent functions. In fact, Carlet [77,78] showed that asymptotically almost all Boolean bent functions are nonnormal.

For more on (weakly) normal bent functions we refer to [104,158,160,271].

5.14 COUNTING BENT FUNCTIONS

In [371] it was observed that the concatenation of the rows of the Sylvester–Hadamard matrix H_n yields a bent function. Moreover, the following result is true (see Preneel et al. [371, Theorem 6, p. 168]).

Lemma 5.50. *For $n = 2k$, the concatenation of the 2^k rows of H_n or their complements in arbitrary order results in $(2^k)!\,2^{2^k}$ different bent functions of n variables.*

Using the above lemma, we can deduce some 'naive' bounds on the number of bent functions, namely we have

$$(2^{n/2})!\,2^{2^{n/2}} \le \#\text{bent} \le 2^{2^{n-1}+\frac{1}{2}\binom{n}{n/2}}. \tag{5.45}$$

The upper bound is obtained from the fact that the degree of the bent function is less than $n/2$, which implies that the number of coefficients that appear in the polynomial expression of f is exactly $\sum_{i=0}^{n/2}\binom{n}{i}$. Since $\sum_{i=0}^{n}\binom{n}{i} = 2^n$, and $\binom{n}{i} = \binom{n}{n-i}$, we get that

$$\sum_{i=0}^{n/2}\binom{n}{i} = 2^{n-1} + \frac{1}{2}\binom{n}{n/2}.$$

From this we deduce the truth of our claim, since the coefficients could take only the values 0, 1.

Considering divisibility by powers of 2 of affine distances of a Boolean function, Carlet and Klapper [92] were able to find bounds on the number of bent, correlation immune and resilient functions. Denote by $B(k,n)$ the number of Boolean functions in n variables, whose algebraic degree is at most k; by $D(k,n)$ the number of Boolean functions in n variables, whose distances to the affine functions are all divisible by 2^k; $C(k,n)$ the number of Boolean functions in n variables, whose distances to the affine functions are all congruent to 2^k modulo 2^{k+1}. The number of bent functions is bounded by $C(n/2-1,n)$, the number $CI(n,m)$ of correlation immune functions of order m is bounded by $D(m,n)$, and the number of resilient functions of order m is bounded by $D(m+1,n)$. Let $\epsilon = 2^{-\left(\binom{n-1}{d-1}-\binom{n-1-d}{d-1}-1\right)}$ where $d = n/2$.

Theorem 5.51. *For $n \ge 6$ even, the number of bent functions in n variables B_n is at most*

$$\frac{B(\frac{n}{2},n)(1+\epsilon)}{2^{2^{n/2}-n/2-1}} + B\left(\frac{n}{2}-1,n\right) = 2^{2^{n-1}+\binom{n}{n/2}/2-2^{n/2}+n/2+1}(1+\epsilon) + 2^{2^{n-1}+\binom{n}{n/2}/2}.$$

Both $C(m-1,n)$ $(m>2)$ and $CI(n,m)$ are at most

$$\frac{2^{m+1}B(n-m,n)(1+\epsilon)}{B(n-m,m)} + B(n-m-1,n), \quad \text{if } 2 \le n/2 \le m < n,$$
$$\frac{B(n-m,n)-B(n-m-1,n)}{2^{2^m-m-1}} + B(n-m-1,n), \quad \text{if } 2 \le m < n/2.$$

The number of $(m-1)$-resilient functions in n variables is at most

$$\frac{2B(n-m,n)(1+\epsilon)}{B(n-m,m)} + B(n-m-1,n), \quad \text{if } n/2 \le m < n,$$
$$\frac{B(n-m,n)-B(n-m-1,n)}{2^{2^m-1}} + B(n-m-1,n), \quad \text{if } 2 \le m < n/2.$$

In the next table, taken from Preneel's thesis [368, p. 256] (except for the upper bound for $n = 6$ which is derived from the previous theorem, the lower and upper bounds for $n = 8$, and the exact count for $n = 8$, which come from Langevin et al. [270,269]), we give the known results concerning the number of bent functions and bounds on it.

n	**Lower bound by** (5.45)	# bent	**Upper bound**	# **Boolean**
2	8	8	8	16
4	384	896	2,048	65,536
6	$2^{23.3}$	$2^{32.3}$	2^{38} [92]	2^{64}
8	$2^{95.6}$ [269]	$2^{106.291}$ [269]	$2^{129.2}$ [270]	2^{256}

Preneel [368, pp. 255–256] described a computation which exactly determined the number B_6 of bent functions in 6 variables; the result is

$$B_6 = 5,425,430,528 \approx 2^{32.3}.$$

The value of B_6 is calculated in a different way by Meng et al. [323, p. 7]; they were apparently unaware of Preneel's computation. Dobbertin and Leander [159] explain a method for exactly determining the number B_8 of bent functions in 8 variables. Dobbertin's untimely death in 2006 delayed their actual computation. However, after extensive computations, Langevin et al. [269] announced on December 31, 2007, that the number of bent Boolean functions is $n_8 = 193,887,869,660,028,067,003,488,010,240 \approx 2^{97.291}$, up to affine terms, which gives the count $B_8 = 2^9 \cdot n_8$ from our table.

We want to mention here an interesting approach of Agievich [6] who introduced the notion of bent rectangles. Let n, m, k be positive integers with $n = m + k$. For a Boolean function f in n variables, we define

$$W'(f)(\mathbf{u},\mathbf{v}) = \sum_{\mathbf{y}\in\mathbb{V}_k} (-1)^{f(\mathbf{u},\mathbf{v})\oplus\mathbf{v}\cdot\mathbf{y}}, \quad \mathbf{u}\in\mathbb{V}_m, \mathbf{v}\in\mathbb{V}_k.$$

If each row of $W'(f)$ and each column (multiplied by $2^{(m-k)/2}$) are spectral sequences (occur as images under the Walsh–Hadamard transform) then we call $W'(f)$ a *bent rectangle*. Note that by definition, the rows of $W'(f)$ are always spectral sequences. Agievich [6] proved that a function is bent if and only if the corresponding rectangle is a bent rectangle. Working with bent squares, that is $m = k = n/2$, the author was able to construct $\sum_{d=2}^{2^n-2} \chi(n,d)$ distinct bent functions on any even dimension, where $\chi(n,d)$ is given by

$$\left(\left(\frac{(2^n-1)(2^{n-1}!)}{(2^{n-1}-d)!}\right)^2 \cdot 2^{4d} \sum_{i=0}^{d} \frac{(-1)^i}{i!} - \delta_n(d)\right) \cdot (2^n - 2d)! \cdot 2^{2^n-2d},$$

for any $d = 2, \ldots, 2^{n-1}$, where

$$\delta_n(d) = (d/2)! \cdot 2^{9d/2} \cdot \left(\frac{4}{3} s_n(d) r_n(d) - \frac{4}{9} r_n^2(d) \right),$$

if d is even and 0, otherwise, and

$$s_n(d) = \frac{(2^n - 1)(2^{n-1})!}{2^{d/2}(2^{n-1} - d)!(d/2)!}$$

$$r_n(d) = (2^n - 1)(2^{n-1} - 1) \binom{2^{n-2}}{d/2}.$$

If $n = 8$, then we obtain the lower bound $2^{70.4}$ for the number of bent functions, which is quite important since it does not use any computer work.

The counting and construction of all quadratic bent functions is possible as one can see in the next result.

Theorem 5.52. *Let f be a Boolean function of degree 2. Then f is bent if and only if the symplectic matrix M_f, associated to f, has full rank. It follows that the number of quadratic bent functions equals*

$$\prod_{i=0}^{\frac{n}{2}-1} \left(2^{2i+1} - 1\right) 2^{2i}.$$

Proof. See the proof of Theorem 9 in Preneel et al. [371, p. 169]. ■

5.15 HIGHLY NONLINEAR BALANCED FUNCTIONS

Bent functions have maximum nonlinearity and satisfy the SAC; however, they are not balanced and hence cannot be directly used in many cryptosystems, where the balance property is needed. We recall that a bent function on $\mathbb{V}_n$ contains $2^{2k-1} \pm 2^{k-1}$ zeros and $2^{2k-1} \mp 2^{k-1}$ ones. Meier and Staffelbach observed in [318] that by changing 2^{k-1} positions in a bent function we get a balanced function having a nonlinearity of at least $2^{2k-1} - 2^k$. It turns out that we can get the same nonlinearity by concatenating bent functions.

Using Theorems 5.43 and 5.44 we can construct highly nonlinear SAC balanced functions (see Theorems 5.57 and 5.58). Seberry and Zhang, and Seberry, Zhang and Zheng [407,409,410,412] also have done extensive work on the subject.

Remark 5.53. *If either f is balanced on $\mathbb{V}_n$ or g is balanced on $\mathbb{V}_m$, then $f(\mathbf{x}) \oplus g(\mathbf{y})$ is balanced on $\mathbb{V}_{m+n}$.*

Lemma 5.54. *Let f_i, $i = 1,2,3,4$ be bent functions on $\mathbb{V}_{2k}$. Set*

$$g_1(u,x_1,x_2,\ldots,x_{2k}) = (1\oplus u)f_1(\mathbf{x})\oplus u f_2(\mathbf{x})$$
$$g_2(u,v,x_1,\ldots,x_{2k}) = (1\oplus u)(1\oplus v)f_1(\mathbf{x})\oplus(1\oplus u)vf_2(\mathbf{x}) \oplus u(1\oplus v)f_3(\mathbf{x})\oplus uvf_4(\mathbf{x}). \quad (5.46)$$

Then we must have

$$\mathcal{N}_{g_1} \geq 2^{2k}-2^k, \quad \mathcal{N}_{g_2} \geq 2^{2k+1}-2^{k+1}. \quad (5.47)$$

Proof. Let η_{f_i} be the sequence of $f_i, 1\leq i\leq 4$. It is easy to see that $\eta_{g_1} = \eta_{f_1} \,||\, \eta_{f_2}$ is the sequence of g_1, of length 2^{2k+1}. Let η_ℓ be the sequence of an affine function, say ℓ. We know that η_ℓ is a row of the Sylvester–Hadamard matrix $\pm H_{2k+1}$. Since $H_{2k+1} = H_1\otimes H_{2k}$, η_ℓ can be expressed as a concatenation of two copies of $\pm L$, an affine sequence on $\mathbb{V}_{2k}$, that is $\eta_\ell = L\,||\,L$ or $\eta_\ell = L\,||\,(-L)$.

Obviously, $\eta_{g_1}\cdot\eta_\ell = \eta_{f_1}\cdot L\pm\eta_{f_2}\cdot L$, and since f_1,f_2 are bent functions, one must have $\eta_{f_i}\cdot L = \pm 2^k$, $i=1,2$. Thus $|\eta_{g_1}\cdot\eta_\ell|\leq 2^{k+1}$, and since $d(f,g) = 2^{n-1}-\frac{1}{2}\eta_f\cdot\eta_g$, for any two Boolean functions f,g, we deduce $d(g_1,\ell)\geq 2^{2k}-2^k$ which implies a similar inequality for the nonlinearity.

The function g_2 is defined in such a way that $\eta_{g_2} = \eta_{f_1}\,||\,\eta_{f_2}\,||\,\eta_{f_3}\,||\,\eta_{f_4}$. Now we use the fact that $H_{2k+2} = H_2\otimes H_{2k}$ and basically the same calculations imply $\mathcal{N}_{g_2}\geq 2^{2k+1}-2^{k+1}$. ■

Now we are able to construct SAC balanced functions with high nonlinearity. Consider a bent function f on $\mathbb{V}_{2k}$, $\mathbf{x} = (x_1,\ldots,x_{2k})$ and let ℓ be an affine function on $\mathbb{V}_{2k}$. From Theorem 5.29 we know that $f\oplus\ell$ is also bent. Since f takes the value zero $2^{2k-1}\pm 2^{k-1}$ times and the value one $2^{2k-1}\mp 2^{k-1}$ times, we may assume that f takes the value zero $2^{2k-1}+2^{k-1}$ times (otherwise, we replace f by $f\oplus 1$). By the same reasoning we may assume that $f\oplus\ell$ takes the value zero $2^{2k-1}-2^{k-1}$ times.

In [412, Theorem 3, Section 5], Seberry et al. proved the following result.

Theorem 5.55. *Under the previous assumptions, we define*

$$g_3(u,x_1,\ldots,x_{2k}) = f(\mathbf{x})\oplus u\,\ell(\mathbf{x}).$$

Then g_3 is balanced, satisfies the SAC and $\mathcal{N}_{g_3}\geq 2^{2k}-2^k$.

Proof. We prove first that g_3 is balanced. Note that $g_3(0,\mathbf{x}) = f(\mathbf{x})$ has $2^{2k-1}+2^{k-1}$ zeros and $g_3(1,\mathbf{x}) = f(\mathbf{x})\oplus\ell(\mathbf{x})$ has $2^{2k-1}-2^{k-1}$ zeros. Thus, g_3 has 2^{2k} zeros and 2^{2k} ones.

Now we prove that g_3 satisfies the SAC. Let $\mathbf{e} = (a_0,a_1,\ldots,a_{2k})$ be a vector of Hamming weight $wt(\mathbf{e}) = 1$, $\mathbf{a} = (a_1,\ldots,a_{2k})$, $\mathbf{z} = (u,x_1,\ldots,x_{2k})$ and $\mathbf{x} = (x_1,\ldots,x_{2k})$. Then

$$g(\mathbf{z}\oplus\mathbf{e}) = f(\mathbf{x}\oplus\mathbf{a})\oplus(u\oplus a_0)\ell(\mathbf{x}\oplus\mathbf{a}),$$

so

$$g(\mathbf{z}) \oplus g(\mathbf{z} \oplus \mathbf{e}) = f(\mathbf{x}) \oplus f(\mathbf{x} \oplus \mathbf{e}) \oplus u\,(\ell(\mathbf{x} \oplus \mathbf{e}) \oplus \ell(\mathbf{x})) \oplus a_0 \ell(\mathbf{x} \oplus \mathbf{a}). \tag{5.48}$$

If $a_0 = 0$, so $wt(\mathbf{a}) = 1$, then

$$g(\mathbf{z}) \oplus g(\mathbf{z} \oplus \mathbf{e}) = f(\mathbf{x}) \oplus f(\mathbf{x} \oplus \mathbf{a}) \oplus u\,(\ell(\mathbf{x} \oplus \mathbf{a}) \oplus \ell(\mathbf{x})).$$

But ℓ is a nonconstant affine function, so the last term in the above sum is a constant c. Thus

$$g(\mathbf{z}) \oplus g(\mathbf{z} \oplus \mathbf{e}) = f(\mathbf{x}) \oplus f(\mathbf{x} \oplus \mathbf{a}) \oplus cu,$$

and

$$\sum_{\mathbf{z} \in \mathbb{V}_{2k+1}} g(\mathbf{z}) \oplus g(\mathbf{z} \oplus \mathbf{e}) = 0. \tag{5.49}$$

If $a_0 = 1$, so $wt(\mathbf{a}) = 0$, then $g(\mathbf{z}) \oplus g(\mathbf{z} \oplus \mathbf{e}) = \ell(\mathbf{x})$, which is balanced. Thus the Equation (5.49) is true in this case as well.

Since $g_3 = f \oplus u\ell = (1 \oplus u)f \oplus u(f \oplus \ell)$, with f and $f \oplus \ell$ bent functions, we deduce, using the previous lemma, that $\mathcal{N}_{g_3} \geq 2^{2k} - 2^k$. ■

Seberry et al. [412, Lemma 17] also constructed balanced Boolean SAC functions with high nonlinearity on even dimensional spaces. Let $k \geq 2$, $f(x_1, \ldots, x_{2k})$ be a bent function. Consider three affine functions $\ell_i, i = 1,2,3$ such that $\ell_i \oplus \ell_j$ is nonconstant for any $i \neq j$. As before, without loss of generality we may assume that both f and $f \oplus \ell_1$ take the value zero $2^{2k-1} + 2^{k-1}$ times and both $f \oplus \ell_2$ and $f \oplus \ell_3$ assume the value zero $2^{2k-1} - 2^{k-1}$ times.

Theorem 5.56. *Set*

$$g_4(u, v, x_1, \ldots, x_{2k}) = f(\mathbf{x}) \oplus v\ell_1(\mathbf{x}) \oplus u\ell_2(\mathbf{x}) \oplus uv\,(\ell_1(\mathbf{x}) \oplus \ell_2(\mathbf{x}) \oplus \ell_3(\mathbf{x})).$$

Then g_4 is balanced, satisfies the SAC and $\mathcal{N}_{g_4} \geq 2^{2k+1} - 2^{k+1}$.

Proof. We can prove in a similar way as above that g_4 is balanced and satisfies the SAC. Now, for the nonlinearity, we use the identity

$$\begin{aligned} g_4 &= f \oplus v\ell_1 \oplus u\ell_2 \oplus uv(\ell_1 \oplus \ell_2 \oplus \ell_3) \\ &= (1 \oplus u)(1 \oplus v)f \oplus (1 \oplus u)v(f \oplus \ell_1) \oplus u(1 \oplus v)(f \oplus \ell_2) \oplus uv(f \oplus \ell_3), \end{aligned}$$

and using Lemma 5.54 we deduce that $\mathcal{N}_{g_4} \geq 2^{2k+1} - 2^{k+1}$. ■

Another application for Theorem 5.43 involves the modification of the bent functions constructed there in order to produce balanced Boolean functions with high nonlinearity (see [434]).

Theorem 5.57. *Suppose $n = 2k \geq 8$ and, for $i = 1, 2, \ldots, 2^{k-1}$, let M_i, N_i denote arbitrary affine functions made up of 2^{k-4} 4-bit blocks. Define g to be the concatenation of 2^{k-1} segments S_i, where for each $i \leq 2^{k-2}$ the segment S_i is of the form*

$$(M_i N_i \cdots M_i \bar{N}_i | \bar{N}_i M_i \cdots N_i M_i)$$

or

$$(M_i N_i \cdots M_i \bar{N}_i | N_i \bar{M}_i \cdots \bar{N}_i \bar{M}_i)$$

and the segment $S_{i+2^{k-2}}$ is of the form

$$(\bar{N}_i \bar{M}_i \cdots \bar{N}_i \bar{M}_i | \bar{M}_i \bar{N}_i \cdots \bar{M}_i N_i)$$

or

$$(N_i \bar{M}_i \cdots \bar{N}_i \bar{M}_i | \bar{M}_i \bar{N}_i \cdots \bar{M}_i N_i),$$

respectively. Furthermore, we impose the following conditions:

(i) *Exactly 2^{k-1} strings M_i and 2^{k-1} strings N_j are based on each of the blocks A, B, C, D.*
(ii) *For any i, j with $i \neq j$ and $1 \leq i, j \leq 2^{k-2}$, the strings $M_i \oplus M_j$ and $N_i \oplus N_j$ are balanced.*

Then the function g is balanced, satisfies the SAC and has nonlinearity $\mathcal{N}_g \geq 2^{2k-1} - 2^k$ (a similar result holds for $n = 2k + 1$).

Modifying Theorem 5.44 in the same way produces the following result.

Theorem 5.58. *Construct functions g, for $n = 2k$ (or $2k + 1$) ≥ 8, by concatenation of elements of T (see Equation* (5.34)*) such that we have the conditions* (i) *and* (iii) *of Theorem* 5.44 *and the condition*

(ii$'$) *Same as condition* (ii) *of Theorem* 5.44 *(for n odd we have 2^{k-2} subsegments), but with the special subsegment containing 2^{k-3} pairs $(M|M)$ and 2^{k-3} pairs $(\bar{M}|\bar{M})$.*

Then the function g is balanced, satisfies the SAC and has the nonlinearity $\mathcal{N}_g \geq 2^{2k-1} - 2^k$ ($\mathcal{N}_g \geq 2^{2k} - 2^k$).

Proof. Obviously g is balanced in either case. It follows directly from the definition of SAC that if the truth table of the Boolean function f is $(v_1, v_2, \ldots, v_{2^n})$ (we shift the indices for easy writing of our next argument), then f satisfies the SAC if and only if

$$\begin{aligned}
&(w_1 w_{2^{i-1}+1} \oplus w_2 w_{2^{i-1}+2} \oplus \cdots \oplus w_{2^{i-1}} w_{2^i}) \oplus \\
&(w_{2^i+1} w_{2^i+2^{i-1}+1} \oplus \cdots \oplus w_{2^i+2^{i-1}} w_{2^{i+1}}) \oplus \cdots \oplus \\
&(w_{2^n-2^i+1} w_{2^n-2^{i-1}+1} \oplus \cdots \oplus w_{2^n-2^{i-1}} w_{2^n}) = 0,
\end{aligned} \tag{5.50}$$

for each $i = 1,2,\ldots,n$, where $w_i = (-1)^{v_i}$. We introduce a map denoted $\odot$ from the Cartesian product $T \times T$ (T as in (5.34)) to $\{-4,-2,0,2,4\}$ defined by

$$M \odot N = 2^n - 2wt(M \oplus N) = \#0\text{s} - \#1\text{s in the sum } M \oplus N. \tag{5.51}$$

Then the relation (5.50) reads as follows, for an arbitrary $i \geq 3$:

$$(X_1 \odot X_{2^{i-3}+1} \oplus X_2 \odot X_{2^{i-3}+2} \oplus \cdots \oplus X_{2^{i-3}} \odot X_{2^{i-2}}) \oplus \cdots = 0. \tag{5.52}$$

To show that g satisfies the SAC we employ the Equation (5.52).

The case $i \leq 2$ is handled easily by associating the blocks in the following way: $D \Leftrightarrow C, \bar{D} \Leftrightarrow \bar{C}, A \Leftrightarrow B, \bar{A} \Leftrightarrow \bar{B}$. Next, we compute the nonlinearity of g for $n = 2k$. Consider an affine function ℓ formed with A or/and $\bar{A}$. From the part of g that does not contain $A, \bar{A}$ we get $2^{2k-1} - 2^{2k-3}$ ones. Without loss of generality we assume that the first segment in g contains the 2^{k-3} pairs $(A|A)$ and 2^{k-3} pairs $(A|\bar{A})$. Since each segment in ℓ is similar to one of the previous segments or is the complement of one and from Lemma 5.45, which applies here as well, we can cancel at most all pairs from the first segment, and from each of the $2^{k-3} - 1$ remaining segments we have left at least half of the blocks uncanceled, from the first and from the second half of g. Consequently, we get $8 \cdot 2^{k-3} \cdot (2^{k-3} - 1) = 2^{2k-3} - 2^k$ units. Thus $\mathcal{N}_g \geq 2^{2k-1} - 2^k$. The case $n = 2k+1$ is similar. ■

5.16 PARTIALLY BENT FUNCTIONS

To resolve a conjecture proposed in [370], Carlet [64] was led to the notion of a partially bent function (we already mentioned these in Section 3.7). Let $W(\hat{f})$ and $\hat{r}$ be the Walsh transform, respectively, the autocorrelation function, as defined in Sections 2.2 and 2.3. Preneel et al. [370] proposed that the number of zeros of $\hat{r}$, respectively, $W(\hat{f})$, say $N_{\hat{r}}$, respectively, $N_{W(\hat{f})}$, satisfy

$$(2^n - N_{\hat{r}})(2^n - N_{W(\hat{f})}) \geq 2^n,$$

which was proved by Carlet [64]. Although it was conjectured that the functions satisfying the equality either will have degree 2 or satisfy $PC(n)$, or $PC(n-1)$, this was shown by Carlet to be incorrect. He dubbed the functions for which equality holds *partially bent*.

Connections between this notion and some already existing notions were also investigated. Let f be a partially bent function, let $\phi_f(\mathbf{u},\mathbf{v}) = f(0) \oplus f(\mathbf{u}) \oplus f(\mathbf{v}) \oplus f(\mathbf{u} \oplus \mathbf{v})$ defined on $\mathbb{V}_n \times \mathbb{V}_n$, and let $E = \{\mathbf{u} \in \mathbb{V}_n \mid \phi_f(\mathbf{u},\mathbf{v}) = 0, \forall \mathbf{v} \in \mathbb{V}_n\}$, called the *kernel associated to* f.

Proposition 5.59. *A partially bent function f on $\mathbb{V}_n$ satisfies $PC(k)$ if and only if its associated kernel E only contains* 0, *or elements of Hamming weight $> k$. A partially bent* *function f is balanced if and only if its restriction to its associated kernel is nonconstant, that is, if and only if there exists an element $\mathbf{u} \in \mathbb{V}_n$ such that for any $\mathbf{x} \in \mathbb{V}_n$, $f(\mathbf{x} \oplus \mathbf{u}) = f(\mathbf{x}) \oplus 1$. Otherwise, its weight is equal to $2^{n-1} \pm 2^{n-1-h}$.*

Proof. We follow Carlet [64]. The first part is immediate observing that $\hat{r}(\mathbf{x}) = 0$ if and only if $\mathbf{x} \notin E$.

Next, let f be a partially bent function, E its associated kernel, and so $\mathbb{V}_n = E \oplus E^{\perp}$. Now,

$$W(\hat{f})(0) = \sum_{\mathbf{u} \in \mathbb{V}_n} (-1)^{f(\mathbf{u})} = \sum_{\mathbf{x} \in E} (-1)^{\mathbf{t} \cdot \mathbf{x}} \sum_{\mathbf{y} \in E^{\perp}} (-1)^{f(\mathbf{y})}.$$

Further,

$$\sum_{\mathbf{y} \in E^{\perp}} (-1)^{\mathbf{t}(\mathbf{y})} = \pm \sqrt{|E^{\perp}|},$$

since $f|_{E^{\perp}}$ is bent, and

$$\sum_{\mathbf{x} \in E} (-1)^{f \cdot \mathbf{x}} = \begin{cases} 0 & \text{if } \mathbf{t} \notin E^{\perp} \\ |E| & \text{otherwise.} \end{cases}$$

Therefore, we infer that f is balanced if and only if $\mathbf{t}$ does not belong to $E^{\perp}$, and so, if and only if f is nonconstant on E. If that is so, by taking any element $\mathbf{u} \in E \setminus \mathbf{t}^{\perp}$, where $\mathbf{t}^{\perp} = \{\mathbf{x} \in \mathbb{V}_n \mid \mathbf{x} \cdot \mathbf{t} = 0\}$, we have

$$f(\mathbf{x} \oplus \mathbf{u}) = f(\mathbf{x}) \oplus \mathbf{t} \cdot \mathbf{u} = f(\mathbf{x}) \oplus 1.$$

The converse is immediate.

Finally, assume that f is unbalanced. Suppose E has dimension $n - 2h$. Then

$$2^n - 2wt(f) = \sum_{\mathbf{u} \in \mathbb{V}_n} (-1)^{f(\mathbf{u})} = \pm |E| \sqrt{|E^{\perp}|} = \pm 2^{n-2h} 2^h = \pm 2^{n-h},$$

from which we derive $wt(f) = 2^{n-1} \pm 2^{n-h-1}$. ∎

Partially bent functions can also be used to construct correlation immune functions.

Proposition 5.60. *A partially bent function defined by*

$$f(\mathbf{x} \oplus \mathbf{y}) = f(\mathbf{y}) \oplus \mathbf{t} \cdot \mathbf{x}, \quad \mathbf{x} \in E, \mathbf{y} \in E^{\perp},$$

is correlation immune of order k (respectively, resilient of order k) if and only if $\mathbf{t} \oplus E^{\perp}$ contains elements of Hamming weight greater than k or equal to 0 (respectively, greater than k).

Proof. We can easily find that the Walsh transform

$$W(\hat{f})(\mathbf{s}) = \sum_{\mathbf{x}\in E,\mathbf{y}\in E^{\perp}} (-1)^{\mathbf{x}\cdot\mathbf{t}\oplus\mathbf{x}\cdot\mathbf{s}\oplus f(\mathbf{y})\oplus\mathbf{y}\cdot\mathbf{s}}$$
$$= \sum_{\mathbf{x}\in E}(-1)^{\mathbf{x}\cdot(\mathbf{t}\oplus\mathbf{s})} \sum_{\mathbf{y}\in E^{\perp}}(-1)^{f(\mathbf{y})\oplus\mathbf{y}\cdot\mathbf{s}}.$$

The second sum $\sum_{\mathbf{y}\in E^{\perp}}(-1)^{f(\mathbf{y})\oplus\mathbf{y}\cdot\mathbf{s}} \neq 0$, since f is bent on $E^{\perp}$, from which we derive that $W(\hat{f})(\mathbf{s}) \neq 0$ if and only if $\mathbf{s}\oplus\mathbf{t}\in E^{\perp}$. ■

Although we saw a wealth of nice (arithmetical) properties of this class of partially bent functions, we are not able to characterize the algebraic normal form of these, nor to count or even give a good bound on the size of this class (cf. [64]).

5.17 SEMI-BENT FUNCTIONS

At Asiacrypt 1994, Chee et al. [106] introduced a new (quite useful) notion, namely *semi-bent* Boolean functions on odd-dimensional vector spaces over $\mathbb{F}_2$.

Definition 5.61. *Let f_0 be a bent function on $\mathbb{V}_n$, $n = 2k$, $\mathbf{a} \in \mathbb{V}_n$, and $f_1(\mathbf{x}) = f_0(A\mathbf{x} \oplus \mathbf{a}) \oplus 1$, A is a nonsingular $n \times n$ matrix over $\mathbb{F}_2$ and $\mathbf{a}$ is any vector in $\mathbb{V}_n$. The concatenation $g = f_0 \,||\, f_1$ on $\mathbb{V}_{2k+1}$ is called a* semi-bent function.

It is rather straightforward to write explicitly the action of g. If $\mathbf{x} \in \mathbb{V}_{2k}$ and $\mathbf{x}^* = (\mathbf{x}, x_{2n+1}) \in \mathbb{V}_{2k+1}$, then

$$g(\mathbf{x}^*) = (1\oplus x_{2n+1})f_0(\mathbf{x}) \oplus x_{2n+1}f_1(\mathbf{x})$$
$$= f_0(\mathbf{x}) \oplus x_{2n+1}f_0(\mathbf{x}) \oplus x_{2n+1} \oplus x_{2n+1}f_0(A\mathbf{x}\oplus\mathbf{a}).$$

Chee et al. [106, Theorems 16 and 18, and Corollary 21] prove the next two theorems.

Theorem 5.62. *Any semi-bent function g on $\mathbb{V}_{2k+1}$ is balanced, and $\mathcal{N}_g = 2^{2k} - 2^k$. Further, for any $\mathbf{w}^* \in \mathbb{V}_{2k+1}$, the correlation value between g and the linear function $\ell_{\mathbf{w}^*}(\mathbf{x}) = \mathbf{w}^* \cdot \mathbf{x}$ is 0 or $\pm 2^{-k}$, and*

$$\#\{\mathbf{w}^* \in \mathbb{V}_{2k+1} \,|\, c(g,\ell_{\mathbf{w}^*}) = 0\} = 2^{2k} = \#\{\mathbf{w}^* \in \mathbb{V}_{2k+1} \,|\, c(g,\ell_{\mathbf{w}^*}) = \pm 2^{-k}\}.$$

Recall that the crosscorrelation function between $f_0, f_1 : \mathbb{V}_n \to \mathbb{F}_2$ is defined as

$$c(\hat{f}_0,\hat{f}_1)(\mathbf{s}) = \sum_{\mathbf{x}\in\mathbb{V}_n} \hat{f}_0(\mathbf{x})\cdot\hat{f}_1(\mathbf{x}\oplus\mathbf{s}).$$

Proposition 5.63. *Let g on $\mathbb{V}_{2k+1}$ be a semi-bent function with $A = I$ (the identity matrix). Then g satisfies the PC with respect to all nonzero vectors $\mathbf{s}^* = (\mathbf{s}, s_{2n+1}) \in \mathbb{V}_{2k+1}$ with $\mathbf{s} \neq \mathbf{a}$. If further $\mathbf{a} = (1,1,\ldots,1)$, then g satisfies $PC(2k)$.*

Proof. We follow Chee et al. [106]. Let $\mathbf{s}^* = (\mathbf{s}, s_{2n+1}) \in \mathbb{V}_{2k+1}$ with $\mathbf{s} \neq \mathbf{a}$. If $s_{2n+1} = 0$, by Theorem 2.16 and since a bent function satisfies $PC(n)$, then the autocorrelation function satisfies

$$\hat{r}_{\hat{f}_0}(\mathbf{s}) = \hat{r}_{\hat{f}_1}(\mathbf{s}) = 0,$$

and $\hat{r}_{\hat{g}}(\mathbf{s}^*) = 0$, so the first claim of the proposition is true.

Further, if $s_{2n+1} = 1$, then

$$\begin{aligned}
\hat{r}_{\hat{g}}(\mathbf{s}^*) &= 2c(\hat{f}_0, \hat{f}_1)(\mathbf{s}) \\
&= \frac{1}{2^{2k-1}} \sum_{\mathbf{y} \in \mathbb{V}_{2k}} W(\hat{f}_0)(\mathbf{y}) W(\hat{f}_1)(\mathbf{y}) \cdot (-1)^{\mathbf{s} \cdot \mathbf{y}} \\
&= \frac{-1}{2^{2k-1}} \sum_{\mathbf{y} \in \mathbb{V}_{2k}} (W(\hat{f}_0)(\mathbf{y}))^2 \cdot (-1)^{(\mathbf{a} \oplus \mathbf{s}) \cdot \mathbf{y}} \\
&= -2 \sum_{\mathbf{y} \in \mathbb{V}_{2k}} (-1)^{(\mathbf{a} \oplus \mathbf{s}) \cdot \mathbf{y}} = 0.
\end{aligned}$$

The last claim follows immediately. ■

5.18 SYMMETRIC BENT FUNCTIONS

In this section we consider Boolean functions that are invariant under the action of some subgroup of permutations on their algebraic normal forms.

We first consider the action of the entire symmetric group S_n. A function invariant under the full symmetric group S_n is called *symmetric*. The value of any symmetric function depends on the Hamming weight of the input only, and so it can be expressed as $f(\mathbf{x}) = c_k$, where $wt(\mathbf{x}) = k$ and c_k is a constant that depends on k only. Further, the Walsh transform of f can be computed easily by

$$W(f)(\mathbf{t}) = \sum_{k=0}^{n} (-1)^{c_k} \sum_{wt(\mathbf{x})=k} (-1)^{\mathbf{t} \cdot \mathbf{x}},$$

where the second sum can be expressed as $P_k(wt(\mathbf{t}), n)$, where $P_k(y, n)$ is the Krawtchouk polynomial [287, pp. 130 and 150–153]

$$P_k(y, n) = \sum_{j=0}^{k} (-1)^j \binom{y}{j} \binom{n-y}{k-j}$$

of even degree k in y. It is known (see [287, p. 129] and [400]) that the generating function of these polynomials is

$$(1-z)^y(1+z)^{n-y} = \sum_{k=0}^{n} P_k(y,n)z^k. \tag{5.53}$$

In 1994, Savicky [400] proved that the only Boolean functions in n variables that are symmetric and bent have degree 2, therefore, there are only four distinct symmetric bent functions in n (even) variables (see [296,481] for alternate proofs of this result).

Theorem 5.64. *If f is a symmetric bent function in n variables, then*

$$f(x_1,x_2,\ldots,x_n) = \sum_{1\le i<j\le n} x_i x_j \oplus c\sum_{i=1}^{n} x_i \oplus d,$$

with $c,d \in \mathbb{F}_2$.

Proof. Since f is bent, then for every $\mathbf{t} \in \mathbb{V}_n$ such that $wt(\mathbf{t}) = n/2$, then the Walsh transform $|W(f)(\mathbf{t})| = 2^{n/2}$. Using the generating function (5.53) we easily infer that

$$\sum_{\mathbf{x},wt(\mathbf{x})=\mathbf{k}} (-1)^{\mathbf{x}\cdot\mathbf{w}} = \begin{cases} 0 & \text{if } k \text{ odd} \\ (-1)^m \binom{n/2}{m} & \text{if } k = 2m, \end{cases}$$

and so the Walsh transform of f becomes $W(f)(\mathbf{t}) = \sum_{m=0}^{n/2}(-1)^{c_{2m}}(-1)^m\binom{n/2}{m}$.

If there exists $0 \le m \le n/2-1$ such that $c_{2m+2} = c_{2m}$, then the sum would contain nonzero terms of opposite sign and the absolute value of the Walsh transform would be strictly less than $\sum_{m=0}^{n/2}\binom{n/2}{m} = 2^{n/2}$, which is obviously a contradiction. Thus, for every $0 \le m \le n/2-1$, $c_{2m+2} = c_{2m} \oplus 1$.

In a similar fashion we treat the case of $wt(\mathbf{t}) = n/2-1$ with $|W(f)(\mathbf{t})| = 2^{n/2}$, and infer that $c_{2m+3} = c_{2m+1} \oplus 1$, for $0 \le m \le n/2-2$. Therefore, $c_{k+2} = c_k \oplus 1$, for all $0 \le k \le n-2$. Set $c = c_0 \oplus c_1, d = c_0$. It is immediate that c_k will satisfy the congruence

$$c_k \equiv \binom{k}{2} + ck + d \pmod 2,$$

for any k, and so the identity $f(x_1,x_2,\ldots,x_n) = \sum_{i<j} x_i x_j \oplus c\sum_i x_i \oplus d$ holds, since $f(\mathbf{x}) = c_k$ when $wt(\mathbf{x}) = k$. ■

5.19 ROTATION SYMMETRIC FUNCTIONS

Hash functions are used to map a large collection of *messages* into a small set of *message digests* and can be used to generate efficiently both signatures and message authentication codes, and they can be also used as one-way functions in key agreement and key establishment protocols. There are two approaches to the study of hash functions: *information theory* and *complexity theory*. The first method provides unconditional security – an enemy cannot attack such systems even if he/she has unlimited computing power. Unfortunately, this is still a theoretical approach and is generally impractical [17]. In the second method based on complexity theory, some assumptions are made on the computing power of the enemy or the weaknesses of the existing systems and algorithms. The best we can hope for is to estimate the computing power necessary for the attacker to break the algorithm. Interpolation cryptanalysis [237] and high order differential cryptanalysis [338] have shown that the algebraic degree is an important factor in the design of cryptographic primitives. In fact, in [338] the algebraic degree is the crucial parameter in determining how secure certain cryptosystems are against higher order differential attacks. Together with propagation criteria, differential and nonlinearity profile, resiliency, correlation immunity, local and global avalanche characteristics they form a class of design criteria which we have to consider in the design of such primitives.

In [365], Pieprzyk and Qu studied some functions, which they called *rotation symmetric*, as components in the rounds of a hashing algorithm. This is a highly desirable property when efficient evaluation of the function is important, for instance in the implementation of MD4, MD5 or HAVAL, since we can reuse evaluations from previous iterations. It turns out that a quadratic rotation symmetric function takes $\frac{3n-1}{2}+6(m-1)$ operations (additions and multiplications) to evaluate in m consecutive rounds of a hashing algorithm. In [338] the authors showed how to break in less than 20 milliseconds a block cipher that employs quadratic Boolean functions as its S-boxes and is provably secure against linear and differential attacks. Therefore, it is necessary to employ higher degree rotation symmetric functions in our algorithms. To protect from differential attack, we need rotation symmetric functions with high nonlinearity.

In this section we follow Cusick and Stănică [131], Stănică and Maitra [438,439].

Let $(x_1,x_2,\ldots,x_n)\in\mathbb{V}_n$. For $1\le k\le n$, we define

$$\rho_n^k(x_i)=\begin{cases}x_{i+k} & \text{if } i+k\le n,\\ x_{i+k-n} & \text{if } i+k>n.\end{cases}$$

Definition 5.65. *A Boolean function f is* rotation symmetric *(RotS) if and only if for any $(x_1,\ldots,x_n)\in\mathbb{V}_n$,*

$$f(\rho_n^k(x_1,\ldots,x_n))=f(x_1,\ldots,x_n)$$

for any $1\le k\le n$.

In other words, rotation symmetric functions are invariant under shifting indices.

We extend the definition of ρ to tuples by $\rho_n^k(x_1,\ldots,x_n)=(\rho_n^k(x_1),\ldots,\rho_n^k(x_n))$, and to monomials by $\rho^k(x_{i_1}\cdots x_{i_\ell})=\rho^k(x_{i_1})\cdots\rho^k(x_{i_\ell})$. If, for instance, $n=4$, and the term $x_1x_2x_3$ is present in the algebraic normal form (ANF) of a rotation symmetric function f, then the terms $x_2x_3x_4, x_3x_4x_1, x_4x_1x_2$ must also be present in the ANF of f.

Pieprzyk and Qu [365] showed that the nonlinearity of a homogeneous of degree k rotation symmetric function f_k satisfies $\mathcal{N}_{f_k} \geq 2^{n-k}$, $k \geq 2$. They concentrated on quadratic rotation symmetric functions and proved the next result.

Theorem 5.66. *Let $f_2 : \mathbb{V}_n \to \mathbb{F}_2$ be a homogeneous rotation symmetric Boolean function of degree 2 which is generated from a term of degree 2 using the rotation symmetry property. Then the weight of f_2 satisfies $2^{n-2} \leq wt(f_2) \leq 2^n + 2^{n-2}$; the nonlinearity $\mathcal{N}_{f_2} \geq 2^{n-2}$; if n is odd, then f_2 is balanced; and f_2 will satisfy the propagation criterion with respect to all vectors $\alpha \in \mathbb{V}_n$ of weight strictly between 0 and n (in particular, f_2 will satisfy the SAC).*

In fact, rotation symmetric functions appeared in the literature before 1999, precisely in 1994, when Pieprzyk [364] was investigating bent permutations in $\mathbb{F}_{2^n}$, which are permutations whose nonlinearity attains the nonlinearity of bent functions in $\mathbb{F}_{2^{n-1}}$. He showed that if n is odd, then $p(x) = x^3 : \mathbb{F}_{2^n} \to \mathbb{F}_{2^n}$ is such a bent permutation and in the course of that proof he showed that the trace of p is the following rotation symmetric function

$$\mathrm{Tr}_{\mathbb{F}_{2^n}/\mathbb{F}_2}(p(x)) = f_2^{(n)} = x_1x_2 \oplus x_2x_3 \oplus \cdots \oplus x_{n-1}x_n \oplus x_nx_1,$$

whose nonlinearity is in fact $\mathcal{N}_{f_2^{(n)}} = 2^{n-1} - 2^{(n-1)/2}$, for n odd. If n is even, the nonlinearity is then $\mathcal{N}_{f_2^{(n)}} = 2^{n-1} - 2^{n/2}$. Regardless of parity, the weight is $wt(f_2^{(n)}) = 2^{n-1} - 2^{n/2-1}(1+(-1)^n)$ (cf. Cusick and Stănică [131]).

For the cubic rotation symmetric function

$$f_3^{(n)} = x_1x_2x_3 \oplus x_2x_3x_4 \oplus \cdots \oplus x_nx_1x_2, \tag{5.54}$$

it was shown in [131] that the weight of $f_3^{(n)}$ satisfies

$$wt\left(f_3^{(n)}\right) = 2\left(wt\left(f_3^{(n-2)}\right) + wt\left(f_3^{(n-3)}\right)\right) + 2^{n-3}.$$

It was also conjectured in [131] that the weight and nonlinearity of any cubic (homogeneous) rotation symmetric function are equal.

The general case seems to be quite hard to investigate. As is the case with any cryptographic property, a count of various rotation symmetric functions was desirable. This was done in [438,439].

From Definition 5.65, it is clear that any rotation symmetric function f will have the same value on each of the subsets generated from the rotational symmetry. As an example, for $n = 4$, one gets the following partitions:

$$\{(0,0,0,0)\},$$
$$\{(0,0,0,1),(0,0,1,0),(0,1,0,0),(1,0,0,0)\},$$
$$\{(0,0,1,1),(0,1,1,0),(1,0,0,1),(1,1,0,0)\},$$

Table 5.1 The values of g_n, $1 \le n \le 14$

n	1	2	3	4	5	6	7	8	9	10	11	12	13	14
g_n	2	3	4	6	8	14	20	36	60	108	188	352	632	1182

$$\{(0,1,0,1),(1,0,1,0)\},$$
$$\{(0,1,1,1),(1,0,1,1),(1,1,0,1),(1,1,1,0)\},$$
$$\{(1,1,1,1)\}.$$

Therefore, there are six different subsets which partition the 16 input patterns and any 4-variable rotation symmetric Boolean function can have a specific value corresponding to each subset. Thus there are $2^6 = 64$ rotation symmetric functions in 4 variables.

We let

$$G_n(x_1,\ldots,x_n) = \{\rho_n^k(x_1,\ldots,x_n), \text{for } 1 \le k \le n\},$$

that is, the orbit of $(x_1,\ldots,x_n)$ under the action of ρ_n^k, $1 \le k \le n$. Certainly, $G_n(x_1,\ldots,x_n)$ generates a partition of cardinality g_n, of the set $\mathbb{V}_n$. For $(x_1,\ldots,x_n)$, a function is rotation symmetric if it takes the same value for all the elements in $G_n(x_1,\ldots,x_n)$. It is clear that there are 2^{g_n} n-variable rotation symmetric Boolean functions. From Burnside's lemma, we will get in Section 5.20 that $g_n = \frac{1}{n}\sum_{t|n}\phi(t)2^{\frac{n}{t}}$. In Table 5.1, we list the first few values of g_n.

For binary strings S_1, S_2 of the same length λ, we denote by $\#(S_1 = S_2)$ (respectively, $\#(S_1 \neq S_2)$), the number of places where S_1 and S_2 are equal (respectively, unequal). Certainly, $d(S_1,S_2) = \#(S_1 \neq S_2)$. We will also use the notation $wd(S_1,S_2) = \#(S_1 = S_2) - \#(S_1 \neq S_2)$. Note that $wd(S_1,S_2) = \lambda - 2d(S_1,S_2)$.

We select the representative element of $G_n(x_{i_1}x_{i_2}\ldots x_{i_\ell})$ as the lexicographically first element. For example, the representative element of $\{x_1x_2x_3,\ x_2x_3x_4,\ x_3x_4x_1,\ x_4x_1x_2\}$ is $x_1x_2x_3$. Note that it is also clear that the variable x_1 will always exist in the lexicographically first element (the representative element).

We now define the *short algebraic normal form* (SANF) of a rotation symmetric function. A rotation symmetric function $f(x_1,\ldots,x_n)$ can be written as

$$a_0 \oplus a_1x_1 \oplus \sum_{j=1}^{n} a_{1j}x_1x_j \oplus \cdots \oplus a_{12\ldots n}x_1x_2\cdots x_n,$$

where the coefficients $a_0, a_1, a_{1j}, \ldots, a_{12\ldots n} \in \mathbb{F}_2$, and the existence of a representative term $x_1x_{i_2}\ldots x_{i_\ell}$ implies the existence of all the terms from $G_n(x_1x_{i_2}\ldots x_{i_l})$ in the algebraic normal form. This representation of f is called the *short algebraic normal form* (SANF) of f.

As an example, consider the ANF of a 4-variable rotation symmetric Boolean function $x_1 \oplus x_2 \oplus x_3 \oplus x_4 \oplus x_1x_2x_3 \oplus x_2x_3x_4 \oplus x_3x_4x_1 \oplus x_4x_1x_2$. Its SANF is $x_1 \oplus x_1x_2x_3$.

As we have already mentioned, a Boolean function is said to be homogeneous if its algebraic normal form contains terms of same degree only. Homogeneous bent functions have recently received quite a bit of attention in the literature [102,103,324,374,373,406, 437,465]. Using a computer search in a reduced space, in [439] the authors found the exact counts of 8, 48, and 15104, rotation symmetric bent functions in 4, 6, and 8 variables, respectively.

Filiol and Fontaine [169] discussed the set of idempotent Boolean functions in an experimental setting. Let $\mathcal{B} = (\mathbf{b}_1,\ldots,\mathbf{b}_n)$ be a basis of $\mathbb{V}_n$ (which is identified with $\mathbb{F}_{2^n}$). An *idempotent* f is a Boolean function on $\mathbb{F}_{2^n}$ that satisfies $f^2 = f$. Define the *Mattson-Solomon (MS) polynomial* by $MS_f(Z) = \sum_{j=0}^{2^n-2} A_j Z^{2^n-j-1}$, where $A_j = \sum_{i=0}^{2^n-1} f(\alpha^i)\alpha^{ij}$ (α is a primitive element of $\mathbb{F}_{2^n}$). Using the representation $f = \sum_{g\in\mathbb{F}_{2^n}^*} f(g)(g)$ (in the multiplicative algebra $\mathbb{F}_2[\mathbb{F}_{2^n},\times]$), we get that f is an idempotent iff $f(g)=f(g^2)$, for all g; the coefficients of the MS polynomial belong to $\mathbb{F}_2$; $A_j = A_k$ for all k in the 2-cyclotomic class of j ($\{j,2j,\ldots,2^{n-1}j\}$); the algebraic normal form of f, using a normal basis $(\gamma,\gamma^2,\ldots,\gamma^{2^{n-1}})$ remains invariant under circular shift. This gives that the set of idempotents is the same as the class of rotation symmetric Boolean functions. For $n=5,7$, they found idempotents of highest nonlinearity (12, respectively 56) of degrees 2, 3 (for $n=5$), and degrees 2, 3, 4, 5, 6 (for $n=7$). For $n=6,8$ they found all idempotents of highest nonlinearity (28, respectively 120), of degrees 2, 3, respectively, 2, 3, 4. They were not able to find all idempotent functions for $n=8$, though. Finally, for $n=9$, they found 1142395 functions (up to equivalence) with nonlinearity 240, some of which are balanced, of degrees 2 to 7.

While the search of [169] among rotation symmetric functions considers nonlinearity only, in [438,439] other cryptographic properties were considered, like resiliency, propagation characteristics, and autocorrelation spectra, and the authors found some optimized examples for 6 and 7 variables. Maximov et al. [306] showed the nonexistence of a Boolean function in 9 variables, which is resilient of order 3, with nonlinearity 240 in the rotation symmetric class (the question was posed at the Fast Software Encryption conference in 2004). They also found 16, 812 rotation symmetric functions that are correlation immune of order 3, have algebraic degree 5 and nonlinearity 240.

It is interesting to note that the famous Patterson–Wiedemann functions [360] that achieve nonlinearity 16276 (strictly greater than nonlinearity $2^{15-1}-2^{(15-1)/2}$ obtained by bent functions concatenation) in 15 variables are in fact rotation symmetric. Moreover, Kavut et al. [249,250] proved that there exist rotation symmetric functions in 9 variables having nonlinearity 241 (which is also strictly greater than the nonlinearity obtained by bent functions concatenation, $2^{9-1}-2^{(9-1)/2}$), and recently in [251], Kavut et al. found in an extended class of rotation symmetric functions examples of Boolean functions in 9 variables having nonlinearity 242, which was rather surprising and gives further motivation for the rotation symmetric Boolean functions investigation. As a side comment, we note that this discovery helped also with the covering radius problem for Reed-Muller code $R(1,9)$, which is now reduced to two possible values 242, 244 (see [228] for further clarifications).

5.20 ENUMERATION OF ROTATION SYMMETRIC FUNCTIONS

If $|G_n(x_1,\ldots,x_n)| = n$, we say that the elements of $G_n(x_1,\ldots,x_n)$ form a *long cycle* (of length n). If $|G_n(x_1,\ldots,x_n)| < n$, we call it a *short cycle* (length $<n$).

It turns out that the sequence g_n (the cardinality of the partition of $\mathbb{V}_n$ generated by $G_n(\mathbf{x})$ with $\mathbf{x} \in \mathbb{V}_n$) counts also the number of n-bead necklaces with two colors when turning over is not allowed, output sequences from a simple n-stage cycling shift register, and binary irreducible polynomials whose degree divides n (see Sloane [431]).

We shall be using the Burnside's lemma [386] (apparently, the lemma was discovered by Cauchy about 50 years earlier [341]).

Lemma 5.67 (*Burnside's lemma*). *Let G be a group of permutations acting on a set S. Then the number of orbits induced on S is given by $\frac{1}{|G|}\sum_{\pi\in G}|fix_S(\pi)|$, where $fix_S(\pi) = \{x \in S \mid \pi(x) = x\}$.*

Theorem 5.68. $g_n = \frac{1}{n}\sum_{t|n}\phi(t)2^{n/t}$, *where $\phi(t)$ is Euler's phi-function.*

Proof. We provide here a proof (see also [431,439]). Here $G = \{\rho_n^1,\ldots,\rho_n^n\}$ and $S = \{0,1\}^n$. To use Burnside's lemma we need to find the number of fixed points of ρ_n^i, $i = 1,\ldots,n$. The number of permutation cycles of ρ_n^i is $\gcd(n,i)$, each of them of length $\frac{n}{\gcd(n,i)}$. Observe that ρ_n^i has order $\frac{n}{\gcd(n,i)}$. Since, to be fixed by ρ_n^i, each input cycle must consist of all 0s or all 1s, we get that the number of fixed points of ρ_n^i is $2^{\gcd(n,i)}$. Applying Burnside's lemma we obtain,

$$\begin{aligned} g_n &= \frac{1}{n}\sum_{i=1}^{n} 2^{\gcd(n,i)} = \frac{1}{n}\sum_{k|n}\ \sum_{i,\gcd(n,i)=k}^{n} 2^k \\ &= \frac{1}{n}\sum_{k|n}2^k \sum_{j,\gcd(n/k,j)=1} 1 = \frac{1}{n}\sum_{k|n}\phi\left(\frac{n}{k}\right)2^k = \frac{1}{n}\sum_{t|n}\phi(t)2^{\frac{n}{t}}, \end{aligned}$$

which proves the theorem. ■

Oftentimes, a more compact expression for g_n can be found. Since $\phi(p^i) = p^i - p^{i-1}$, by Theorem 5.68 we obtain the next corollary.

Corollary 5.69. *For a prime p, $g_{p^m} = p^{-m}\left(2^{p^m} + \sum_{i=1}^{m}(p^i - p^{i-1})2^{p^{m-i}}\right)$.*

Let h_n be the number of long cycles, that is, the number of monomials whose period under ρ is exactly n. Let ω_n be the number of prime factors of n, and write the prime power factorization of n as $n = p_1^{a_1}\cdots p_{\omega_n}^{a_{\omega_n}}$, $p_i \neq p_j$.

Theorem 5.70. *We have* (i) $h_1 = 2$,

(ii) *If $n = p^a$, p prime, then $h_{p^a} = \frac{1}{n}\sum_{d|n}\phi(d)2^{n/d} - \sum_{i=1}^{a-1}\frac{2^{p^i}-2^{p^{i-1}}}{p^i} - 2$. In particular, if $a = 1$, $h_p = \frac{2^p-2}{p}$.*

(iii) *Let $n = p_1^{a_1} \cdots p_{\omega_n}^{a_{\omega_n}}$ be the prime power factorization of n. Then*

$$h_n = \frac{1}{n}\sum_{d|n}\phi(d)2^{n/d} - \sum_{i=1}^{\omega_n}\sum_{j=1}^{a_i}\frac{2^{p_i^j} - 2^{p_i^{j-1}}}{p_i^j} - 2,$$

if $\omega_n \geq 2$.

Proof. The complete proof can be found in [439] and investigates the fixed points of ρ^i and their relation to the fixed points of ρ^d, where $d = \gcd(n, i)$. ■

We next consider $G_n(x_1, \ldots, x_n)$, where $wt(x_1, \ldots, x_n) = w$ is fixed. Note that in this way we get a partition over the $\binom{n}{w}$ n-bit binary strings of weight w. Denote the number of such partitions by $g_{n,w}$, and let $h_{n,w}$ be the number of distinct sets $G_n(x_1, \ldots, x_n)$, where $wt(x_1, \ldots, x_n) = w$ and $|G_n(x_1, \ldots, x_n)| = n$, that is, the number of long cycles of weight w. Clearly, $h_{n,w} < g_{n,w}$. We will write $k\,|'\,m$, if $1 < k \leq m$ is a proper divisor of m.

Theorem 5.71. *We have*

(i) $g_{n,w} = \frac{1}{n}\binom{n}{w}$, *if* $\gcd(n, w) = 1$. *Also,* $g_{n,0} = g_{n,n} = 1$.

(ii) $g_{n,w} = \frac{1}{n}\left(\binom{n}{w} - \sum_{k|'\gcd(n,w)} \frac{n}{k} \cdot h_{\frac{n}{k},\frac{w}{k}}\right) + \sum_{k|'\gcd(n,w)} h_{\frac{n}{k},\frac{w}{k}}$, *if* $w < n$.

Proof. First, note that $g_{n,w}$ is the sum of the number of long and short cycles. Obviously, $x = (x_1, \ldots, x_n)$ is part of a short cycle if and only if there is a minimal block $b = [x_1, x_2, \ldots,]$ such that $x = \overbrace{bbb\ldots b}^{k\ times}$. Furthermore, k divides w, so the weight of b is w/k. Since x is covered by concatenating k copies of b, it follows that k divides n, as well. This implies that there cannot be any short cycle if $\gcd(n, w) = 1$ and hence we obtain the first claim of (i). If $w = 0$ (respectively $w = n$), then the only element x of such a weight is $(0, \ldots, 0)$ (respectively $(1, \ldots, 1)$), so $g_{n,0} = g_{n,n} = 1$. The proof of (i) is completed.

Assume $1 < w < n$. Using the same observation as above, we note that $(x_1, \ldots, x_n)$ is part of a short cycle under g_n, if and only if there is a minimal block b, of length n/k, where $k\,|'\gcd(n, w)$, which renders x by concatenation of k copies of b. Since b is minimal, then it must be a full cycle under $g_{n/k}$, of weight w/k. Thus,

$$\#\text{ short cycles} = \sum_{k\,|'\gcd(n,w)} h_{\frac{n}{k},\frac{w}{k}}. \tag{5.55}$$

Let L (respectively S) be the sets of elements in $\mathbb{V}_n$ of weight w, which are part of long (respectively short) cycles. Recall that the total number of elements of weight w is $\binom{n}{w}$. Therefore, $|L| = \binom{n}{w} - |S|$. The number of long cycles is $\frac{1}{n}|L|$. Moreover, each short cycle under g_n of weight w is the concatenation of k copies (for some value of $k\,|'\gcd(n, w)$)

of a long cycle under $g_{\frac{n}{k}}$ of weight w/k. Since in each long cycle under $g_{\frac{n}{k}}$ of weight w/k there are n/k elements, it follows that

$$\#\text{ long cycles} = h_{n,w} = \frac{1}{n}\binom{n}{w} - \frac{1}{n}\sum_{k\,|'\gcd(n,w)} \frac{n}{k}\cdot h_{\frac{n}{k},\frac{w}{k}}. \quad (5.56)$$

Putting together (5.55) and (5.56), we obtain (ii). ■

Recall that $g_{n,w}$ is the number of distinct cycles of weight w. This means that the degree w monomials can be partitioned in $g_{n,w}$ different cycles. We obtain:

Corollary 5.72. *Consider n-variable rotation symmetric Boolean functions. The number of* (i) *degree w homogeneous functions is $2^{g_{n,w}} - 1$,* (ii) *the number of degree w functions is $(2^{g_{n,w}} - 1)2^{\sum_{i=0}^{w-1} g_{n,i}}$, and* (iii) *the number of functions with degree at most w is $2^{\sum_{i=0}^{w} g_{n,i}}$.*

As an example, take $w = 2$ first. If n is odd, then $g_{n,2} = \frac{n-1}{2}$. If n is even, $g_{n,2} = \frac{1}{n}\left(\binom{n}{2} - \frac{n}{2}\cdot h_{\frac{n}{2},1}\right) + h_{\frac{n}{2},1}$. Since $h_{\frac{n}{2},1} = 1$, we get $g_{n,2} = \frac{n}{2}$. Thus there are $2^{\lfloor\frac{n}{2}\rfloor}$ homogeneous quadratic rotation symmetric Boolean functions.

Further, take $w = 3$. If $3 \nmid n$, then $g_{n,3} = \frac{1}{n}\binom{n}{3} = \frac{(n-1)(n-2)}{6}$. If $3|n$, then $g_{n,3} = \frac{1}{n}\left(\binom{n}{3} - \frac{n}{3}\cdot h_{\frac{n}{3},1}\right) + h_{\frac{n}{3},1}$. Now $h_{\frac{n}{3},1} = 1$. Hence, $g_{n,3} = \frac{1}{n}\left(\binom{n}{3} - \frac{n}{3}\right) + 1 = \frac{n(n-3)}{6} + 1$. The number of homogeneous cubic rotation symmetric functions is $2^{g_{n,3}}$.

Since $g_{n,w}$ depends on values of $h_{*,*}$, it is desirable to have an exact formula for these values. In [439] the recurrence Equation (5.56) is solved giving the next result.

Let n, w be such that $\gcd(n,w) = 1$ and $d = \prod_{j=1}^{t} p_j^{a_j}$, p_j primes. With n, w, d fixed, let the binomial coefficients $b_{\alpha_1,\ldots,\alpha_t} = \binom{n\prod_{j=1}^{t} p_j^{\alpha_j}}{w\prod_{j=1}^{t} p_j^{\alpha_j}}$.

Theorem 5.73. *We have*

$$h_{nd,wd} = \frac{1}{nd}\left(\sum_{0\le i_1,\ldots,i_t\le 1} (-1)^{\sum_{j=1}^{t} i_j}\, b_{a_1-i_1,\ldots,a_t-i_t}\right). \quad (5.57)$$

Proof. See the combinatorial proof of [439]. ■

Since the class of rotation symmetric functions is much smaller than the class of Boolean functions, a search was now possible. In [438,439] Stănică and Maitra found among the 2^{21} 8-variable rotation symmetric functions (the space cardinality was reduced from 2^{36} using various equivalence tricks) that exactly $4\cdot 3776$ of them are bent functions. The following 8 are also homogeneous, expressed in SANF:

x_1x_5	$x_1x_4 \oplus x_1x_5$
$x_1x_3 \oplus x_1x_5$	$x_1x_3 \oplus x_1x_4 \oplus x_1x_5$
$x_1x_2 \oplus x_1x_5$	$x_1x_2 \oplus x_1x_4 \oplus x_1x_5$
$x_1x_2 \oplus x_1x_3 \oplus x_1x_5$	$x_1x_2 \oplus x_1x_3 \oplus x_1x_4 \oplus x_1x_5$

The search in 10 variables is hard, since $g_{10} = 108$ and one needs to consider functions of up to degree 5, and since there are $g_{10,2} + g_{10,3} + g_{10,4} + g_{10,5} = 65$ groups for searching bent functions, the search space is of size 2^{65}. However, if we throw in homogeneity, the search can be accomplished efficiently [439]. The following functions are the only 10 variable homogeneous bent functions of degree 2 (in SANF):

x_1x_6	$x_1x_5 \oplus x_1x_6$
$x_1x_4 \oplus x_1x_6$	$x_1x_3 \oplus x_1x_6$
$x_1x_3 \oplus x_1x_4 \oplus x_1x_6$	$x_1x_3 \oplus x_1x_4 \oplus x_1x_5 \oplus x_1x_6$
$x_1x_2 \oplus x_1x_6$	$x_1x_2 \oplus x_1x_5 \oplus x_1x_6$
$x_1x_2 \oplus x_1x_4 \oplus x_1x_5 \oplus x_1x_6$	$x_1x_2 \oplus x_1x_3 \oplus x_1x_5 \oplus x_1x_6$
$x_1x_2 \oplus x_1x_3 \oplus x_1x_4 \oplus x_1x_6$	$x_1x_2 \oplus x_1x_3 \oplus x_1x_4 \oplus x_1x_5 \oplus x_1x_6$

Based on experimental data, Stănică and Maitra [439] proposed a conjecture.

Conjecture 5.74. *There are no homogeneous rotation symmetric bent functions of degree greater than* 2.

Construction and enumeration of bent rotation symmetric functions have been studied in [169,438–440,116].

A partial but ultimately unsuccessful attempt to prove the above conjecture was presented in [440, Theorem 5]. Further attempts were made by Stănică in [437], which gives more insight to the problem than [440, Theorem 5], and Theorem 5.76(iii) supports the mentioned conjecture for a large class of homogeneous rotation symmetric Boolean functions.

First, we recall a result from Zheng et al. [486, Theorem 30].

Theorem 5.75. *Let f be a function on $\mathbb{V}_n$ and $J \subseteq \{1,2,\ldots,n\}$ such that f does not contain any term $x_{j_1} \cdots x_{j_t}$ where $t > 1$ and $j_1,\ldots,j_t \in J$. Then the nonlinearity of $f \leq 2^{n-1} - 2^{s-1}$, where $s = |J|$.*

As an example, take the 8-variable rotation symmetric Boolean function f having SANF $x_1x_2x_3$, that is, its algebraic normal form is $x_1x_2x_3 \oplus x_2x_3x_4 \oplus x_3x_4x_5 \oplus x_4x_5x_6 \oplus x_5x_6x_7 \oplus x_6x_7x_8 \oplus x_7x_8x_1 \oplus x_8x_1x_2$. Let $J = \{1,2,4,5,7\}$ as in the previous theorem. It is easy to see that there is no term in f with *all* indices from J. Since $|J| = 5$, it follows that the nonlinearity $\leq 2^7 - 2^4 = 128 - 16 = 112$; in reality, the nonlinearity is 80.

For a homogeneous degree d rotation symmetric Boolean function f with its SANF given by $\sum_{i=1}^{s} \beta_i$, where $\beta_i = x_{k_1^{(i)}} x_{k_2^{(i)}} \cdots x_{k_d^{(i)}}$ (note that $k_1^{(i)} = 1$, for all i), we define a sequence $d_j^{(i)}$, $j = 1,2,\ldots,k_{i-1}^{(i)}$, by $d_j^{(i)} = k_{j+1}^{(i)} - k_j^{(i)}$. Let $d_f = \max_{i,j}\{d_j^{(i)}\}$, that is, the largest distance between two consecutive indices in all monomials of f. The next theorem generalizes in some direction the results of [364] and [131].

Theorem 5.76. *The following hold for a homogeneous rotation symmetric Boolean function f of degree $d \geq 3$ in $n \geq 6$ variables:*

(i) *If the SANF of f is $x_1 \dots x_d$, then f is not bent.*
(ii) *If the SANF of f is $x_1x_2 \dots x_{d-1}x_d \oplus x_1x_2 \dots x_{d-1}x_{d+1}$, then f is not bent, assuming: $\frac{n-2}{4} > \lfloor \frac{n}{d} \rfloor$, if $n \not\equiv 1 \pmod d$; $\frac{n}{4} > \lfloor \frac{n}{d} \rfloor$, if $n \equiv 1 \pmod d$.*
(iii) *In general, if $d_f < \frac{n/2-1}{\lfloor n/d \rfloor}$, then f is not bent.*

Proof. We follow Stănică [437]. It is easy to check the claim for $n = 6$. Now we consider $d \ge 3$ and $n \ge 8$.

Take the rotation symmetric Boolean function f with SANF $x_1x_2 \dots x_d$. Assume first that $n \not\equiv 0 \pmod d$. Let $J = \{1,2,\dots,d-1,d+1,d+2,\dots,2d-1,2d+1,\dots,\lfloor n/d \rfloor d - 1, \lfloor n/d \rfloor d + 1,\dots,n-1\}$. Since f is homogeneous and there are no d consecutive indices (assume $x_{n+1} := x_1$, etc.), as required by the terms of f, it follows that the set J satisfies the conditions of Theorem 5.75. To find the number of elements of J, we count the missing indices, obtaining $|J| = n - \lfloor n/d \rfloor - 1$. Thus, $\mathcal{N}_f \le 2^{n-1} - 2^{n-\lfloor n/d \rfloor - 2}$. Since $d \ge 3$ and $n \ge 8$, then $\lfloor n/d \rfloor + 1 \le \lfloor n/3 \rfloor + 1 \le n/3 + 1 < n/2$. Therefore, $n - \lfloor n/d \rfloor - 2 > n/2 - 1$, which implies $\mathcal{N}_f < 2^{n-1} - 2^{n/2-1}$, so f is not bent.

If $n \equiv 0 \pmod d$, take $J = \{1,2,\dots,d-1,d+1,d+2,\dots,2d-1,2d+1,\dots,\lfloor n/d \rfloor d - 1 = n-1\}$, with $|J| = n - n/d$. Thus, $\mathcal{N}_f \le 2^{n-1} - 2^{n-\lfloor n/d \rfloor - 1} < 2^{n-1} - 2^{n/2-1}$, so f is not bent, in this case, as well.

We prove next claim (ii) for the homogeneous rotation symmetric Boolean function f with SANF $x_1x_2 \dots x_d \oplus x_1x_2 \dots x_{d-1}$. Assume that $n \not\equiv 0, 1 \pmod d$. Take $J = \{1,2,\dots,d-1,d+2,\dots,\lfloor n/d \rfloor d - 1, \lfloor n/d \rfloor d + 2,\dots,n-2\}$, which satisfies Theorem 5.75, since there are no d consecutive indices with a gap of length 2. By counting missing indices, we obtain $|J| = n - 2\lfloor n/d \rfloor - 1$, therefore $\mathcal{N}_f \le 2^{n-1} - 2^{n-2\lfloor n/d \rfloor - 2} < 2^{n-1} - 2^{n/2-1}$, if $n/2 - 1 < n - 2\lfloor n/d \rfloor - 2$, which is equivalent to $n > 4\lfloor n/d \rfloor + 2$.

Next, assume that $n \equiv 0 \pmod d$, respectively, $n \equiv 1 \pmod d$. In these cases, take $J_0 = \{2,\dots,d-1,d+2,\dots,\lfloor n/d \rfloor d - 1 = n-1\}$, respectively, $J_1 = \{1,2,\dots,d-1,d+2,\dots,\lfloor n/d \rfloor d - 1 = n-2\}$. Both J_0,J_1 satisfy Theorem 5.75 and as before, counting missing indices, we obtain $|J_0| = n - 2\lfloor n/d \rfloor - 1$ and $J_1 = n - 2\lfloor n/d \rfloor$. It follows that, under $n \equiv 0 \pmod d$, $\mathcal{N}_f \le 2^{n-1} - 2^{n-2\lfloor n/d \rfloor - 2} < 2^{n-1} - 2^{n/2-1}$, if $n/2 - 1 < n - 2\lfloor n/d \rfloor - 2$, which is equivalent to $n > 4\lfloor n/d \rfloor + 2$. Also, under $n \equiv 1 \pmod d$, $\mathcal{N}_f \le 2^{n-1} - 2^{n-2\lfloor n/d \rfloor - 1} < 2^{n-1} - 2^{n/2-1}$, if $n/2 - 1 < n - 2\lfloor n/d \rfloor - 1$, which is equivalent to $n > 4\lfloor n/d \rfloor$.

We prove now claim (iii). If $d_f = 1$, it follows that f is generated by $x_1x_2 \cdots x_d$, and the result follows from part (i). Assume that $d_f \ge 2$.

Case 1. $n \equiv k_0 \pmod d$, $k_0 \ge d_f$. We use once again Theorem 5.75. Take $J_1 = \{d_f, d_f+1,\dots,d-1,d+d_f,d+d_f+1,\dots,d\lfloor n/d \rfloor - 1 = n - k_0 - 1, d\lfloor n/d \rfloor + d_f,\dots,n-1\}$.

Case 2. $n \equiv k_0 \pmod d$, $0 \le k_0 < d_f$. Take $J_2 = \{d_f - k_0, d_f - k_0 + 1,\dots,d-1,d+d_f,d+d_f+1,\dots,d\lfloor n/d \rfloor - 1 = n - k_0 - 1\}$.

Both J_1,J_2 satisfy the conditions of Theorem 5.75 and $|J_1| = n - d_f\lfloor n/d \rfloor - 1$, $|J_2| = n - d_f\lfloor n/d \rfloor$. Therefore, in Case 1, $\mathcal{N}_f \le 2^{n-1} - 2^{n-d_f\lfloor n/d \rfloor - 2} < 2^{n-1} - 2^{n/2-1}$, with the last inequality holding if and only if $n/2 - 1 < n - d_f\lfloor n/d \rfloor - 2$. The last inequality follows from our imposed condition $d_f < \frac{n/2-1}{\lfloor n/d \rfloor}$.

In Case 2, $\mathcal{N}_f \leq 2^{n-1} - 2^{n-d_f\lfloor n/d \rfloor - 1} < 2^{n-1} - 2^{n/2-1}$, with the last inequality holding if and only if $n/2 - 1 < n - d_f\lfloor n/d \rfloor - 1$. The last inequality follows from $d_f < \frac{n/2-1}{\lfloor n/d \rfloor} < \frac{n/2}{\lfloor n/d \rfloor}$. ■

Certainly, if the rotation symmetry property is removed, there are plenty of, say cubic, homogeneous bent functions [406]. Using a matrix representation method for Boolean functions of Hou [229], Xia et al. [465] showed that homogeneous cubic bent functions exist on any even dimensional vector $\mathbb{V}_n$ space over $\mathbb{F}_2$, where $n \geq 6, n \neq 8$. In fact, Qu et al. [374] found that in any space $\mathbb{V}_{6k}$ there are more than $30^k \binom{6k}{6}$ cubic homogeneous bent functions. An interesting attempt was made by Charnes et al. [102,103] to find invariants of bent functions. They showed that some cubic homogeneous bent functions in 6 variables that appeared previously in Qu et al. [373] arise as invariants under an action of the symmetric group on four objects. They used the developed technique to construct cubic homogeneous bent functions in 8, 10, and 12 variables.

At the opposite pole, Xia et al. [465] showed that there are no homogeneous bent functions of degree k in $2k$ variables, for $k > 3$. This was generalized recently by Meng et al. [324] who showed that for any nonnegative integer ℓ, there exists a positive integer N such that for $k \geq N$ there exist no $2k$ variable homogeneous bent functions having degree $k-\ell$ or more, where N is the least integer such that $2^{N-1} > \binom{N+1}{0} + \binom{N+1}{1} + \cdots + \binom{N+1}{\ell+1}$.

The problem of constructing quartic, quintic, etc., homogeneous bent functions is still open. We would like to mention here the following conjecture of Meng et al. [324].

Conjecture 5.77. *For any $k > 1$, there exists N such that for any $n > N$, there exist homogeneous bent functions of degree k on $\mathbb{V}_{2n}$.*

The previous conjecture seems hard, but perhaps one can answer the following 'easier' existence problem: for any k find a homogeneous bent function of degree k (in some dimension).

CHAPTER

Stream cipher design

6

6.1 INTRODUCTION

He who loves practice without theory is like the sailor who boards ship without a rudder and compass and never knows where he may cast.

Leonardo da Vinci (1452–1519)

We consider in this chapter only the simplest type of stream cipher, in which the plaintext is given as a string of bits $\{p_i\}$ and the ciphertext string $\{c_i\}$ is produced by adding this stream mod 2 to a keystream $\{k_i\}$, that is, another string of bits produced by some process. Thus

$$c_i \equiv p_i + k_i \pmod 2. \tag{6.1}$$

This is often called a *binary additive stream cipher*.

The archetype for all stream ciphers of this kind is the famous Vernam cipher, proposed by Gilbert Vernam in 1917 (the history is recounted in the book of Kahn [248, pp. 394–403]) for the enciphering of telegraph messages. In this stream cipher, the keystream is simply a bitstream of the same length as the plaintext bitstream; the keystream is never reused (the reason for this is explained in the next paragraph), so each new plaintext requires the production of a new keystream of the same length. If the keystream bits are always chosen at random (it is a fundamental fact of cryptography that it is not easy to generate long strings of genuinely random bits!), then this cipher is called the one-time pad. In his classic paper [418, pp. 679–682], Shannon proved that the one-time pad has perfect security, in the sense that given any amount of ciphertext, an attacker cannot deduce any information whatsoever about the corresponding plaintext, other than the number of bits it contains. This fact does not put the designers of new stream ciphers out of business. The one-time pad is an impractical cipher because it requires so much keystream. If you could send large amounts of keystream securely to your intended recipients of messages, why not do away with the encryption and simply send the messages themselves securely?

A possible solution to the problem of distributing large amounts of key is to have the two people who wish to communicate purchase identical copies of some book. This seems to provide suitable long bitstrings by having the letters (and spaces, numbers and punctuation also, if this is desired) converted into bits by some agreed upon method. Now the two parties can obtain a keystream for a one-time pad by one of them sending a precise starting point in the book to the other. Historically, this method was actually used by various espionage agents, and there are references to it in some novels about spies. Alas, this system is not a true one-time pad because the keystream, since it corresponds to sensible text in some language, is not random. This cryptosystem is often called a *running key cipher*. The system is broken by exploiting the redundancy of the language in which the book is written. Two recent detailed descriptions of computer methods for doing this are given in articles in the cryptography journal *Cryptologia* [143,18]. Note that the attacks on the running key cipher show why it is not safe to reuse a portion of the keystream for any one-time pad: reusing keystream simply converts the one-time pad into a running key cipher.

There are several methods to solve the problem of generating large amounts of keystream. One way is to make the keystream periodic, so that it repeats itself after d bits for some fixed (large) d. In this case we say that we have a periodic stream cipher. If the periodic keystream can be reproduced by using a short generating key (for example, if the keystream is the output of a linear feedback shift register, or LFSR; see the book of Golomb [211] for a readable account of the basic theory of LFSRs), then a long keystream can be obtained from a short bitstring; thus the key generation problem is solved. In this chapter we will examine various keystream generators other than the simple LFSR.

If the bits in the keystream do not depend on the plaintext or ciphertext, then we say that the stream cipher is *synchronous*. Synchronous stream ciphers require that the keystreams used for encryption and decryption be exactly synchronized in order to recover the plaintext from the ciphertext. If synchronization is lost at some bit in the keystream, then the deciphered plaintext will be garbled from that point on. This is a disadvantage, because after loss of synchronization it may be necessary to retransmit some or all of the original plaintext, enciphered again. There is also an advantage, in that enemy interception of transmitted ciphertext, which results in loss of synchronization, will be detected.

In a *self-synchronous stream cipher*, each keystream bit is derived from a fixed number n of the preceding ciphertext bits. Thus, when a ciphertext bit is lost or altered during transmission, the error propagates forward for n bits, but synchronization is regained after n correct ciphertext bits have been received. This eliminates the disadvantage of needing to retransmit after a loss of synchronization. Of course a self-synchronous stream cipher is not periodic, since each key bit depends on the preceding plaintext and/or keystream bits. Keystream for a stream cipher can be produced by using a block cipher in output feedback (OFB) or counter mode; also, both the cipher block chaining (CBC) and cipher feedback (CFB) modes of operation of block ciphers can be regarded as stream ciphers (see [322, pp. 230–233], and our Chapter 7). In this chapter we focus on stream ciphers whose design involves Boolean functions.

6.2 BOOLEAN FUNCTIONS IN PSEUDORANDOM BIT GENERATORS

It is not a simple matter to define precisely what is meant by saying that a string of bits is ‘random’. Our intuition tells us that repeatedly tossing a fair coin will give a random string of bits if, say, we associate 0 with 'heads' and 1 with 'tails'. If our intuition has some acquaintance with probability, then it will also tell us that a random string of bits cannot be predicted with any better odds of success than we would have by simple guessing. We will not attempt any precise definition of randomness here, since an intuitive understanding is sufficient to appreciate almost all of the discussion in this chapter. Knuth [261, pp. 142–166] gives an excellent discussion of various precise answers to the question 'What is a random sequence?'

We already remarked that genuinely random long bitstrings are not easily obtained. Various physical phenomena, such as static on a radio channel, are believed to be random, but it is difficult to be sure of this and also difficult to design a way to actually use such phenomena to produce long random bitstrings. Some of the problems here are discussed in [322, pp. 171–173], and some actual implementations from the 1950s are recounted in [415, p. 146]. Since random long bitstrings are hard to obtain, in most cryptographic applications we must be content with 'pseudorandom' bitstrings. A pseudorandom bit generator (PRBG) is an efficient algorithm which, given as input a short truly random bitstring (called a 'seed' for the PRBG), produces as output a long bitstring which an adversary ignorant of the seed cannot efficiently distinguish from a truly random bitstring of the same length.

Many PRBGs give periodic outputs with long periods. In the early days of the study of PRBGs, various necessary conditions (or statistical tests to be passed) for a pseudorandom sequence to appear random were proposed. Among the first of these were the three randomness postulates of Golomb [211, pp. 43–45], [322, pp. 180–181]. A few more conditions are enumerated in [322, pp. 179–185] and a very extensive account (which applies also to sequences that are not bitstrings) of statistical tests is in [261, pp. 38–113]. These tests are all concerned with various statistics computed from the given sequence, so we say that a sequence which satisfies many of these conditions has good statistical properties.

Each statistical test passed by a PRBG shows that the generator does not have a certain statistical weakness, but no matter how many tests we try we cannot be certain that there is not some new statistical weakness that has been overlooked. The following definition enables us to consider all possible statistical tests.

Definition 6.1 (*Cryptographic security of a PRBG*). *We say that a PRBG passes all polynomial-time statistical tests if no polynomial-time algorithm can distinguish the output of the generator from a truly random bitstring of the same length with a probability significantly greater than* $1/2$.

Note that if we are willing to spend an exponential amount of time, we can always distinguish the output of a generator from a truly random bitstring: we simply input all possible seeds into the generator and then search to see if the bitstring we have is among the output bitstrings.

The next definition describes a very special polynomial-time statistical test (first defined formally by Blum and Micali [44,45]), but the theorem which follows it shows that passing this special test is equivalent to passing all polynomial-time statistical tests. This establishes the 'universality' of the next bit test.

Definition 6.2. *We say that a PRBG passes the next bit test if there exists no polynomial-time algorithm which, given as input the first $n-1$ bits of some string produced from a random seed by the generator, can predict the nth bit of the string with a probability significantly greater than 1/2.*

Theorem 6.3. *A PRBG passes the next bit test if and only if it passes all polynomial-time statistical tests.*

Theorem 6.3 is proved in the seminal paper of Yao [469], which also contains precise versions of Definitions 6.1 and 6.2. We do not need the precise definitions or the details of the proof of Theorem 6.3, but they can be found in [445, Section 8.2]. A leisurely exposition (aimed at statisticians, not specialists in cryptography) of this theorem and the cryptographic concepts behind it is given in [48]. In view of Yao's theorem, we can define a (cryptographically) secure PRBG as follows:

Definition 6.4. *If a PRBG satisfies Definition* 6.1 *or* 6.2, *we will say that the generator is* secure.

The m-sequences produced by maximum period length LFSRs certainly have good statistical properties (for example, they satisfy Golomb's randomness postulates, as shown in [211, pp. 43–45]) and are therefore frequently used in the design of PRBGs for stream ciphers. However, an LFSR by itself is completely unsuitable for cryptographic purposes, because the Berlekamp–Massey algorithm [300] (see Section 2.11) gives an efficient means for determining the feedback connections for any LFSR of length L if any subsequence of the output of length $2L$ is known. This algorithm shows that any PRBG must have a large linear complexity (see Section 2.11), because there is always the possibility that an attacker could calculate the shortest LFSR which duplicates the output of a PRBG, and then use the Berlekamp–Massey algorithm to determine the PRBG.

The simplest way to get around the defects in using an LFSR as a PRBG is to use a nonlinear Boolean function to generate the bitstring. Methods to do this using nonlinear combinations of LFSRs or using nonlinear feedback shift registers are discussed in the next two sections. In this section we only consider PRBGs not based on feedback shift registers.

We first consider the possibility of actually constructing PRBGs which are secure (in the sense of Definition 6.4). It is not known if secure generators exist, but if we assume the intractability of any one of several problems widely believed to be hard, then we can construct some generators that are provably secure.

Motivated by [469], Blum and Micali [45, Section 4] gave the first explicit example of a secure PRBG; it is based on the discrete log problem.

Theorem 6.5 (*Blum–Micali generator*). *Let p be a large prime and let g be a primitive root mod p, so the powers of g mod p give the cyclic group $G = \mathbb{F}_p^* = \{1,2,\ldots,p-1\}$. Define $f : G \to G$ by $f(x) \equiv g^x \pmod{p}$. We define a PRBG with input an n-bit string and output an $L(n)$-bit string as follows: randomly choose an n-bit seed x and define the function $B : G \to \{0,1\}$ by $B(x) = 1$ if $0 \le \log_g x \le (p-1)/2$ and $B(x) = 0$ if $\log_g x > (p-1)/2$.*

Let $f^i(x)$ denote the ith iteration of $f(x)$ and define the output bitstring z_i, $1 \le i \le L(n)$, by $z_i = B(f^i(x))$. If we assume the intractability of the discrete logarithm problem in G, then this PRBG is secure.

Proof. See [45]. ■

Note that in the PRBG of Theorem 6.5, we must perform one modular exponentiation to obtain one pseudorandom bit. This was improved in [285] and [362], where it was shown that each modular exponentiation can give $O(\log\log p)$ pseudorandom bits. If p is an n-bit prime and $c < n$ is a constant, then each modular exponentiation can give up to $n - c - 1$ bits provided a further assumption about the discrete log problem mod p is made. This is explained in [359] and [184]. If one uses a composite integer m for the discrete log modulus, then [221] shows that each modular exponentiation can provide $O(m)$ bits.

Another explicit example of a secure PRBG was given by Blum, Blum and Shub [43].

Theorem 6.6 *(Blum–Blum–Shub, $x^2 \pmod{n}$ generator). Let $n = pq$, where p and q are two large secret primes, both $\equiv 3 \pmod{4}$. Randomly choose an integer seed x_0, which is a quadratic residue mod n (that is, x_0 is relatively prime to n and $x_0 \equiv a^2 \pmod{n}$ for some a). Define*

$$x_i = x_{i-1}^2 \text{ Mod } n \quad \textit{for } i = 1, 2, \ldots \tag{6.2}$$

(here x Mod n means the least nonnegative integer which is in the residue class $x \pmod{n}$). Define the output bitstring $z_i, i = 1, 2, \ldots$, by z_i = the parity bit of x_i (that is, the least significant bit of x_i). If we assume that the problem of factoring the modulus n is intractable, then this PRBG is secure.

Proof comments. The paper [43] contains a proof that the PRBG is secure if the problem of deciding whether a given integer a is a quadratic residue mod n (which we will call the *quadratic residuosity problem*) is intractable when the factorization of n is not known. In fact, it is widely believed that the quadratic residuosity problem is equivalent in difficulty to the problem of factoring n. In one direction, this is very easy in the setup here: for a prime $p \equiv 3 \pmod{4}$, one solution of $x^2 \equiv a \pmod{p}$ is simply $x = a^{(p+1)/4}$. Thus if we know the factors p and q of n, we solve the quadratic congruences for the two prime moduli and then combine the solutions using the Chinese remainder theorem to obtain a solution of $x^2 \equiv a \pmod{n}$, if such a solution exists. There is no known proof in the other direction, but Vazirani and Vazirani ([455] or [456] – the two papers are the same) were able to complete the proof of Theorem 6.6 by proving that if we assume the problem of factoring the modulus n is intractable, then the $x^2 \pmod{n}$ generator is secure.

The $x^2 \pmod{n}$ generator in Theorem 6.6 is computationally more efficient than the Blum–Micali generator in Theorem 6.5 because in the former we need only perform a modular squaring to obtain one pseudorandom bit. In [455] or [456] it was shown that in fact one can obtain $O(\log\log n)$ pseudorandom bits for each modular squaring; the

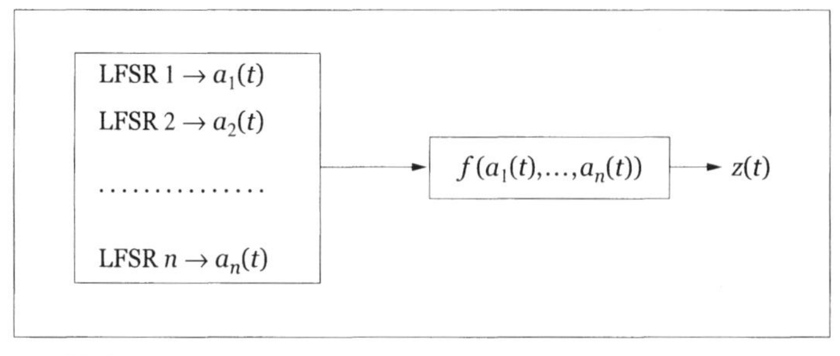

FIGURE 6.1

A nonlinear combination generator with output bitstring $z(t) = f(a_1(t),\ldots,a_n(t))$, where f is a nonlinear Boolean function.

same result, with a different proof, was found by Alexi et al. [7]. In both [455] and [7] it is shown that Theorem 6.6 is also true if (6.2) is replaced by

$$x_i = x_{i-1}^2 \text{ Mod } N = M, \quad \text{say, if } M < n/2; \text{ otherwise } x_i = n - M;$$

this is sometimes called the *modified Rabin generator*. These papers also both show, by different methods, that each modular squaring in this generator can give $O(\log\log n)$ pseudorandom bits.

There are other problems whose assumed intractability can be used to design provably secure PRBGs. Impagliazzo and Naor [233] designed a provably secure PRBG, assuming the intractability of the subset sum problem. Fischer and Stern [171] designed a provably secure PRBG, assuming the intractability of the syndrome decoding problem. We will not explore any details of these approaches.

6.3 NONLINEAR COMBINATION GENERATORS

We mentioned in the previous section that using an LFSR as a PRBG is cryptographically very weak because of the Berlekamp-Massey algorithm. However, if we use the outputs from n LFSRs as the input n-vector to a nonlinear Boolean function $f(x_1,\ldots,x_n)$, then the output bitstring made up of the values of f might be a good PRBG. This is called a *nonlinear combination generator*, as shown in Figure 6.1.

We already saw in Section 4.3 that a nonlinear combination generator is subject to correlation attacks unless the function f is correlation immune of sufficiently high order. For example, the Geffe generator [183] discussed in Section 4.3 is weak because of deficiencies in correlation immunity. Another property that the combining function must have is a sufficiently high linear complexity, as defined in Section 2.11.

Given some natural conditions, it is possible to easily compute the linear complexity of the output function f for a nonlinear combination generator, if the linear complexities of the input LFSRs are known. One fundamental result of this kind is the following theorem:

Theorem 6.7. *Suppose we have a nonlinear combination generator of the form in Figure 6.1. Assume that LFSR i has length $r(i)$ and maximum period $2^{r(i)} - 1$, and assume that the lengths $r(1),\ldots,r(n)$ are relatively prime in pairs. Let $F(x_1,\ldots,x_n)$ denote the real polynomial corresponding to the algebraic normal form of $f(x_1,\ldots,x_n)$ (that is, all xors in the algebraic normal form are replaced by ordinary real number addition). Then the linear complexity of the output sequence $z(t)$ is $F(r(1),\ldots,r(n))$.*

Proof. We let $P(i) = 2^{r(i)} - 1$ denote the period of LFSR i. One can easily show (see also [384, p. 89] for a more general result) that for positive integers a, b we have

$$\gcd(2^a - 1, 2^b - 1) = 2^{\gcd(a,b)} - 1, \tag{6.3}$$

so by our hypotheses the periods $P(i)$ are relatively prime in pairs.

First, consider the sequence $L(1,2) = \{a_1(t)a_2(t) : t = 1,2,\ldots\}$ defined by the termwise product of the output sequences from the first two LFSRs. Let $P(1,2)$ denote the least period of this sequence. Clearly, $P(1,2)$ divides $P(1)P(2)$, and $\gcd(P(1),P(2)) = 1$ implies that $P(1,2) = d(1)d(2)$, where $d(i)$ divides $P(i)$ for $i = 1,2$. Thus $d(1)P(2)$ must be a period of $L(1,2)$, so we have

$$a_1(t + d(1)P(2))a_2(t) = a_1(t + d(1)P(2))a_2(t + d(1)P(2)) = a_1(t)a_2(t)$$

for all large t. There must exist some integer b such that $a_2(t)$ is nonzero for all sufficiently large $t \equiv b \pmod{P(2)}$, so $a_1(t + d(1)P(2)) = a_1(t)$ for all such t. Fix such a t sufficiently large; by the Chinese remainder theorem, there exists an integer $m \geq t$ with $m \equiv t \pmod{P(1)}$ and $m \equiv b \pmod{P(2)}$. Now

$$a_1(t) = a_1(m) = a_1(m + d(1)P(2)) = a_1(t + d(1)P(2)),$$

so $d(1)P(2)$ is a period of LFSR 1. Hence $P(1)$ divides $d(1)P(2)$, and so it divides $d(1)$. This gives $d(1) = P(1)$ and a symmetrical argument shows $d(2) = P(2)$. Thus $P(1,2) = P(1)P(2)$.

By induction, the argument in the previous paragraph can be extended to show that the period $P(i(1),\ldots,i(k))$ of the termwise product sequence

$$L(i(1),\ldots,i(k)) = \{a_{i(1)}(t)\cdots a_{i(k)}(t) : t = 1,2,\ldots\} \tag{6.4}$$

formed from the output sequences of any $k > 2$ of the LFSRs is equal to the product $P(i(1))\cdots P(i(k))$ of the periods. Also, a version of the same argument shows that any sequence made up of sums of product sequences (6.4) with distinct sets $\{i(1),\ldots,i(k)\}$ has period equal to the sum of the corresponding periods $P(i(1))\cdots P(i(k))$. The output sequence $z(t)$ is such a sequence, so Theorem 6.7 is proved. ■

The ingredients in the proof of Theorem 6.7 have been known for a long time. The first detailed account of results on the period of sums of linear recurrent sequences was given in the seminal paper of Zierler [487]. The general study of products of linear recurrent sequences was begun by Selmer [414, Chapter 4] and more detailed results on this topic are in Zierler and Mills [488]. Herlestam [225] gives another general method of deriving results on combinations of LFSR outputs.

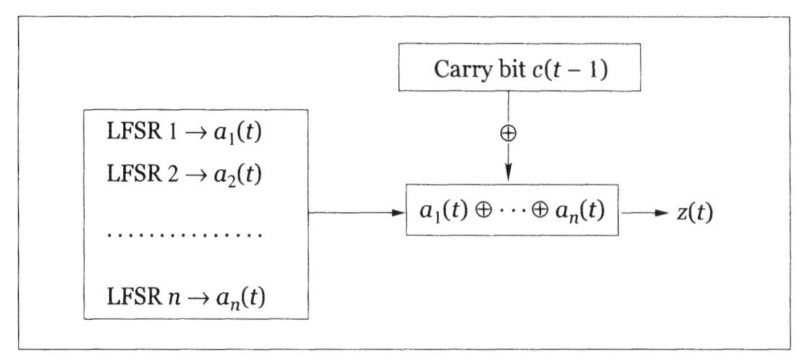

FIGURE 6.2

The summation generator with output bitstring
$z(t) = a_1(t) \oplus \cdots \oplus a_n(t) \oplus c(t-1)$.

Since a good PRBG should have a long period, designers of nonlinear combination generators generally insist that the component LFSRs have maximum periods, so one of the hypotheses of Theorem 6.7 is always satisfied. Rueppel [388, pp. 92–113] and Rueppel and Staffelbach [390] give a detailed discussion of the results that can be achieved if either the maximum period hypothesis or the relative primality assumption on the lengths in Theorem 6.7 is weakened. The latest improvements of results on the linear complexity of products of linear recurrent sequences are given in [214].

Theorem 6.7 shows that it is straightforward to design a nonlinear combination generator with high linear complexity; we simply choose a combining function of high degree. Unfortunately, Theorem 4.12 shows that there is a tradeoff between high degree and the other desirable property of high order correlation immunity for f.

Rueppel [387], [388, Chapter 9 and pp. 117, 140] pointed out that introducing even 1 bit of memory into the nonlinear combination generator allows one to avoid the tradeoff between degree and correlation immunity. One way to do this is the summation generator [387], [388, pp. 217–226] shown in Figure 6.2.

The summation generator starts with $c(0) = 0$ and $z(1) = a_1(1) \oplus \cdots \oplus a_n(1)$. Then for $t \geq 1$

$$c(t) = [(a_1(t) + \cdots + a_n(t))/2] \tag{6.5}$$

and

$$z(t) = a_1(t) \oplus \cdots \oplus a_n(t) \oplus c(t-1). \tag{6.6}$$

The formula for $z(t)$ in (6.6) looks like a linear function but, as explained in detail in [388, pp. 182–187], ordinary integer addition is actually a nonlinear operation when carried out (no pun intended!) in $\mathbb{F}_2$ with the extra bit $c(t-1)$, as in the summation generator. The nonlinearity comes from the carry bits used in the integer addition.

If we let $P(i)$ denote the period of LFSR i in the summation generator, then it is not difficult to see [387, pp. 268–270], [388, pp. 220–225] that the period of the output is the product $P(1)P(2)\cdots P(n)$ and the linear complexity of the output is close to this product.

Further, by a natural extension of the definition of correlation immunity (Section 4.1) and modifying the combining function in Figure 6.2 (in particular, the new function can have any desired degree) [387, pp. 262–264], [388, pp. 209–216], we can produce a generalized summation generator which has the maximum possible correlation immunity. Thus the tradeoff between high degree and high correlation immunity is eliminated.

A straightforward extension of the definition of correlation immunity for Boolean functions to correlation immunity for the summation generator would be to say that the generator is correlation immune of order $k(<n)$ if the information obtained about $z(t)$ given the values of the inputs to any k of the LFSRs is zero. This is not satisfactory from a cryptographic point of view because a PRBG must also have the property that information about a given output bit cannot be reliably deduced from knowledge of previous output bits. This property will not necessarily hold if we use the straightforward extension just stated (as pointed out by Siegenthaler [425] soon after his original definition of correlation immunity); for example, if we define

$$z(t) = (1 \oplus z(t-1))(a_1(t) \oplus \cdots \oplus a_n(t)), \tag{6.7}$$

then we have correlation immunity of order $n-1$ according to the straightforward extension, but (6.7) gives $z(t) = 0$ whenever $z(t-1) = 1$ and therefore defines a worthless PRBG. Thus the definition we want is: the generator is correlation immune of order k if the information obtained about $z(t)$ given $z(1), \ldots, z(t-1)$ and the values of the inputs to any k of the LFSRs is zero. To obtain a combining function of any desired high degree, we simply replace (6.5) by

$$c(t) = g(a_1(t), \ldots, a_n(t), c(t-1)), \tag{6.8}$$

where $g(x_1, \ldots, x_{n+1})$ is any Boolean function. Now the simple nonlinear function (6.6) provides the maximum correlation immunity (order $n-1$) for the generator and, independently, the possibly complicated function g in (6.8) provides the large linear complexity. The information-theoretic proof that the definitions (6.6) and (6.8) achieve order $n-1$ correlation immunity is given in [387, pp. 262–264] and [388, pp. 212–214].

Although the summation generator may have large period and linear complexity plus maximum correlation immunity, it can still be attacked with correlation methods and is completely broken by these when $n = 2$ [320] (earlier conference version [319]), [141, 332]. There is also a divide and conquer attack on the summation generator when $n = 2$ [142], and a known plaintext attack on the summation generator based on the fact that it has a relatively low 2-adic span [259]. Fast correlation attacks (see Section 4.3 for a discussion of these) on the generator can also sometimes succeed [204].

We know that the output of any nonlinear combination generator has a nonzero correlation to certain linear functions of the inputs, because if we let

$$c(f, \ell_{\mathbf{u}}) = c(f(\mathbf{x}), \ell_{\mathbf{u}}(\mathbf{x})), \quad \mathbf{u} = (u_1, \ldots, u_n), \mathbf{x} = (x_1, \ldots, x_n) \tag{6.9}$$

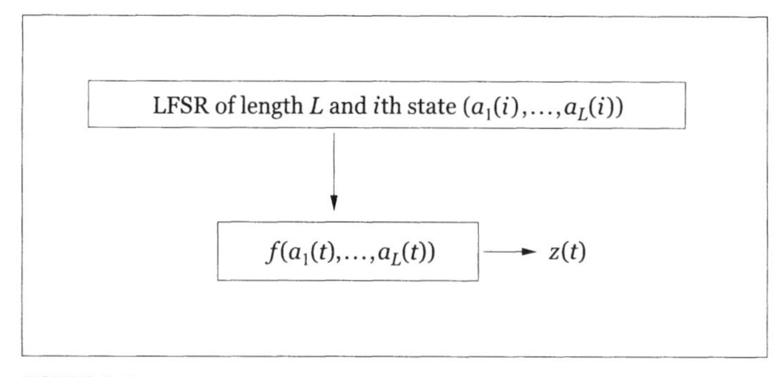

FIGURE 6.3

A nonlinear filter generator with output $z(t) = f(a_1(t),\ldots,a_L(t))$, where f is a nonlinear Boolean function.

denote the correlation value (defined by (2.4)) between the Boolean function f and the linear function

$$\ell_{\mathbf{u}}(\mathbf{x}) = u_1 x_1 \oplus \cdots \oplus u_n x_n,$$

then Parseval's equation (2.16) implies

$$\sum_{\mathbf{u} \in \mathbb{V}_n} c(f, \ell_{\mathbf{u}})^2 = 1. \tag{6.10}$$

(We have already seen (6.10) in Lemma 4.4, where the details of the proof are given.) Now (6.10) shows that not all of the correlation values (6.9) can be zero, so there is some linear combination of the n inputs which has a nonzero correlation with the output $f(x_1,\ldots,x_n)$. It was shown in [320] that a result very similar to (6.10) (namely, the sum of the squares of the correlations to linear functions is a nonzero constant which does not depend on the combining function) holds for any nonlinear combination generator with a single bit of memory, in particular for the summation generator of Figure 6.2.

6.4 NONLINEAR FILTER GENERATORS

Another method of using an LFSR in a PRBG without immediately allowing an attack via the Berlekamp–Massey algorithm involves applying a nonlinear function f to the outputs at a fixed number of stages in a single LFSR. This is called a nonlinear filter generator (since the LFSR outputs are filtered through the function f), as shown in Figure 6.3.

Of course, a nonlinear filter generator can be regarded as a special case of nonlinear combination generator (in which the n component LFSRs are all the same), but this case has traditionally been analyzed separately in the literature.

If we assume the LFSR in Figure 6.3 has length L and maximum period length $2^L - 1$, then the outputs $z(1), \ldots, z(2^L - 1)$ can be regarded as successive values of the function $f(\mathbf{s}_i)$ in L variables ($i = 1, 2, \ldots, 2^L - 1$), where $\mathbf{s}_i$ is the L-vector which gives the state of the LFSR after i clock ticks. To uniquely specify $f(\mathbf{s}_i)$ (and so uniquely specify the function f in the generator), we must assign to every input L-vector its corresponding value of f. Since f is uniquely determined by the values of the 2^L coefficients in its algebraic normal form, there is a one-to-one correspondence between f and the initial state $\mathbf{s}_1$ of the LFSR. Because the zero state $\mathbf{0}$ never occurs in the output of the LFSR, this means that the two functions f and $f+1$ will give the same function f and so the same generator. Thus there is no loss of generality if we assume $f(\mathbf{0}) = 0$.

We need to analyze the linear complexity of the nonlinear filter generator. A result of Key [253] gives a nice upper bound.

Theorem 6.8. *If the function $f(x_1, \ldots, x_L)$ in the nonlinear filter generator of Figure 6.3 has degree k, then the linear complexity of the output sequence $z(t)$ is at most*

$$\sum_{j=1}^{k} \binom{L}{j}.$$

Proof. First, we assume that the output sequence $\{b(t) : t = 1, 2, \ldots\}$ of the LFSR for any particular initial state has the form

$$b(t) = \sum_{i=1}^{L} A_i \alpha^{2^{i-1} t}, \tag{6.11}$$

where α is a root of the characteristic polynomial

$$g(x) = x^L + \sum_{i=1}^{L} c(L-i) x^{L-i} \quad (c(i) \in \mathbb{F}_2 \text{ for each } i)$$

of the LFSR in the extension field $K = \mathbb{F}_{2^L}$ and each A_i is an element, depending on the initial state, of K. It is easy to verify that in fact any sequence of form (6.11) must satisfy the linear recursion of order L, namely

$$b(t) + \sum_{i=1}^{L} c(L-i) b(t-i) = 0, \tag{6.12}$$

which corresponds to the LFSR.

First, consider the case where the polynomial f is of degree 2, and suppose that the quadratic term in this polynomial multiplies together the two sequences $s_{n,1}$ and $s_{n,2}$ generated by the LFSR with two different initial states. By (6.11), these sequences have the form

$$s_{n,k}(t) = \sum_{i=1}^{L} A_{i,k} \alpha^{2^{i-1} t}, \quad \text{for } k = 1 \text{ or } 2$$

and therefore their product sequence is

$$\sum_{i=1}^{L}\sum_{j=1}^{L} A_{i,1}A_{j,2}\alpha^{2^{i-1}+2^{j-1}t}. \tag{6.13}$$

There are $\binom{L}{2}$ ways of choosing unequal integers i and j in the double sum and L ways of choosing $i = j$, so the total number of nonzero terms in the double sum is at most (since some of the $A_{i,k}$ might be zero) $L + \binom{L}{2}$. A computation like the one showing that (6.11) must satisfy the recursion (6.12) shows that the sequence (6.13) satisfies a recursion of order at most $L + \binom{L}{2}$ and an exactly analogous argument applies to the output sequence $f(a_1(t),\ldots,a_L(t))$, where f is of degree 2, in the nonlinear filter generator. This gives the desired upper bound on the linear complexity when $\deg f = 2$. It is straightforward but notationally messy to extend this reasoning to get the upper bound in the theorem when $\deg f = k$.

Now we have the theorem whenever the LFSR output is of form (6.11). By [278, Theorem 8.24, p. 406] this is certainly true if the characteristic polynomial $g(x)$ is irreducible over $\mathbb{F}_2$, for then we have $b(t) = \mathrm{Tr}(\theta\alpha^t)$ for some θ in K, where Tr is the usual trace function of K over $\mathbb{F}_2$. If $g(x)$ is reducible over $\mathbb{F}_2$, then $b(t)$ is a sum of trace functions [278, Exercise 8.41, p. 467]. This complicates the notation even more in writing out a detailed proof of Theorem 6.8 using the reasoning above, but there is no new idea. This gives the theorem in full generality. ■

Of course from a cryptographic point of view, we are interested in lower bounds on the linear complexity, rather than upper bounds such as the one given in Theorem 6.8. Massey and Serconek [301] show that in the special case where L is a prime and f is a product of two distinct variables, there is always equality in the bound of Theorem 6.8; we omit the proof, which is based on the discrete Fourier transform.

Theorem 6.9. *If L is prime and the function $f(x_1,\ldots,x_L)$ in the nonlinear filter generator of Figure 6.3 has the form x_ix_j for $i < j$, then the linear complexity of the output sequence $z(t)$ is $L + \binom{L}{2}$.*

If L is prime and $\deg f > 2$, then it is not hard to find examples where the linear complexity is smaller than the upper bound in Theorem 6.8. Some examples for $L = 7$ are given in [388, p. 87]. When L is large, these examples must be rare, because the next theorem [388, pp. 89–91] shows that when L is prime and $\deg f = k$ is fixed, the fraction of all possible Boolean functions f which give the maximum possible linear complexity in Theorem 6.8 has limit 1 as $L \to \infty$. Our proof corrects some small errors in [388].

Theorem 6.10. *Suppose the nonlinear filter generator of Figure 6.3 has L equal to a large prime. Let $L_k = \sum_{j=1}^{k}\binom{L}{j}$ denote the maximum possible linear complexity of the output sequence of the generator when the function f has degree k. Then in the set of all output sequences produced when f varies over all polynomials of degree k, the fraction of sequences with linear complexity L_k is at least $e^{-1/L}$.*

Proof. We shall say that an output sequence $\{z(t) : t = 1,2,\ldots\}$ of the generator with $\deg f = k$ is *nondegenerate* if the sequence has the maximum linear complexity L_k. We define a formal power series (called the D-transform of the sequence) by

$$Z(D) = z(1) + z(2)D + z(3)D^2 + \cdots \tag{6.14}$$

It follows from the Berlekamp–Massey algorithm (see Section 2.12) that for some polynomials $P(D)$ and $C(D)$ we have

$$Z(D) = \frac{P(D)}{C(D)} \quad \text{with } \deg P(D) < \deg C(D) \text{ and } \deg C(D) = L_k. \tag{6.15}$$

We can expand $Z(D)$ in (6.15) into partial fractions, so

$$Z(D) = \sum_{i=1}^{r} \frac{P_i(D)}{C_i(D)},$$

where the $C_i(D)$ are the irreducible factors of $C(D)$ and $\deg P_i(D) < \deg C_i(D)$ for $i = 1,2,\ldots,r$. By Berlekamp–Massey, the linear complexity of $\{z(t)\}$ is L_k only if $\deg C(D) = L_k$ and the numerators $P_i(D)$ are all different from zero. Thus the number of nondegenerate output sequences is the number of choices of $P_i(D), 1 \le i \le r$, such that all of these polynomials are nonzero.

Since L is prime, every element of $\mathbb{F}_{2^L}$ other than 0 and 1 has a minimal polynomial of degree L. Thus all the irreducible factors $C_i(D)$ of $C(D)$ have degree L, and there are $r = \deg C(D)/L$ such factors. For each of the r factors, there are $2^L - 1$ different nonzero choices of the corresponding numerator polynomial $P_i(D)$ of degree $< L$. Thus the fraction of nondegenerate sequences is

$$\left(\frac{2^L - 1}{2^L}\right)^r = (1 - 2^{-L})^{2^L(L_k/(2^L L))}. \tag{6.16}$$

We have $L_k \le 2^L - 1$, so as $L \to \infty$ the right side of (6.16) has a limit at least $e^{-1/L}$, and this gives the theorem. ∎

A general method for studying the linear complexity of periodic sequences was given by Massey and Serconek [302]; the techniques there can readily be applied to computations for the special case in Theorem 6.8.

Correlation attacks on nonlinear filter generators are possible. If the cryptanalyst knows the characteristic polynomial of the LFSR but does not know anything else about the generator in Figure 6.3, then one way to proceed would be to try to determine an equivalent nonlinear combination generator (Figure 6.1 with $n = L$ and all of the LFSRs with the same taps, but different initial states). A ciphertext only attack which sometimes succeeds in finding this equivalent generator is described by Siegenthaler [424]. In this attack, the filter function f is determined by analyzing the crosscorrelation between ciphertext and output from the LFSR. The basic idea behind this attack and the numerous later variants of it is to break the generator by finding good correlations between the generator output and some linear combinations of the inputs to the filter function f. Such correlations always exist because the squares of the various correlation coefficients have sum equal to 1 (Lemma 4.4). A detailed, but now dated, account of one attack of this kind, the *best affine approximation attack*, is given in the book of Ding, Xiao and Shan [156, pp. 17–27].

The so-called 'fast correlation attacks' on nonlinear filter generators are discussed in Section 4.3; in particular the attack in [110] is explained in some detail. Some new approaches in these correlation attacks are given in [10]. The idea there is to determine how much information a given filter function $f(\mathbf{x})$ 'leaks' about inputs into it. Since the goal is to find the maximum leakage of $f(g(\mathbf{x}))$ over all linear functions $g(\mathbf{x})$, we shall call this the *maximum leakage attack*; in the literature, it has also been called the *conditional correlation attack*. This attack seems to have been partly inspired by the *differential cryptanalysis* attack on multi-round Feistel-type ciphers of Biham and Shamir [37,38]. The maximum leakage attack has been refined and extended in various papers, such as [274,284], and we shall use the term *conditional correlation attack* for any of these extensions.

Another type of correlation attack is discussed by Golić [190,191]. This 'linear model approach' exploits the fact that the generator output, with probability different from one half, satisfies the same linear recurrence as the LFSR sequence. One more attack, the *inversion attack*, is introduced in Golić [192]. This attack assumes that the filter function f has the (cryptographically desirable) property that the generator output is a random sequence whenever the input sequence is, and assumes that the only unknown in Figure 6.3 is the initial state of the LFSR. Then that initial state is reconstructed from a sufficiently long known string of output bits. The paper [192, p. 184] also gives a list of design criteria that a nonlinear filter generator should satisfy in order to be secure against a variety of attacks.

A survey of the various attacks on all kinds of shift register-based generators up to 1993 is given by Golić [189].

6.5 MULTIPLEXER GENERATOR

The idea of using multiplexers in a PRBG was proposed by Jennings [238] (the multiplexer generator actually first appeared in her 1980 University of London Ph.D. thesis). The Jennings design is given in Figure 6.4.

The vector $\mathbf{v}(t)$ is defined from the output sequence $\{a(t)\}$ of LFSR 1 by

$$\mathbf{v}(t) = (a(t+i_0), a(t+i_1), \ldots, a(t+i_{h-1})),$$

where the h *address inputs* $i_0, \ldots, i_{h-1}$ in $\{1,2,\ldots,m\}$ are given in increasing order and h is chosen to satisfy $h \le \min(m, \log_2 n)$. The vector $\mathbf{v}(t)$ is transformed into an h-bit integer

$$u(t) = a(t+i_0) + 2a(t+i_1) + \cdots + 2^{h-1}a(t+i_{h-1}). \tag{6.17}$$

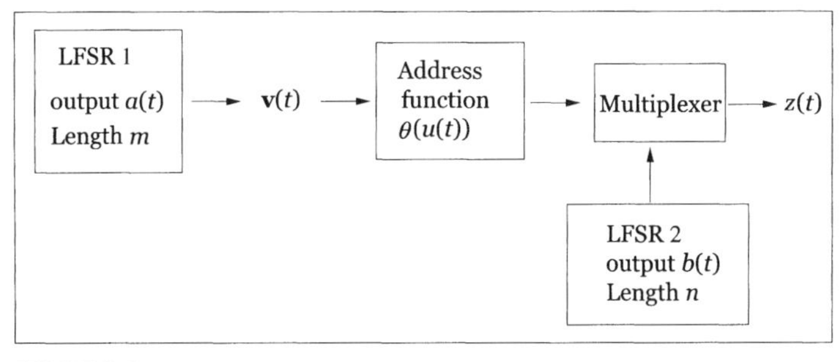

FIGURE 6.4

Multiplexer generator with output bitstring $z(t) = b(t + \theta(u(t)))$, $t = 0, 1, \ldots$.

The fixed function θ is any one-to-one map from $\{0, 1, \ldots, 2^h - 1\}$ to $\{0, 1, \ldots, n-1\}$ (this is where the upper bound on h is used). Now $\theta(u(t))$ specifies an address for the LFSR 2 and the output bit $z(t)$ is simply the bit $b(t + \theta(u(t)))$ from state t of LFSR 2.

Given natural assumptions on the two LFSRs (cf. Theorem 6.7), we can analyze the period and linear complexity of the output sequence $z(t)$.

Theorem 6.11. *Suppose we have a multiplexer generator of the form in Figure* 6.4. *Assume that the LFSRs have their maximum period lengths and that* $\gcd(m, n) = 1$. *Then the period of the output sequence* $z(t)$ *has its maximum value* $(2^m - 1)(2^n - 1)$ *and an upper bound on the linear complexity of* $z(t)$ *is* $n \sum_{j=0}^{r} \binom{m}{j}$. *If the h taps for LFSR 1 satisfy* $i_k = (k+1)d$ *for* $k = 0, 1, \ldots, h-1$, *d fixed, then the linear complexity is equal to this upper bound.*

Proof. See [238, Theorems 9 and 11]. She does not refer to linear complexity (not yet defined in 1982!) but to the degree of the minimum polynomial for $z(t)$, which is the same thing. ■

The multiplexer generator was first successfully attacked by Anderson [8] and Zeng, Yang and Rao [476], using two different methods. The method of [8], which we shall call the *inconsistency attack*, is particularly simple to implement, so we discuss an example in detail. The example assumes that we know the taps for the two LFSRs, the address inputs and the function θ. Suppose we have $m = 7$ with taps at bits 1 and 7, so the output of LFSR 1 is $a(t)$ with $a(t) \oplus a(t+6) \oplus a(t+7) = 0$; and $n = 8$ with taps at bits 1, 5, 6 and 7, so the output of LFSR 2 is $b(t)$ with $b(t) \oplus b(t+4) \oplus b(t+5) \oplus b(t+6) \oplus b(t+8) = 0$. Suppose the address inputs are bits 3, 4, 5 in each state of LFSR 1. Suppose the initial state for LFSR 1 is the bitstring 0100110 and for LFSR 2 is the bitstring 01101011. Since we have $2^h = n = 8$, the map θ must be a bijection. We define

$$\theta(u(t)) = 7 - u(t),$$

where $u(t)$ is the 3-bit integer given by (6.17). This address position identifies the bit in state t of LFSR 2 which is output as $z(t)$. Table 6.1 gives the first 15 states of the two LFSRs and the first 15 addresses for the output $z(t)$.

Table 6.1 LFSR register states and addresses for a multiplexer generator example. The addresses are $7-u$, where u is the 3-bit integer formed by bits 3, 4, 5 in the states of LFSR 1

LFSR 1	Address	LFSR 2
		01234567 ← Addresses
0100110	6	01101011
1001100	4	11010110
0011001	1	10101101
0110011	3	01011011
1100111	6	10111110
1001110	4	01101101
0011101	0	11011010
0111011	1	10110101
1110111	2	01101010
1101110	4	11010100
1011101	0	10101000
0111010	1	01010000
1110100	2	10100000
1101001	5	01000001
1010010	3	10000010

We will be able to break the system with just 15 bits of output, so we assume that we have intercepted the first 15 bits of output, which we see from the table are

$$\{z(t) : t = 1, 2, \ldots, 15\} = 1001\ 1110\ 1011\ 100 \tag{6.18}$$

(here and later we group bitstrings by fours to make reading easier). We assume in our attack that we know the LFSR lengths and taps, the map θ and the three address inputs. Thus we need only to find the key for the generator, that is the initial states of the two LFSRs.

The first step in the attack is to guess an initial state for LFSR 1; we begin by guessing the initial state is 0000 001. Now the address for the first bit to be taken from LFSR 2 is 7, so the least significant bit of the initial state of LFSR 2 must be the first ouput bit, which we know is 1 from (6.18). Moving to the next state of LFSR 1, we see that the least significant bit in the next state of LFSR 2 is 0. Proceeding in this way through the first nine states of LFSR 1, we get the results shown in Table 6.2.

We see from Table 6.2 that there is an inconsistency in state 8: the known output bit is 0, but the computed address for LFSR 2 gives the output bit as 0. This shows that our guess for the initial state of LFSR 1 must be wrong (another inconsistency at state 9 also shows this). If we now guess the initial state of LFSR 1 to be 0000 010, 0000 011, etc., we

Table 6.2 Some information on LFSR 2 deduced from guessing the initial state of LFSR 1 is 0000 001. Not all determined bits in LFSR 2 are filled in.

LFSR 1	Address	Known	L	F	S	R	2				
		output	0	1	2	3	4	5	6	7	← Addresses
0000001	7	1								1	
0000011	7	0							1	0	
0000111	6	0						1	0		
0001111	4	1					1	0			
0011111	0	1	1			1	0				
0111111	0	1	1		1	0					
1111111	0	1	1	1	0						
1111110	0	0	1	0							
1111101	0	1	0								

find inconsistency until we reach the 38th guess 0100 110, which is the correct one. Also the initial state of LFSR 2 is determined and the system is completely broken.

Anderson [8, p. 287] refers to one 'commercially available system' with $m = 31$ and $n = 32$ in Figure 6.4. He states that in 1990 an attack on this system could be carried out successfully in about three hours, using no more than 63 bits of output.

The inconsistency attack requires knowledge of the map θ, but the linear consistency attack of [476] does not. In fact, despite the large number of possible choices for θ, the linear consistency attack shows that this function makes a negligible contribution to the security of the multiplexer generator.

The idea behind the linear consistency attack is this: for some PRBGs, one can find a relatively small subset K_1 of the generator key K such that there exists a system of linear equations whose matrix form is

$$A(K_1)\mathbf{x} = \mathbf{b}, \tag{6.19}$$

where the coefficient matrix $A(K_1)$ depends only on the subkey K_1 and the vector $\mathbf{b}$ is determined by a certain segment of the generator output. If the subkey K_1 is chosen to be the one actually used in generating the segment, then (6.19) is necessarily a consistent linear system. If K_1 is not the actual subkey that was used, then (provided the segment is long enough) the system (6.19) will be inconsistent with probability very close to 1. When (6.19) is consistent, then one can usually deduce the remaining part $K - K_1$ of the key from the solution of (6.19). Therefore, with probability close to 1, one can recover the entire key by an exhaustive search of all possibilities for the subkey K_1; the consistency of (6.19) serves to identify the correct subkey. The success of this attack means that the bits of the key K which are outside of the subkey K_1 do not contribute much to the cryptographic strength of the system, that is, the key bits outside of K_1 are essentially redundant.

We need two lemmas [476, Lemmas 1 and 2] to prove the main theorem of the linear consistency attack.

Lemma 6.12. *Suppose $A = [a_{ij}]$ is an $m \times n$ matrix, $m \geq n$, with entries chosen independently and randomly from $\mathbb{F}_2$. Then for any integer r, $0 < r \leq n$,*

$$\Pr(\text{rank}(A) = r) = \binom{n}{r} 2^{-m(n-r)} \prod_{i=m-r+1}^{m} (1 - 2^{-i}).$$

Proof. This is a straightforward but complicated application of Polya-style counting using group actions; the details are in [476, pp. 165–166]. ■

Lemma 6.13. *Let $\mathbf{b}$ be any nonzero vector in $\mathbb{V}_m$ and let r be any integer such that $0 < r \leq m$. If the r-dimensional subspaces $\mathbb{W}$ of $\mathbb{V}_m$ are generated with equal probability, then* $\Pr(\mathbf{b} \in \mathbb{W}) = (2^r - 1)/(2^m - 1)$.

Proof. Each r-dimensional subspace of $\mathbb{V}_m$ which contains $\mathbf{b}$ can be spanned by a basis of the form $\{\mathbf{b}, \mathbf{v}_1, \ldots, \mathbf{v}_{r-1}\}$. There are

$$B_r = 2^{m(r-1)} \prod_{i=m-r+1}^{m-1} (1 - 2^{-i})$$

such bases of $\mathbb{V}_m$ and the basis $\{\mathbf{b}, \mathbf{v}'_1, \ldots, \mathbf{v}'_{r-1}\}$ will span the same space as $\{\mathbf{b}, \mathbf{v}_1, \ldots, \mathbf{v}_{r-1}\}$ if and only if

$$\{\mathbf{b}, \mathbf{v}'_1, \ldots, \mathbf{v}'_{r-1}\} = \{\mathbf{b}, \mathbf{v}_1, \ldots, \mathbf{v}_{r-1}\} \begin{bmatrix} 1 & \mathbf{c} \\ 0 & Q \end{bmatrix},$$

where $\mathbf{c}$ is an arbitrary $(r-1)$-vector and Q is in the general linear group $\mathrm{GL}(r-1, \mathbb{F}_2)$. So the number $N_{r,\mathbf{b}}$ of r-dimensional subspaces in $\mathbb{V}_m$ which contain $\mathbf{b}$ is given by

$$N_{r,\mathbf{b}} = \frac{B_r}{2^{r-1}|\mathrm{GL}(r-1, \mathbb{F}_2)|}.$$

In a similar way, the number N_r of arbitrary r-dimensional subspaces of $\mathbb{V}_m$ is seen to be

$$N_r = \frac{2^{mr} \prod_{i=m-r+1}^{m-1} (1 - 2^{-i})}{|\mathrm{GL}(r, \mathbb{F}_2)|}.$$

Using the well-known formula

$$|\mathrm{GL}(r, \mathbb{F}_2)| = 2^{m^2} \prod_{i=1}^{m} (1 - 2^{-i}),$$

we obtain

$$\Pr(\mathbf{b} \in \mathbb{W}) = N_{r,b}/N_r = (2^r - 1)/(2^m - 1).$$

■

Theorem 6.14. *Suppose $A = [a_{ij}]$ is an $m \times n$ matrix, $m > n$, with entries chosen independently and randomly from $\mathbb{F}_2$. Suppose $\mathbf{b}$ is any nonzero vector in $\mathbb{V}_m$. Then*

$$\Pr(A\mathbf{x} = \mathbf{b} \text{ is consistent}) < 2^{n-m}(1 + 2^{-m-1})^n.$$

Proof. We give the proof of [476, Theorem 1]. If $L(A)$ denotes the subspace of $\mathbb{V}_m$ spanned by the n columns of A, then the system is consistent if and only if $\mathbf{b} \in L(A)$. Thus the probability in the theorem is

$$\begin{aligned}
\Pr(\mathbf{b} \in L(A)) &= \sum_{r=0}^{n} \Pr(\text{rank}(A) = r)\Pr(\mathbf{b} \in L(A) | \dim L(A) = r) \\
&= \sum_{r=0}^{n} \binom{n}{r} 2^{-m(n-r)} \prod_{i=m-r-1}^{m} (1 - 2^{-i})(2^r - 1)/(2^m - 1) \\
&= 2^{-m} \sum_{r=0}^{n} \binom{n}{r} 2^{-m(n-r)} \prod_{i=m-r-1}^{m-1} (1 - 2^{-i})(2^r - 1) \\
&< 2^{-m} \sum_{r=0}^{n} \binom{n}{r} 2^{-m(n-r)} (2^r - 1) \\
&= 2^{-m} \left(\sum_{r=0}^{n} \binom{n}{r} 2^{-m(n-r)} 2^r - \sum_{r=0}^{n} \binom{n}{r} 2^{-m(n-r)} \right) \\
&= 2^{-m}(2 + 2^{-m})^n - 2^{-m}(1 + 2^{-m})^n < 2^{n-m}(1 + 2^{-m-1})^n;
\end{aligned}$$

we used Lemmas 6.12 and 6.13 in the second equality. ■

Now we can prove the theorem at the heart of the linear consistency attack. There is little loss of generality in assuming that the taps of both LFSRs are known (equivalent to knowing the two characteristic polynomials) and that the address inputs $i_0, \ldots, i_{h-1}$ are known (especially since Theorem 6.11 shows that a simple way to guarantee maximum linear complexity highly restricts the choices for these inputs in LFSR 1).

Theorem 6.15. *Suppose we have a multiplexer generator of the form in Figure 6.4, with known characteristic polynomials $f(x)$ and $g(x)$ for LFSRs 1 and 2, respectively, and known address inputs $i_0, \ldots, i_{h-1}$. If an output segment of length $N \geq m + 2^h n$ is known, then the generator can be broken by 2^{m+h} linear consistency tests.*

Proof. We give a step-by-step description of the attack. There is no loss of generality in assuming that the intercepted output segment of length N is $z(0), \ldots, z(N-1)$. The first goal of the attack is to recover the initial state $\mathbf{a}_0 = (a(0), \ldots, a(m-1))$ of LFSR 1.

Step 1. Suppose

$$g(x) = x^n + c(n-1)x^{n-1} + \cdots + c(1)x + c(0)$$

is the characteristic polynomial of LFSR 2, so the output $b(t)$ satisfies the recursion

$$b(t) = c(n-1)b(t-1) + \cdots + c(1)b(t-(n-1)) + c(0)b(t-n), \quad t \geq n. \quad (6.20)$$

We compute and store the coefficients $g_{t,j}$ such that

$$b(t) = g_{t,n-1}b(n-1) + \cdots + g_{t,1}b(1) + g_{t,0}b(0), \quad 0 \leq t \leq N-1.$$

We can compute the $g_{t,j}$ from the recursion (6.20) or from the polynomial congruence

$$x^t \equiv g_{t,n-1}x^{n-1} + \cdots + g_{t,1}x + g_{t,0} \pmod{g(x)}, t \geq 0,$$

which follows from (6.20).

Step 2. For every nonzero vector $\mathbf{a}$ in $\mathbb{V}_m$, start LFSR 1 with $\mathbf{a}$ as its initial state. Construct 2^h linear systems S_k, where

$$S_k \text{ is } A_k\mathbf{x} = \mathbf{z}_k, \quad 0 \leq k \leq 2^h - 1, \quad (6.21)$$

by defining, say, $\mathbf{x} = (x_0, x_1, \ldots, x_{n-1})$ and

$$\mathbf{g}(t) = (g_{t,n-1}, \ldots, g_{t,1}, g_{t,0}),$$

and then putting the equation $\mathbf{g}(t) \cdot \mathbf{x} = z(t)$ into S_k whenever $u(t) = k, 0 \leq t \leq N-1$. Note that because $u(t) = k$, each entry in z_k has the form $b(t + \theta(k)) = z(t) =$ one of the intercepted output bits for some value of t. The matrix A_k will have $m(k)$ rows, one for each value of $t, 0 \leq t \leq N-1$, such that $u(t) = k$.

Now test the consistency of system S_k for each $k, 0 \leq k \leq 2^h - 1$. We discard the vector $\mathbf{a}$ whenever an inconsistent system is found. We keep every vector $\mathbf{a}$ for which all of the systems S_k are consistent, and these vectors are the candidates for the vector $\mathbf{a}_0$, which must be one of the kept ones. We can estimate the probability p for an arbitrary vector $\mathbf{a}$ to be kept, as follows: suppose $m(k) > n$ (so Theorem 6.14 applies to A_k) for $k < q$ and $m(k) \leq n$ for $k \geq q$. By Theorem 6.14, the probability p_k for system S_k to be consistent satisfies

$$p_k < 2^{n-m(k)}(1 + 2^{-m(k)-1})^n, \quad 0 \leq k \leq q-1,$$

so we have

$$\begin{aligned} p &< 2^{qn-m(0)-\cdots-m(q-1)} \prod_{k=0}^{q-1}(1 + 2^{-m(k)-1})^n \\ &\leq 2^{2^h n - N}(1 + 2^{-n-2})^{qn} < 2^{-m} e^{n^2/2^{n+2}}; \end{aligned}$$

we use our hypothesis $N \geq m + 2^h n$ and the fact $q < 2^h < n$ in the last inequality. Note that this step involves 2^{m+h} linear consistency tests.

Step 3. Let $\mathbf{a}$ be any of the candidates for $\mathbf{a}_0$. Consider any of the systems (6.21) in which A_k has its largest rank, and let V denote the set of solutions in $\mathbb{V}_n$ for this system. By Lemma 6.13, V has only one element with probability nearly 1. Choose an arbitrary $\mathbf{v}_0$ in V and consider the set of vectors

$$\mathbf{v}_{-n+1},\ldots,\mathbf{v}_{-1},\mathbf{v}_0,\mathbf{v}_1,\ldots,\mathbf{v}_{n-1} \tag{6.22}$$

which would be generated in order by LFSR 2 starting with $\mathbf{v}_{-n+1}$. Check whether there is in (6.22) a subset of 2^h vectors

$$\mathbf{v}_{i(0)},\mathbf{v}_{i(1)},\ldots,\mathbf{v}_{i(2^h-1)} \tag{6.23}$$

such that the conditions (derived from (6.21))

$$A_k\mathbf{v}_{i(k)}^T=\mathbf{z}_k,\quad 0\leq k\leq 2^h-1$$

and

$$\max_k i(k)-\min_k i(k)<n$$

are both satisfied. Discard $\mathbf{v}_0$ if such a subset of (6.22) does not exist, and discard candidate $\mathbf{a}$ if all vectors in V are discarded.

Since the probability that a random set of 2^h vectors in $\mathbb{V}_n$ will have the property that LFSR 2, starting with one of these vectors, will generate the remaining vectors within $n-1$ steps has order $O(2^{-n(2^h-1)})$, all of the candidate vectors $\mathbf{a}$ except $\mathbf{a}_0$ will be discarded, and $\mathbf{v}_0$ in V will be uniquely determined.

Step 4. Given the 2^h indices $i(k)$ in (6.23), define $\sigma=\min_k i(k), \tau=\max_k i(k)$ and $\rho=n+\sigma-\tau$. Suppose that LFSR 2, starting from $\mathbf{v}_\sigma$, arrives at the vector $\mathbf{v}_{i(k)}$ in (6.23) after $n(k)$ steps. Then it follows that

$$\theta(k)=n(k)+\nu,\quad 0\leq k\leq 2^h-1,$$

where ν (independent of k) may be any nonnegative integer $\leq\rho$, and the corresponding initial state of LFSR 2 will be the vector produced by LFSR 2 after ν steps, starting at $\mathbf{v}_\sigma$.

Now we have recovered θ and the initial states of the two LFSRs, so the generator is broken. ■

A fact not mentioned explicitly in the literature but worth noting is that in Step 3 of the attack in the proof of Theorem 6.15, the number of subsets to be checked is exponential in n. It is here that the length n of LFSR 2 contributes to the security of the generator; elsewhere the size of the attack only depends linearly on n. On the other hand, the proof of Theorem 6.15 shows that the function θ makes only a minute contribution to the security of the generator.

The inconsistency attack was refined in [9,10]; the method in the latter paper has often been called the *conditional correlation attack*. The approach emphasizes the 'divide and conquer' nature of this attack, a feature it has in common with the linear consistency attack. Another attack on the multiplexer generator is given by Simpson et al. [429].

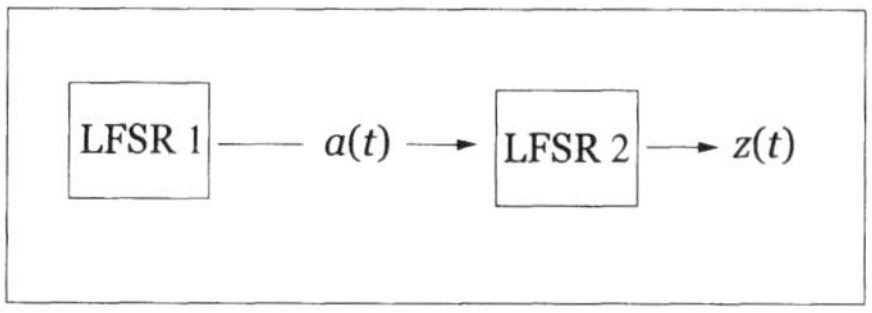

LFSR 1 —— $a(t)$ → LFSR 2 → $z(t)$

FIGURE 6.5

Basic generator using one irregularly clocked LFSR, namely LFSR 2.

6.6 IRREGULARLY CLOCKED LFSRs IN GENERATORS

So far in this chapter, all of the LFSR-based generators which we considered have been 'regularly clocked', that is, one bit of the output sequence is produced each time that the underlying shift registers are clocked. Now we shall consider 'irregularly clocked' LFSRs, where bits of the output sequence are not produced each time that the underlying shift registers are clocked, but rather at irregular clock intervals. This provides a new way to introduce nonlinearity in the operation of the generators.

The simplest design for a generator using an irregularly clocked LFSR is shown in Figure 6.5.

In Figure 6.5, the output sequence $a(t)$ of LFSR 1 serves to control the generator output sequence $z(t)$, which is a subset of the output sequence of LFSR 2. In the basic setup, LFSR 1 is regularly clocked and its output $a(t)$ determines the clocking of LFSR 2. A simple example of this is the step-1/step-2 generator of Gollman and Chambers [209, p. 528], in which LFSR 2 is stepped once at time t if $a(t) = 0$, and is stepped twice if $a(t) = 1$. If the two LFSRs both have length n and the maximum period $2^n - 1$, then it can be shown that the period of the output sequence is $(2^n - 1)^2$ and its linear complexity is $n(2^n - 1)$ (see [209, p. 528], or [100] for a more general result).

Alas, the step-1/step-2 generator is subject to correlation attacks of various kinds. In [489], an attack is given which is very likely to recover the initial state of LFSR 2 if enough consecutive output bits $z(0), z(1), \ldots$ are known and if the length of LFSR 2 is not too large (say <50 with present computer speeds). This gives the first stage of a divide and conquer attack on the generator. A correlation attack which is applicable to more general forms of the generator in Figure 6.5 is explained in [199]. An even more general attack is described in [203].

A more general form for the basic generator of Figure 6.5 has been analyzed by Rueppel [389, pp. 101–104]. We call LFSR 1, say with length L_1 and period P_1, the control register, and we call LFSR 2, say with length L_2 and period P_2, the generating register. We denote the states of the registers 1 and 2 by $\mathbf{a}_i = (a(i), a(i+1), \ldots, a(i+L_1-1))$ and $\mathbf{b}_i = (b(i), b(i+1), \ldots, b(i+L_2-1))$ $(i = 0, 1, 2, \ldots)$, respectively. We define some function f with domain $\{\mathbf{a}_i\}$ and range $\{0, 1, \ldots, L_2 - 1\}$, so f maps a state of LFSR 1 to an address in LFSR 2. The generator is now defined by

$$\sigma(t) = \sum_{k=0}^{t} f(\mathbf{a}_k), \quad z(t) = b(\sigma(t)) \; (t = 0, 1, \ldots). \tag{6.24}$$

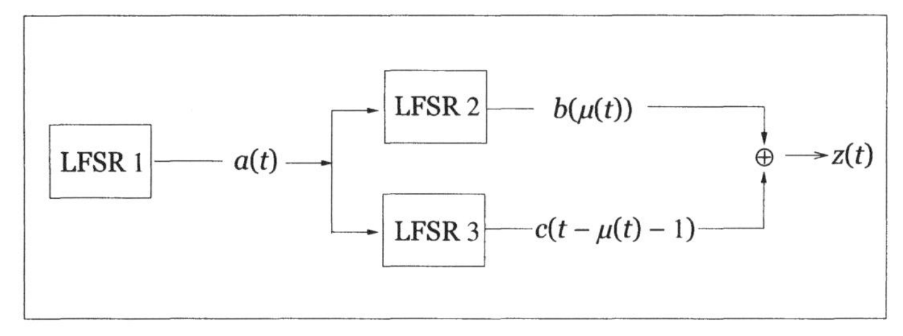

FIGURE 6.6

The alternating step generator.

In other words, for $t = 0, 1, \ldots$ we clock LFSR 1 once and get $f(\mathbf{a}_t)$, then we clock LFSR 2 $f(\mathbf{a}_t)$ times and get the resulting output $b(\sigma(t))$, where $\sigma(t)$ is the sum of the number of steps that LFSR 2 has made at time t.

The step-1/step-2 generator that we discussed above has $f(\mathbf{a}_t) = 1 + a(t)$. The simplest possible choice of f is $f(\mathbf{a}_t) = a(t)$, which was first considered by Beth and Piper [33]. They called the resulting output a 'stop-and-go' sequence, because LFSR 2 stops (is not clocked) whenever $a(t) = 0$ and goes (is clocked once) whenever $a(t) = 1$. Clearly, this sequence is cryptographically weak (an output bit is simply repeated whenever $a(t) = 0$, and we know $a(t) = 1$ if an output bit does not repeat). Beth and Piper were well aware of this and so defined the stop-and-go generator by xor-ing the stop-and-go sequence with the output of a third LFSR, say LFSR 3. The stop-and-go generator may have large linear complexity and period, plus good statistical properties, but as so often happens these are far from enough to provide cryptographic security. The improved linear consistency attack of Zeng et al. [477] (which they call the *linear syndrome attack*) breaks the generator; in fact, [477, Theorem 4] shows that the generator is readily broken if the feedback polynomials of only LFSRs 1 and 3 are known, provided the degree of the unknown feedback polynomial of LFSR 2 is not too large.

Günther [215] introduced the alternating step generator of Figure 6.6. It is based on the stop-and-go generator and retains the advantage of high output speed, but has better cryptographic properties (for example, it turns out to be secure against the linear syndrome attack).

We define

$$b(-1) = c(-1) = 0; \quad \mu(t) = -1 + \sum_{k=0}^{t} a(t).$$

The 'alternating steps' come in part (ii) of the following detailed description of the operation of the alternating step generator.

(i) Get $a(t)$ from LFSR 1 at time t.
(ii) If $a(t) = 1$, then step LFSR 2 from $t-1$ to t, outputting $b(\mu(t))$.
If $a(t) = 0$, then step LFSR 3 from $t-1$ to t, outputting $c(t - \mu(t) - 1)$.
(iii) Define the output $z(t)$ to be $b(\mu(t)) \oplus c(t - \mu(t) - 1)$.

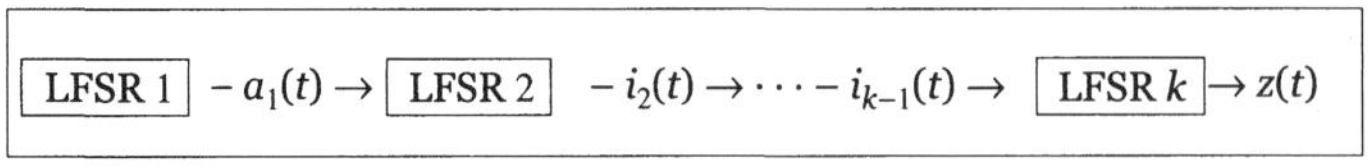

FIGURE 6.7

The cascade generator.

We see that the alternating step generator is essentially two stop-and-go generators with the same control register LFSR 1, with the outputs of these two generators being xor-ed together to produce the alternating step generator output. Günther [215] points out that the alternating step generator is vulnerable to a divide and conquer attack (using a method which is essentially a special case of the linear consistency attack, which was developed some years later) which can recover the initial state of the control register in about 2^L steps, where L is the length of the register. Once that is found, the initial states of LFSRs 2 and 3 are easily recovered and the generator is broken. However, this attack is easily prevented by choosing L large enough to make the computations infeasible, and there was no other attack on the alternating step generator until the work of Golić and Menicocci [196]. They showed that a divide and conquer attack on LFSRs 2 and 3 is sometimes feasible by adapting the method of Golić and Mihaljević [199]. Some empirical work in their paper was later analyzed theoretically by Golić [195]. This attack can be prevented if the lengths of LFSRs 2 and 3 are large enough to prevent a complete search over their possible initial states. More work on the divide and conquer correlation attack is in [197] and a new fast correlation attack on LFSRs 2 and 3 is developed in [198].

An attack on the alternating step generator based on the decoding problem for a certain communication channel is given in [239]. A similar attack on the shrinking generator is described in more detail below, when we define that generator.

We note that in the original paper [215] the alternating step generator is defined with LFSR 1 replaced by a register which outputs a de Bruijn sequence (we know from Section 2.13 that such a sequence can be obtained from the output sequence of a maximum period LFSR of length L simply by adding a 0 at the end of each string of $L-1$ zeros in the output of the LFSR). With this change, it was possible to easily prove good statistical properties and large values for the period and linear complexity of the output of the generator. It is conjectured that all of this can be proved if we have an LFSR 1 with maximum period in Figure 6.6, but this remains unproved.

Another attempt to strengthen the stop-and-go generator is the cascade generator of Figure 6.7. The idea is to extend the basic generator of Figure 6.5 by adding more LFSRs in series, where the output of each LFSR controls the clock of the next register in the series.

The first LFSR, with output sequence $a_1(t)$, is clocked regularly, and its output bit $a_1(t)$ controls the second LFSR: if $a_1(t) = 1$, then the second register is clocked; if $a_1(t) = 0$, then the second register is not clocked, but its output bit is repeated. Now the output bit of LFSR 2 is xor-ed with $a_1(t)$ to give the input bit $i_2(t)$ for LFSR 3. In general, if the input bit $i_{j-1}(t)$ to register j is 1, then register j is clocked; if $i_{j-1}(t) = 0$, then register j is not clocked and its last output bit is repeated. Now the output bit of register j is xor-ed with $i_{j-1}(t)$ to give the input bit $i_j(t)$ to register $j+1$. The output of the final register k is the generator output $z(t)$.

The original paper of Gollmann [206] on cascade generators analyzed the situation in which all of the k registers were cyclic (that is, having a feedback polynomial of form $1+x^L$, so successive states of the generator are cyclically permuted) with the same length L, where $L = p$ is chosen to be a prime; these are sometimes called p-cycle cascades. Given some benign number theoretic assumptions, Gollmann showed that the period and linear complexity of the generator are both p^k. This suggests that for security one should take k large in preference to p large, that is, use long cascades of short registers in preference to short cascades of long registers. Chambers and Gollmann [99] found a property they called lock-in, which allows an attacker to reduce the size of the key space which needs to be searched: if the attacker knows the feedback polynomial but not the initial states of the k LFSRs, then he can reconstruct the initial states one at a time, moving from the last register to the first (a type of divide and conquer attack). This reduces the key space which needs to be searched from size $(2^p-2)^k$ to $((2^p-2)/p)^k$, which is a significant saving for a cryptanalyst if p is not too large.

Example 6.16. *(Three-cycle cascade with initial generator states* 010, 010, 001*) (Recall our convention that registers output and shift to the left)*

t	Register 1	Out	In	Register 2	Out	In	Register 3	$z(t)$
0	010	0	0	010	0	0	001	0
1	100	1	1	010	0	1	001	0
2	001	0	0	100	1	1	010	0
3	010	0	0	100	1	1	100	1
4	100	1	1	100	1	0	001	0

Another version of the cascade generator (sometimes called an m-sequence cascade) uses k maximum period LFSRs, all of the same length, say L. Chambers [98] showed that the output sequence of such a generator has period $(2^L-1)^k$ and linear complexity at least $L(2^L-1)^{k-1}$. Park, Lee and Goh [354] gave an attack on this generator which was successful with $k=9$ and $L=100$, and they indicated the attack might still succeed with $k=10$. Their attack extended earlier work of Menicocci [325]. The lock-in property also still occurs with m-sequence cascades.

More early work on these cascade generators is in [207,210,208].

Another generator using irregular clocking which has received a lot of attention is the shrinking generator of Coppersmith, Krawczyk and Mansour [119]. No subexponential attack on this generator has been found yet.

We see from Figure 6.8 that the output sequence of the shrinking generator is a 'shrunken' version of the output sequence $\{b(t)\}$ of LFSR 2. The bit $b(t)$ is output if and only if the input $a(t)$ to LFSR 2 is 1; if $a(t)=0$ then LFSR 2 is clocked but the bit $b(t)$ is not output. If we define

$$F(t) = \text{the index of the } t\text{th 1 in the sequence } \{a(t)\}, \tag{6.25}$$

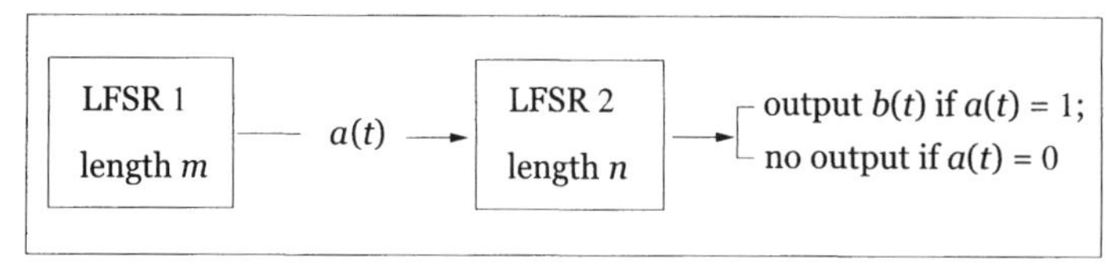

FIGURE 6.8

The shrinking generator.

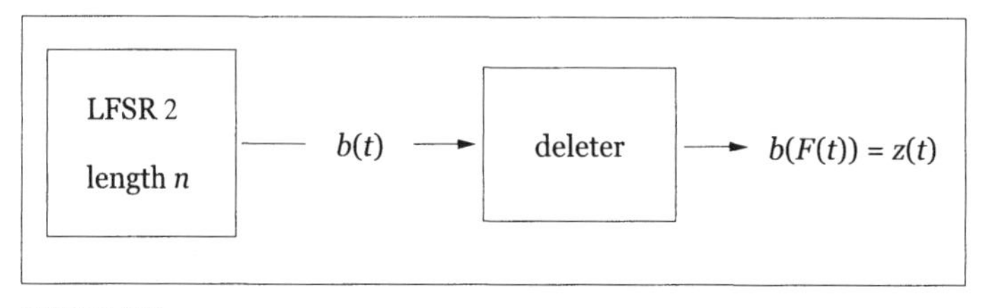

FIGURE 6.9

The shrinking generator as a deletion channel; the deleter is controlled by LFSR 1 in Figure 6.8 and $F(t)$ is defined by (6.25).

then plainly the tth output bit in Figure 6.8 is $b(F(t))$, so we can define the output sequence to be $\{z(t) = b(F(t)) : t = 1, 2, \ldots\}$. The new feature here (unlike the examples of generators of the type in Figure 6.5, which we discussed above) is that long strings of bits from LFSR 2 may be omitted from the output, and it is difficult to tell where these strings occur. We assume that the feedback polynomials for the LFSRs are known, so the secret key for the generator is the initial state for the two LFSRs.

This generator has the virtues of great simplicity, ease of implementation in hardware, good statistics (it is proved in [119] that the period of the generator output is exponential in both m and n, and that the linear complexity is exponential in m; it is also shown that the output $\{z(t)\}$ closely resembles a random bitstring in several ways) and resistance to various attacks (briefly explained in [119, Section 4]). A disadvantage of the shrinking generator is the fact that when LFSR 1 outputs a long string of zeros there is a corresponding long delay before the next output $z(t)$ appears. This could be a serious problem in some applications; [119, Section 5] suggests use of a buffer to solve the problem.

Golić and O'Connor [201] described a *probabilistic correlation attack* on the shrinking generator. This attack is most easily described if we interpret the shrinking generator as a *deletion channel*, as shown in Figure 6.9.

A deletion channel is a communication channel in which bits in the input bitstring to the channel are deleted with a probability p, and the resulting subset of the input bitstring is the output bitstring of the channel. If we take the input string to be $\{b(t)\}$ in Figure 6.9 and the deleter is controlled by LFSR 1 as in Figure 6.8 (we regard the output $\{a(t)\}$ of LFSR 1 as a random string), then the shrinking generator is a deletion channel with $p = 1/2$ and output string $\{z(t)\} = \{b(F(t))\}$, where $F(t)$ is defined by (6.25).

To break the shrinking generator, it suffices to recover the sequence $\{b(t)\}$ from a known segment of the output sequence $\{z(t)\}$. This gives the initial state of LFSR 2 and from this plus the known segment of output we can easily recover the initial state of LFSR 1. Because of Figure 6.9, recovering $\{b(t)\}$ amounts to the decoding problem for a deletion channel: we try to decode the output $\{z(t)\}$ to the correct input $\{b(t)\}$. This problem has been extensively studied, both on its own and in specific cryptographic settings. For example, deletion channels like the one in Figure 6.9 were analyzed from a cryptographic point of view in the papers [199,202,326], all of which were written before the appearance of the shrinking generator. These papers interpret the decoding problem as the problem of finding a match for a given string in a longer string, where various deletions are allowed in the longer string in order to obtain the match. In particular, a probabilistic model for this *approximate string matching*, which was developed by Hall and Dowling [219], is used in [202] and [201].

A refined probabilistic correlation attack on the shrinking generator which has lower complexity than the previous attacks is given by Johansson [239]. He uses maximum a posteriori (MAP) decoding on the deletion channel and gives full details of the complexity estimates. The straightforward attack on the shrinking generator which involves simply trying all 2^m possible states of LFSR 1 and using each one to produce enough generator output to compare with a known output string will break the generator with complexity $O(2^m n^3)$, as explained in [119, Section 4], so this is the complexity estimate to beat. If we choose m and n in Figure 6.8 in the range 61 to 64 (as was done in some implementations of the shrinking generator reported by Krawczyk [263]), then the complexity of this attack is about 2^{80}. According to [239], the method of [201], which involves trying states of LFSR 2, will reduce this to about 2^{77}, and the methods of [239] will get down to 2^{40} to 2^{50}, depending on the length of the known segment of output. We note that much more known output (at least 2^{30} bits with the parameters given) is needed for the methods of [239] than for the method of [201] ($>4n$ bits would be enough).

A detailed account of the application of the probabilistic correlation attack of [201] to the shrinking generator is given by Simpson et al. [428]. As in [201], the attack requires only a short (linear in n) known segment of output. A refined correlation analysis for the attack of [201] is given by Golić [193].

The idea of the shrinking generator led to numerous papers, of which we can only mention a few. Blöcher and Dichtl [42] tried to design a stream cipher which would utilize the shrinking idea but be faster to run in software. Their design was quickly broken by Anderson [11]. A more successful idea was the self-shrinking generator of Meier and Staffelbach [321], which applies the shrinking idea to a single maximum length LFSR. The output of the LFSR is grouped into pairs of consecutive bits. If the bit pair is 10 or 11, then the generator outputs 0 or 1 (the second bit in the pair), respectively. If the bit pair is 00 or 01, then the generator outputs nothing and the next bit pair is considered. Thus on average the self-shrinking generator is expected to discard three quarters of the bits output by the LFSR. The shrinking and self-shrinking generators are related [321, Section 2]:

Lemma 6.17. *Any self-shrinking generator is a particular case of the shrinking generator. Conversely, any shrinking generator is a particular case of the self-shrinking generator.*

Proof. Suppose $\{a(t) : t = 0,1,2,\ldots\}$ is the output sequence of an LFSR, say R, of length n which defines a self-shrinking generator. Then the subsequence $\{a(0),a(2),\ldots\}$ controls the generator output and the subsequence $\{a(1),a(3),\ldots\}$ contains the generator output, plus some discarded bits. Both of these subsequences can be generated by the original LFSR R with initial states $S(1) = \{a(0),a(2),\ldots,a(2n-2)\}$ and $S(2) = \{a(1),a(3),\ldots,a(2n-1)\}$, respectively. Now if we define the shrinking generator in Figure 6.8 with LFSR $1 =$ LFSR $2 = R$, where LFSR i has initial state $S(i)$, then the output of this shrinking generator exactly duplicates the output of the self-shrinking generator.

Conversely, suppose we have a shrinking generator as in Figure 6.8, where LFSRs 1 and 2 have feedback polynomials $f(x)$ and $g(x)$, respectively. If we apply the self-shrinking rule to the sequence $\{s(t) : t = 0,1,2,\ldots\} = \{a(0),b(0),a(1),b(1),\ldots\}$ which interleaves the output sequences of LFSRs 1 and 2, then we reproduce the output of the shrinking generator. It is easy to see that the sequence $\{s(t)\}$ can be produced by an LFSR with feedback polynomial $f(x^2)g(x^2) = f(x)^2 g(x)^2$, so the self-shrinking generator with this LFSR and the indicated initial state duplicates the output of the shrinking generator. ■

In [321, Section 3], exponential lower bounds for the period and linear complexity of the output of a shrinking generator with an LFSR of length n are given, namely $2^{[n/2]}$ and $2^{[n/2]-1}$, respectively. It is also conjectured on the basis of numerical experiments that the linear complexity never exceeds $2^{n-1} - n + 2$, and this conjecture was proved by Blackburn [41]. The further conjecture that for $n > 3$ the period always takes its maximum possible value 2^{n-1} is still unproved.

The divide and conquer attacks on the shrinking generator which we discussed above cannot be applied to the self-shrinking generator, since the two LFSRs in the former are blended into one in the latter. An attack on the self-shrinking generator with LFSR of length n which has complexity $2^{.75n}$ and requires only a 'short' (linear in n) string of known output is explained in [321, Section 5]. This attack is based on information theory and does not even require that the sequence to be shrunk is produced by an LFSR. An improved attack with complexity about $2^{.7n}$ which still only requires a short string of known output is given by Zenner et al. [478]. If the attacker has access to an exponentially long known output string, then there is a better attack due to Mihaljević [327]. If massive memory is available, then the attack based on binary decision diagrams, as introduced by Krause [262], improves the complexity.

Hell and Johansson [224] give two new attacks on the self-shrinking generator. The first attack has the same complexity as the one in [262], requires only a short string of known output, and uses very little memory. The second attack needs an exponentially long known output string, as in [327], but has a significantly lower complexity.

6.7 ALGEBRAIC AND LINEARIZATION ATTACKS

Both nonlinear combination and nonlinear filter generators (Sections 6.3 and 6.4) are subject to *algebraic attacks*, which are unlike any of the previous attacks we have considered because their success depends on solving multivariable *nonlinear* equations of low degree. The idea for this type of attack originated with Courtois [121] and Courtois and Meier [123]. Algebraic attacks have been successful in breaking some PRBGs that seemed resistant to all other known attacks. In other cases, algebraic attacks have greatly reduced the amount of computation needed to break some generators.

In order to give a benchmark with which to compare the amount of computational effort needed to mount an attack, we will describe the *linearization attack*, which is applicable to a very wide class of keystream generators. Suppose the keystream $z(0), z(1), z(2), \ldots$ is computed from some initial secret state (the 'key') given by n bits $a(0), a(1), \ldots, a(n-1)$ in the following way: we have a Boolean function $f(x_1, \ldots, x_n)$ of degree d and another function $L : \mathbb{V}_n \to \mathbb{V}_n$. Let $\mathbf{a} = (a(0), a(1), \ldots, a(n-1))$ denote the initial state vector and define the keystream bits successively by

$$
\begin{aligned}
z(0) &= f(\mathbf{a}) \\
z(1) &= f(L(\mathbf{a})) \\
z(2) &= f(L^2(\mathbf{a})) \\
\ldots &\quad \ldots\ldots \\
z(t) &= f(L^t(\mathbf{a}))
\end{aligned}
\tag{6.26}
$$

Note that this very general model for a PRBG immediately includes the nonlinear filter generators of Section 6.4 (we simply take f to be the filter function and L to be the LFSR in Figure 6.3) and is easily seen to include the nonlinear combination generators of Section 6.3 and many of the other PRBGs we have defined in this chapter. Since $\deg f = d$, every term on the right-hand side of any of the equations in (6.26) is one of the monomials made up of a product of some subset of d or fewer of the unknowns $a_1, \ldots, a_n$. There are $M = \sum_{i=1}^{d} \binom{n}{i}$ of these monomials and we define a variable y_j for each one of them. If a cryptanalyst has access to at least $N \geq M$ keystream bits $z(t)$, then he can solve the linear system of N equations from (6.26) for the values of the variables y_j, and thus recover the values of $a_1, \ldots, a_n$. If d is not large, then the cryptanalyst may well be able to acquire enough keystream bits so that the system of linear equations is highly overdefined (that is, N much larger than M). If we use Gaussian reduction to solve the linear system, then the amount of computation required is

$$
O\left(\binom{n}{d}^{\omega}\right), \tag{6.27}
$$

where ω is the 'exponent of Gaussian reduction'. We can take $\omega = 3$ (Gauss), but the classic paper of Strassen [446] shows that one can take $\omega = \log_2 7 = 2.8\ldots$ with a small O-constant. This result has been successively improved down to $\omega = 2.376$ [120], but the

O-constant for this is very large and so not relevant in this cryptographic setting. The estimate (6.27) for the linearization attack is part of the 'folklore' of cryptanalysis and is mentioned in many places.

The linearization attack is easily defeated if $\binom{n}{d}$ is chosen large enough, but the attack is distinguished by its great generality. It applies to any PRBG whose output can be put in the form (6.26) and it does not matter what functions f and L are used (of course we assume these functions are known to the attacker, so the only secret information is the initial state). Further, the output bits that need to be intercepted for the attack do not need to be consecutive; it is enough to know where they occur in the output sequence.

Algebraic attacks rely on being able to deduce, for many different states of the generator, a multivariable equation of low degree in the state bits. This is done by getting many output bits and using them (for example, with the help of equations like (6.26)) to find the equations of low degree in the state bits. By using the structure of the generator, each such equation gives an equation of low degree in the initial state bits; if we have enough such equations, we get a very overdefined system of equations. Recent work on such overdefined systems of low degree [258,416,121] gives methods much more efficient than the Gaussian reduction for solving the equations to get the unknown initial state bits. In particular, [258] introduces the relinearization method and [416] introduces the XL (standing for 'extended linearization', or 'multiplication and linearization') method, which supersedes relinearization. The XL method does not require as many known keystream bits as are needed for the linearization attack, but this benefit comes at the expense of more computation.

The method of [121] relies on finding a low degree approximation to the function f. If the probability that the approximation holds is near one, then it can be used to derive many equations. Here we describe the improved attack of [123], which is applicable even if there is no good low degree approximation for f. For this we assume that there exists a nonzero polynomial $g(\mathbf{x})$ of low degree in n variables (it does not matter if some of the variables do not actually appear in a term with nonzero coefficient) such that

$$f(\mathbf{x})g(\mathbf{x}) \text{ has low degree } d_1 < d = \deg f. \tag{6.28}$$

Now we derive from the system (6.26) the new system

$$\begin{aligned} z(0)g(\mathbf{a}) &= f(\mathbf{a})g(\mathbf{a}) \\ z(1)g(L(\mathbf{a})) &= f(L(\mathbf{a}))g(L(\mathbf{a})) \\ z(2)g(L^2(\mathbf{a})) &= f(L^2(\mathbf{a}))g(L^2(\mathbf{a})) \\ \cdots &\cdots \cdots \\ z(t)g(L^t(\mathbf{a})) &= f(L^t(\mathbf{a}))g(L^t(\mathbf{a})) \end{aligned} \tag{6.29}$$

Even if $\deg f = d$ is not small, the Equations (6.29) are of low degree in the state bits, so if we have enough known output bits we can obtain the equations needed for the linearization method, or for any other method of solving the overdefined system.

The algebraic attack based on polynomial multiples of low degree (we shall call it the *low degree algebraic attack*) is explicitly carried out against two published stream ciphers in [123]. There is also an extended version [125] with more details. The work [123, p. 356], [125, p. 16] shows that the worst case complexity of the algebraic attack on a PRBG of form (6.26) is $O\left(\binom{n}{d/2}^{\omega}\right)$, which is something like the square root of the complexity (6.27) of the linearization attack.

Three different scenarios ($S3a, S3b, S3c$ below) are given in [125, pp. 4–5] under which low degree relations that can be exploited in an algebraic attack exist:

Low degree algebraic attack scenarios

$S3a$: Assume that there exists a function g of low degree such that $fg = h$, where h is a nonzero function of low degree.

$S3b$: Assume that there exists a function g of low degree such that $fg = 0$.

$S3c$: Assume that there exists a function g of *high* degree such that $fg = h$, where h is a nonzero function of low degree.

Meier, Pasalic and Carlet [315] investigate the existence of low degree multiples of Boolean functions in detail. They begin by condensing the three scenarios above. First, in scenario $S3c$ we have $fg = h \neq 0$. Multiplying by f gives (since $f^2 = f$ over $\mathbb{F}_2$) $f^2g = fg = fh = h$; now we have nonzero h of low degree and $fh = h$, so we are in scenario $S3a$. Thus scenario $S3c$ is redundant and we need only consider scenarios $S3a$ and $S3b$. These last two are related because of the next lemma [315, Proposition 1].

Lemma 6.18. *Suppose we have scenario $S3a$, so $fg = h \neq 0$, where g and h have degrees at most d. Suppose also that $g \neq h$. Then there is a function g' of degree at most d such that $fg' = 0$, and we have scenario $S3b$.*

Proof. We have $f^2g = fg = fh = h$ as before. Now $f(g \oplus h) = 0$ and we take $g' = g \oplus h$. ∎

Lemma 6.18 shows that we need only consider scenario $S3a$ in the case $g = h$ and scenario $S3b$. However, scenario $S3a$ with $g = h$ is equivalent to scenario $S3b$ with f replaced by $f \oplus 1$. Thus we only need scenario $S3b$ for f or $f \oplus 1$ and it is convenient to have the

Definition 6.19. *An* annihilator *of a polynomial f is a nonzero polynomial g such that $fg = 0$.*

We see that in order to prevent an algebraic attack neither f nor $f \oplus 1$ should have an annihilator of low degree. This motivates the definition [315, p. 476] of the *algebraic immunity* $AI(f)$, which is the least value of d such that either f or $f \oplus 1$ has an annihilator of degree d.

In [315, Section 4] there is an algebraic description of the set of all annihilators of a given Boolean function f. In [315, Section 5] there is a description of an efficient algorithm for deciding whether a given Boolean function in n variables has low algebraic immunity. This algorithm is computationally feasible if n is not too large.

Some methods for improving the algebraic immunity of a given Boolean function were given by Carlet [81]. The algebraic immunity of symmetric Boolean functions is studied by Braeken and Preneel [47], with an emphasis on how such functions could be designed for use in a stream cipher. The research area of algebraic attacks is very active at the time of writing of this book. We do not discuss any more papers, but simply list a set of them (not all-inclusive!) for consideration by the reader: [12,273,57,84,134,13,135].

From a complexity theory point of view, the problem of solving nonlinear systems like (6.26) and (6.29) is known to be NP-hard (a basic reference is Garey and Johnson [182]), so we cannot hope for an efficient general algorithm for these systems. This is irrelevant from a cryptographic point of view, since we only need an efficient method (possibly very *ad hoc*) for some particular systems. We now discuss recent work on specific methods for solving such systems of cryptographic interest; we emphasize how these methods can be made practical, in contrast to general methods which work in theory but are impractical computationally.

The classical algorithm for solving a system of multivariable polynomial equations is Buchberger's algorithm for constructing Gröbner bases [51]. This algorithm successively eliminates variables until a single one-variable polynomial equation is reached. This final polynomial is then solved, usually by using Berlekamp's algorithm (this classical algorithm goes back to [25,27]; a detailed exposition of it, including a complexity estimate, is in [261, pp. 420–429] and a good history of factorization algorithms is in [278, pp. 177–183]).

Unfortunately, Buchberger's algorithm is known to have double exponential run time in the worst case (this was proved by Huynh [232], who used a technique by Mayr and Meyer [309]) and seems to have single exponential run time on the average (see [164,413, 443]). A cryptanalyst is usually interested in highly overdefined systems of multivariable polynomial equations. This is because the system variables will represent the secret cryptosystem key and we can typically derive many polynomial equations connecting these variables from the relationships of the key to known samples of plaintext or ciphertext. We saw an example of this in the discussion above of the linearization attack on PRBGs. Since we know a solution of the overdetermined system exists (in contrast to the situation for a random overdetermined system of equations, which is very likely to have no solutions) and we expect the cryptosystem with a given key to produce a unique ciphertext for any given plaintext, there should only be one solution of the system. Thus we need only consider methods that will produce some solution for an overdetermined system. (Alas, some of these methods may introduce extraneous solutions which we then need to eliminate, as we shall see later.) As pointed out in [416, p. 393] it is easy to see that much better algorithms than Buchberger's can exist in the overdefined case: consider a system of $n(n+1)/2$ random homogeneous quadratic equations in n variables $x_1, \ldots, x_n$. Using the linearization method, we can replace each monomial $x_i x_j$ by a new variable y_{ij}. Now the quadratic equations give a system of $n(n+1)/2$ equations in $n(n+1)/2$ variables y_{ij}, which can be solved efficiently by Gaussian elimination. Once we have the y_{ij} values, we can find two possible values for each x_i by taking the square root of y_{ii} in the field, and we can use the values of y_{ij} to combine correctly the roots of y_{ii} and y_{jj}.

The linearization method will fail if it produces a system of linear equations in the new variables which is underdefined (more variables than equations), because then there will be many solutions to the system in the new variables which do not correspond to any solution of the system of equations in the original variables; thus it will be hard to sort out the solutions to the original system. Exactly this situation arises in the cryptanalysis by Kipnis and Shamir [258] of the hidden fields equations (HFE) public key cryptosystem of Patarin [358]. In [258, p. 26] the problem of recovering the secret key of the HFE cryptosystem is reduced to solving a system of ϵm^2 homogeneous quadratic equations in m variables, where $\epsilon < 1/2$. Thus linearization produces a new system of ϵm^2 linear equations in about $m^2/2$ new variables. This linear system is expected to have an exponential number of extraneous solutions (they are called 'parasitic' in [258], because they do not correspond to any solution of the original system of quadratic equations) in the new variables.

A new method called *relinearization* is introduced in [258, pp. 26–27]. This technique allows the extraneous solutions to be disposed of, so the cryptanalysis of HFE can be completed. Suppose the original system has ϵm^2 homogeneous quadratic equations in the m variables $x_1, \ldots, x_m$. We rewrite it (linearization) as a new system of ϵm^2 linear equations in the (approximately) $m^2/2$ new variables $y_{ij} = x_i x_j$, $i \leq j$. The solution space is a linear subspace of expected dimension $(0.5 - \epsilon)m^2$, and each solution can be expressed as a linear function of $(0.5 - \epsilon)m^2$ new variables z_k. Thus there will be many extraneous solutions in the variables y_{ij} which do not correspond to any solutions in the variables x_i.

We will get rid of the extraneous solutions (for ϵ large enough, as we shall see) by producing some new *quadratic* equations in the variables y_{ij}. We do this by considering any four indices a, b, c, d such that $1 \leq a \leq b \leq c \leq d \leq m$. Because $x_i x_j = y_{ij}$, we can group the four-term product $x_a x_b x_c x_d$ into products of pairs in three different ways and we obtain the relations

$$y_{ab} y_{cd} = y_{ac} y_{bd} = y_{ad} y_{bc}. \tag{6.30}$$

There are about $m^4/24$ different ways to choose an ordered four-tuple (a, b, c, d) of distinct indices (we ignore the additional four-tuples which we could obtain from four-tuples of nondistinct indices in this part of the process), and each choice gives rise to two linearly independent equations in the y_{ij} because of (6.30). We thus obtain about $m^4/12$ quadratic equations in the $m^2/2$ variables y_{ij}, and we expect that these equations are linearly independent (in [258, p. 27] it is stated that linear independence 'is not difficult to prove', but this does not seem to be the case; in [416, p. 395], the linear independence of the equations 'was experimentally verified' for a large number of randomly generated systems). We can lower the number of variables to $(0.5 - \epsilon)m^2$ by replacing each one of the y_{ij} by its parametric representation as a linear combination of the new variables z_k.

Now we take the new $m^4/12$ quadratic equations in the $(0.5 - \epsilon)m^2$ variables z_k and linearize again (hence the name relinearization) by replacing each product $z_i z_j$, $i \leq j$ by a new variable v_{ij}. This gives a new system of $m^4/12$ linear equations in the approximately $((0.5 - \epsilon)m^2)^2/2$ variables v_{ij}. We expect this linear system to have a unique solution (no more extraneous solutions!) if $m^4/12 \geq ((0.5 - \epsilon)m^2)^2/2$, which is true when $\epsilon \geq 0.5 - 6^{-1/2} = 0.09\ldots$. Thus in this case relinearization needs one fifth of the number of equations required by simple linearization.

To demonstrate the relinearization technique, we give the toy example from [258, pp. 29–30]. Consider the following five random homogeneous quadratic equations modulo 7 in three variables x_1, x_2, x_3.

$$\begin{aligned}
3x_1^2 + 5x_1x_2 + 5x_1x_3 + 2x_2^2 + 6x_2x_3 + 4x_3^2 &= 5 \\
6x_1^2 + x_1x_2 + 4x_1x_3 + 4x_2^2 + 5x_2x_3 + x_3^2 &= 6 \\
5x_1^2 + 2x_1x_2 + 6x_1x_3 + 2x_2^2 + 3x_2x_3 + 2x_3^2 &= 5 \\
2x_1^2 + 0x_1x_2 + x_1x_3 + 6x_2^2 + 5x_2x_3 + 5x_3^2 &= 0 \\
4x_1^2 + 6x_1x_2 + 2x_1x_3 + 5x_2^2 + x_2x_3 + 4x_3^2 &= 0
\end{aligned} \tag{6.31}$$

We replace each x_ix_j by y_{ij} and solve the system of five equations in six variables to obtain the following parametric solution in one variable z:

$$y_{11} = 2 + 5z, \quad y_{12} = z, \quad y_{13} = 3 + 2z, \quad y_{22} = 6 + 4z, \quad y_{23} = 6 + z, \quad y_{33} = 5 + 3z.$$

This family has seven solutions, but we can see that only two of them solve the original quadratic system in the x_i. To get rid of the extraneous solutions, we form the equations

$$y_{11}y_{23} = y_{12}y_{13}, \quad y_{12}y_{23} = y_{13}y_{22}, \quad y_{12}y_{33} = y_{13}y_{23}.$$

Plugging in the expressions for the y_{ij} in terms of z gives

$$(2+5z)(6+z) = z(3+2z), \quad z(6+z) = (3+2z)(6+4z), \quad z(5+3z) = (3+2z)(6+z),$$

and these equations simplify to the system

$$3z^2 + z + 5 = 0, \quad 4z + 4 = 0, \quad z^2 + 4z + 3 = 0.$$

Relinearization introduces two new variables $v_1 = z$ and $v_2 = z^2$, so we get three linear equations in v_1 and v_2 (overdetermined!) with unique solution $v_1 = 6, v_2 = 1$. Working backwards we obtain $y_{11} = 4$, $y_{22} = 2$, $y_{33} = 2$ and taking square roots modulo 7 gives $x_1 = \pm 2$, $x_2 = \pm 3$, $x_3 = \pm 3$. Finally, we use the values $y_{12} = 6 = x_1x_2$ and $y_{23} = 5 = x_2x_3$ to combine these roots in just two possible ways to obtain

$$(x_1, x_2, x_3) = (2, 3, 4) \text{ or } (5, 4, 3);$$

these are the two solutions of the original quadratic system.

The relinearization method is analyzed in detail in [416]. Much of the work is based on extensive numerical experiments on numerous overdefined systems of m homogeneous quadratic equations in n variables. These experiments involved not only 'degree 4 relinearization' (based on equations like (6.30), which are of degree 4 in the underlying variables x_i), but also higher degree relinearization. For example, degree 6 relinearization is based on equations of the form

$$y_{ab}y_{cd}y_{ef} = y_{gh}y_{ij}y_{kl} \quad \text{with } \{a,b,c,d,e,f\} = \{g,h,i,j,k,l\}.$$

These equations are cubic in the free parameters produced by the method (in the toy example of degree 4 relinearization above, there was one free parameter, namely z, and we obtained quadratic equations in this parameter), even though the original system has only quadratic equations. Thus many more equations are needed for successful degree 6 relinearization. It turns out [416, pp. 396–397] that in higher degree relinearization (unlike the degree 4 case in [258]) most of the new equations are linearly dependent, so the algorithm is much less efficient than initially expected. This disappointment led to the development of the XL method in [416], which is both simpler and more powerful than relinearization.

We describe the XL method applied to a system

$$f_j(x_1, \ldots, x_n) = c_j, \quad 1 \le j \le m \tag{6.32}$$

of quadratic equations, where the goal is to find at least one solution with x_i in a given finite field $\mathbb{F}_q$. We shall use the notation

$$x^{[k]} = \{x_{i(1)}x_{i(2)} \cdots x_{i(k)} : 1 \le i(1) \le i(2) \le \cdots \le i(k) \le n\}$$

for the set of all degree k monomials in the variables x_i (a strikingly inconvenient notation for this set is used in [416, p. 399]). Here is the description in [416, pp. 398–399].

The XL algorithm

Step 1. Multiply: We multiply each of the m equations in (6.32) by all monomials in $x^{[k]}$ for each $k \le D-2$, given some fixed $D \ge 4$.

Step 2. Linearize: Define a new variable for all monomials in $x^{[k]}, k \le D$, and perform Gaussian elimination on the equations in these new variables obtained in Step 1. The ordering of the monomials must be such that all terms containing one given variable (for example, x_1) are eliminated last.

Step 3. Solve: Assume that Step 2 yields at least one single variable equation in the powers of the one given variable. Solve this equation over the given field $\mathbb{F}_q$ (for example, by Berlekamp's algorithm).

Step 4. Repeat: Simplify the equations and repeat the process to find the values of the remaining variables.

Sometimes it is more efficient to work with a proper subset of all of the possible monomials. For example, if the quadratic equations in (6.32) are homogeneous, then it suffices to use only monomials of even degree. This is done in the following toy example of the XL algorithm from [416, p. 400]. Consider the two homogeneous quadratic equations

$$x_1^2 + a x_1 x_2 = c \tag{6.33}$$

$$x_2^2 + bx_1x_2 = d \tag{6.34}$$

with a and b not zero and D chosen to be 4.
Step 1. We get the two initial equations and the six equations obtained by multiplying each of them by the three monomials x_1^2, x_1x_2, x_2^2 in $x^{[2]}$:

$$x_1^4 + ax_1^3x_2 = cx_1^2 \tag{6.35}$$
$$x_1^2x_2^2 + bx_1^3x_2 = dx_1^2 \tag{6.36}$$
$$x_1^2x_2^2 + ax_1x_2^3 = cx_2^2 \tag{6.37}$$
$$x_2^4 + bx_1x_2^3 = dx_2^2 \tag{6.38}$$
$$x_1^3x_2 + ax_1^2x_2^2 = cx_1x_2 \tag{6.39}$$
$$x_1x_2^3 + bx_1^2x_2^2 = dx_1x_2 \tag{6.40}$$

Step 2. By elimination, we find each of the six monomials of degree ≤ 4 in terms of the variable x_1:

$$\text{From (6.33)}, \quad x_1x_2 = (c - x_1^2)/a$$
$$\text{From (6.34)}, \quad x_2^2 = d - (bc/a) + bx_1^2/a$$
$$\text{From (6.35)}, \quad x_1^3x_2 = (cx_1^2 - x_1^4)/a$$
$$\text{From (6.36)}, \quad x_1^2x_2^2 = (d - (bc/a))x_1^2 + bx_1^4/a$$
$$\text{From (6.40)}, \quad x_1x_2^3 = cda^{-1} + (cb^2a^{-1} - bd - da^{-1})x_1^2 - b^2a^{-1}x_1^4$$
$$\text{From (6.38)}, \quad x_2^4 = d^2 - 2bcda^{-1} + (2bda^{-1} + b^2d - b^2ca^{-1})x_1^2 + b^3a^{-1}x_1^4.$$

Step 3. From (6.37) and the equations in Step 2 we get one equation in the variable x_1:

$$c^2 + (abc - a^2d - 2c)x_1^2 + (1 - ab)x_1^4 = 0.$$

We solve this equation over $\mathbb{F}_q$ to find x_1.
Step 4. We find the value of x_2.

The authors of [416] showed (see [416, pp. 404–405] for a sketch of the proof) that the set of equations defined by a successful relinearization of degree D is equivalent to a subset of the equations derived from the XL algorithm with the same choice of D. The extra equations given by the XL algorithm make the overdetermined systems substantially easier to solve. The experimental results in [416] show this in numerous examples and also show that it is sometimes possible to successfully solve a given overdetermined system with the XL algorithm using a smaller value of D than would be required for solution by relinearization.

6.8 THE eSTREAM PROJECT

As an alternative to the NIST process of finding the new encryption standard AES, we would like to mention the European research project NESSIE (New European Schemes for Signatures, Integrity and Encryption), which was active between 2000 and 2003, as an attempt to identify, investigate thoroughly, and promote secure cryptographic algorithms. Unfortunately, all six submissions to NESSIE failed under scrutiny, and in 2004 another project was born, called eCRYPT (European Network of Excellence for Cryptology), 'a 4-year network of excellence funded within the Information Societies Technology (IST) Programme of the European Commission's Sixth Framework Programme (FP6)' [161]. There are multiple goals of the project, and the reader is encouraged to find out more about these in the supplied references. We shall be referring to only one of these goals, namely, the eSTREAM project, meant to identify new stream ciphers, and new techniques in building stream ciphers. There were two profiles that every submission must fall into [161]:

PROFILE 1. Stream ciphers for software applications with high throughput requirements.

PROFILE 2. Stream ciphers for hardware applications with restricted resources such as limited storage, gate count, or power consumption.

There were 34 submissions to eSTREAM, of which 32 were synchronous and two self-synchronizing. Moreover, seven submissions offered authentication along with encryption. The project ended in April of 2008 and the (synchronous) stream ciphers that were chosen were: for Profile 1, HC-128 (submitted by H. Wu), Rabbit (submitted by M. Boesgaard, M. Vesterager, Th. Christensen and E. Zenner), Salsa20/12 (submitted by D.J. Bernstein), SOSEMANUK (submitted by C. Berbain, O. Billet, A. Canteaut, N. Courtois, H. Gilbert, L. Goubin, A. Gouget, L. Granboulan, C. Lauradoux, M. Minier, Th. Pornin and H. Sibert); for Profile 2, F-FCSR-H v.2 (submitted by Th. Berger, F. Arnault and C. Lauradoux), Grain v.1 (submitted by M. Hell, Th. Johansson and W. Meier), MICKEY v.2 (submitted by S. Babbage and M. Dodd), Trivium (submitted by C. de Canniére and B. Preneel).

All of these ciphers must satisfy a plethora of requirements: key-recovery attacks not easier than exhaustive key search; resistance to distinguishing attacks, that is, given a uniform random key, the output is not distinguishable from a uniform random string; better than AES in at least one feature; and, not in the least, have a clear design.

We want to point out a design approach based on nonlinear feedback shift registers (NFSR) in a few of these submissions. As the name suggests, an NFSR uses a nonlinear Boolean function f as a feedback function,

$$s_n = f(s_{n-1}, \dots, s_1, s_1).$$

We do not want to describe all of the finalist stream ciphers, as the reader can easily find them in the mentioned references; however, we shall describe the NFSR feature in Trivium (chosen by us at random). The hardware-oriented stream cipher Trivium [146] generates 2^{64} bits from an 80-bit secret key and an 80-bit IV, with an internal state of

288 bits. After the internal initialization (the 80-bit key and the 80-bit IV are loaded into the 288-bit initial state, setting all remaining bits to 0, except for s_{286}, s_{287}, and s_{288}), the cipher updates the 288 state-bits and generates the stream bits, using a process based on an internal state organized in three interconnected sparse quadratic NFSR relations (explicit below), with keystream bits computed as linear combinations of some internal state bits. Precisely, it extracts the values of 15 specific state bits and uses them both to update 3 bits of the state and to compute 1 bit of keystream z_i. The state bits are then rotated and the process repeats itself until all $N \leq 2^{64}$ of the requested keystream bits have been generated [146]. More details in [146]. We display below the pseudocode used by designers for keystream generation, where the nonlinear Boolean function, which the authors of Trivium use, is explicit:

Keystream Generation
for $i = 1$ to N do

$t_1 \leftarrow s_{66} \oplus s_{93}$
$t_2 \leftarrow s_{162} \oplus s_{177}$
$t_3 \leftarrow s_{243} \oplus s_{288}$
$z_i \leftarrow t_1 \oplus t_2 \oplus t_3$
$t_1 \leftarrow t_1 \oplus s_{91} \cdot s_{92} \oplus s_{171}$
$t_2 \leftarrow t_2 \oplus s_{175} \cdot s_{176} \oplus s_{264}$
$t_3 \leftarrow t_3 \oplus s_{286} \cdot s_{287} \oplus s_{69}$
$(s_1, s_2, \ldots, s_{93}) \leftarrow (t_3, s_1, \ldots, s_{92})$
$(s_{94}, s_{95}, \ldots, s_{177}) \leftarrow (t_1, s_{94}, \ldots, s_{176})$
$(s_{178}, s_{279}, \ldots, s_{288}) \leftarrow (t_2, s_{178}, \ldots, s_{287})$

end for

We recommend the paper by Bernstein [32], which contains an extensive reference list on all stream cipher submissions to eSTREAM and subsequent analysis of all these algorithms.

CHAPTER

Block ciphers

7

7.1 SOME HISTORY

The art of discovering the causes of phenomena, or true hypothesis, is like the art of decyphering, in which an ingenious conjecture greatly shortens the road.

Gottfried Wilhelm Leibniz (1646–1716)

To respond to a need for ensuring security of electronic data, in 1973 (and again in 1974) the US National Bureau of Standards (now National Institute of Standards and Technology — NIST) issued a call for strong encryption primitives that would become a government-wide standard. The first call was unsuccessful since all submissions were unsuitable. However, in 1974, a team from IBM submitted a modified version of their cipher Lucifer (designed in 1971). After a year of intense collaboration between IBM and NSA, Data Encryption Standard (DES) was born, and then published in 1975 in the Federal Register, for comments. Controversy arose since the original Lucifer cipher used a key of length 128, and the key of DES was shortened to 56 bits. Also, the design principles of its substitution and permutation tables were never made public. One of the designers of DES, Alan Konheim, commented: '*We sent the S-boxes off to Washington. They came back and were all different*'. However, another IBM member of the developer team, namely Walter Tuchman, is quoted as saying, '*We developed the DES algorithm entirely within IBM using IBMers. The NSA did not dictate a single wire!*' [462]. The US Senate Committee on Intelligence was charged to investigate the matter, and the conclusion was that all the design decisions were made by the IBM developers: '*NSA did not tamper with the design of the algorithm in any way. IBM invented and designed the algorithm, made all pertinent decisions regarding it, and concurred that the agreed upon key size was more than adequate for all commercial applications for which the DES was intended*' [452,462].

In spite of all these controversies (see also [277, Appendix C, pp. 168–173], [433, Chapter 3], and [452]), the cipher was approved as a federal standard in November 1976 (FIPS PUB 46), and quickly adopted and used in many commercial applications. It was republished as the standard in 1983, 1988 (FIPS-46-1), 1993 (FIPS-46-2), and in 1998 (FIPS-46-3), modified into 'Triple DES'. For about 15 years, exhaustive key search would be the only 'efficient' attack on DES. In 1990, Eli Biham and Adi Shamir discovered *differential cryptanalysis* (see [37,38]), a general method for breaking block ciphers (the IBM team which proposed Lucifer was aware of differential cryptanalysis, and DES was designed to be resistant to this attack [118]). The discovery of *linear cryptanalysis* is attributed to Mitsuru Matsui, who applied the technique to the FEAL cipher (cf. Matsui and Yamagishi [303]). Later, Matsui published an attack on DES, which was the first unclassified experimental cryptanalysis of the cipher [304,305] (this attack on DES is hardly practical, since it requires 2^{43} known plaintexts) (cf. [245,246,462]).

On May 26, 2002, DES was finally superseded by AES (Advanced Encryption Standard), following a public competition, initiated by NIST in 1998. The minimum 'acceptability requirements and evaluation criteria' were [343]:

1. *The algorithm must implement symmetric (secret) key cryptography.*
2. *The algorithm must be a block cipher.*
3. *The candidate algorithm shall be capable of supporting key-block combinations with sizes of* 128-128, 192-128, *and* 256-128 *bits. A submitted algorithm may support other key-block sizes and combinations, and such features will be taken into consideration during analysis and evaluation.*

The evaluation criteria for the candidates were *security*, *design* (ease of software and hardware implementation), *implementation attacks*, *flexibility*. Five finalists were chosen: MARS (proposed by IBM), RC6 (proposed by RSA Laboratories), Rijndael (proposed by Joan Daemen, Vincent Rijmen), Serpent (proposed by Ross Anderson, Eli Biham, Lars Knudsen), and Twofish (proposed by Bruce Schneier and his team at Counterpane Systems). Following a few rounds of discussions, Rijndael was chosen as the new AES. For a look at the comments at the third conference on AES see [5].

7.2 INTRODUCTION

Obviously, the motivation of cryptography is to protect the secrecy of messages that are sent over an insecure channel. Any encryption algorithm contains an encryption function E and a decryption function $D = E^{-1}$. We are concerned here with *symmetric* encryption algorithms, where the encryption transformation must be kept secure, as well, since the knowledge of the encryption function will reveal its inverse, that is, the decryption transformation. In 1976, Diffie and Hellman [150] realized that a system can be devised so that the encryption function would (at least in theory) not reveal the decryption function. Such a system must be based on a *trapdoor one-way* function, which is easy to evaluate, but cannot be efficiently inverted, unless some hidden information is revealed. One year later, Rivest, Shamir and Adleman [380,381] proposed the first such *asymmetric* cipher and *public-key cryptography* was born. We shall not be concerned in this book with this type of cipher.

We will be concerned in this chapter with other types of symmetric encryption algorithms, namely *block ciphers*. Recall that the first type was *stream ciphers*, which were discussed in Chapter 6. A block cipher acts on the plaintext, which is divided into separate blocks of fixed size (for example, 32, 56, 64, 128, etc.). Precisely, a *block cipher* is a pair of functions

$$E : \mathbb{V}_k \times \mathbb{V}_n \to \mathbb{V}_n$$
$$D : \mathbb{V}_k \times \mathbb{V}_n \to \mathbb{V}_n.$$

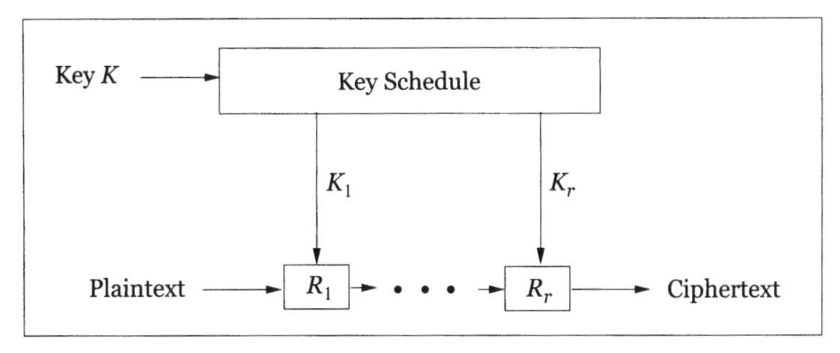

FIGURE 7.1

General structure of a block cipher.

E is called the *encryption* function, which maps an n-bit plaintext into an n-bit ciphertext (block size), using a k-bit (key size) encryption key, and D is the *decryption* function, which maps an n-bit ciphertext into an n-bit plaintext, using a k-bit decryption key. For a key in the *keyspace* $\mathbb{V}_k$ define $E_K(P) = E(K,P), D_K(P) = D(K,P)$. We must have $D_K = E_K^{-1}$, that is, the encryption transformation must be invertible, and its inverse is the decryption function.

Most block ciphers are *iterated block ciphers*: they repeat a sequence of simple operations (called *round functions*) a certain number of times (called *rounds*). Block ciphers input the plaintext in the first round and output the ciphertext in the last round. In each round, we use the same (or slightly changed) operations (except possibly in the first and/or last round, most of the time) (see de Cannière et al. [145]). Each secret subkey K_i that is used in the cipher's ith round is derived (the process is called *key scheduling*) from an initial k-bit secret key K (see Figure 7.1).

7.3 BLOCK CIPHERS' MODES OF OPERATION

7.3.1 Confidentiality modes

Any block cipher can be operated in several modes. There are four modes more often used (ECB, CBC, OFB, CFB). In NIST's Special Publication 800-38A [342], five confidentiality modes are specified for use with any approved block cipher, like the AES algorithm. The modes in SP 800-38A are updated versions of the ECB, CBC, CFB, OFB modes (all specified in FIPS Pub. 81), and CTR mode (specified in SP 800-38A). In addition, there are some modes that are not yet standardized, like the GMC mode proposed by McGrew and Viega [312,313].

We do not attempt to be thorough in our description, as there are various materials (which will be mentioned throughout) devoted to this subject. To help with the analysis of AES, two NIST workshops were dedicated to the study of various block cipher modes (see [335,336]). During these workshops a few other modes were proposed. Some were proven faulty, and some were considered for further investigation.

The *electronic codebook* (ECB) mode (see Figure 7.2) uses the simplest idea, that is, to have the message partitioned into several blocks of n-bit length (recent work is being

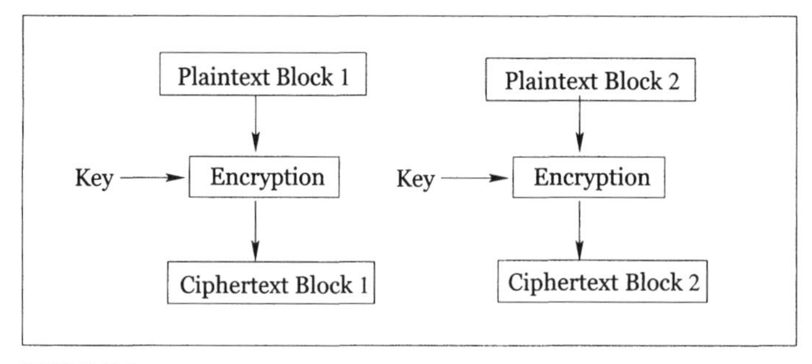

FIGURE 7.2

Electronic codebook (ECB) mode.

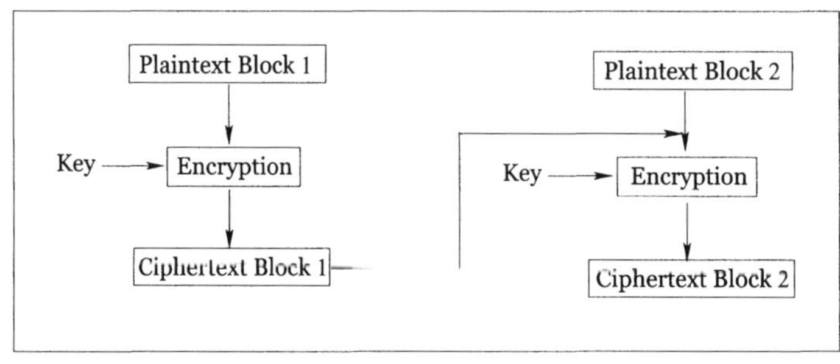

FIGURE 7.3

Cipher block chaining (CBC) mode.

done on devising block ciphers of flexible length - more on that later in the chapter), padding if necessary, and then to apply the cipher to each block. The advantage of this mode is that one can decrypt the n-bit blocks independently, in parallel. Thus, if one error occurs in transmission or encryption, that error affects only that particular block, where the error occurred. However, that is also a disadvantage, since this mode does not hide data patterns, if one uses a key for more than one block, since multiple copies of the same plaintext block map to the same ciphertext block.

The *cipher block chaining* (CBC) mode encrypts the first block using the ECB method applied to the xor of the first block with a block IV. The ith ciphertext block is obtained by applying the block cipher to the ith plaintext block and the $(i-1)$th ciphertext block (see Figure 7.3). Thus, the ciphertext C_i is defined by $C_0 = IV$ (stands for *initialization vector*), $C_i = E_K(P_i \oplus C_{i-1})$. The decryption is easily done: to obtain the plaintext we simply use $P_i = D_K(C_i) \oplus C_{i-1}$, $C_0 = IV$. CBC requires message blocks to have the same length (typically, 8 or 16 bytes), so some blocks may have to be padded to bring them to this length. The main disadvantage of this mode is that one cannot work on blocks independently, and that an error in C_i propagates into C_{i+1}, as well. However, this mode is *self-synchronizing* since if the same error occurs in C_i, but not C_{i+1} (could be that the

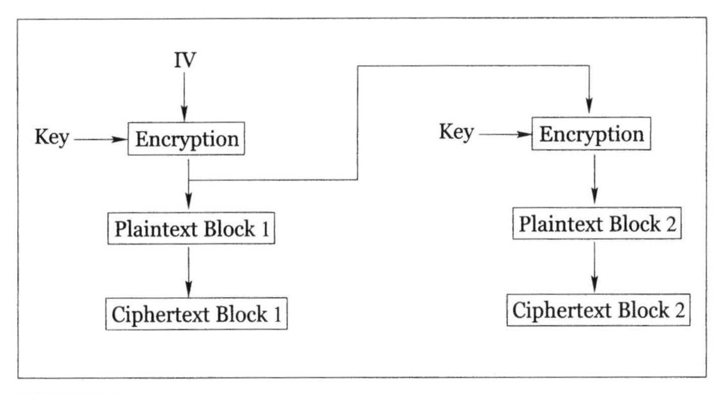

FIGURE 7.4

Output feedback (OFB) mode.

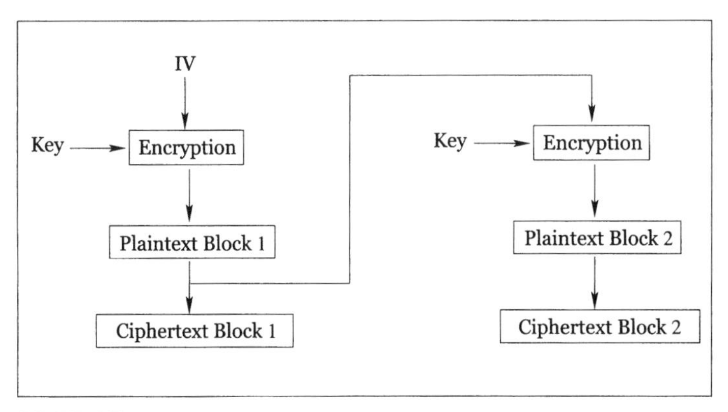

FIGURE 7.5

Cipher feedback (CFB) mode.

ciphertext C_i was changed after its usage in C_{i+1}), then C_{i+2} will be decrypted correctly (see Menezes et al. [322, Chapter 7]).

A slight modification of CBC that propagates more is the *propagating cipher block chaining* mode (PCBC) (not used too often). The encryption uses $P_0 = IV, C_0 = 0$, $C_i = E_K(P_i \oplus P_{i-1} \oplus C_{i-1})$, and of course, decryption is done by $P_0 = IV, C_0 = 0$, $P_i = D_K(C_i) \oplus P_{i-1} \oplus C_{i-1}$.

The *output feedback* (OFB) mode (see Figure 7.4), which is an alternative to CBC to avoid error propagation, produces a unique stream independent from other streams produced by the same encryption key, without rekeying. It uses the output of the encryption function as a feedback, as opposed to the *cipher feedback* (CFB) mode, where the output of the ciphertext is used as a feedback (see Figure 7.5). The ciphertext for OFB is obtained by using the mathematical formula $C_i = P_i \oplus O_i$, where the feedback $O_i = E_K(O_{i-1})$, $O_0 = IV$. The decryption is done in reverse, $P_i = C_i \oplus O_i$. Since

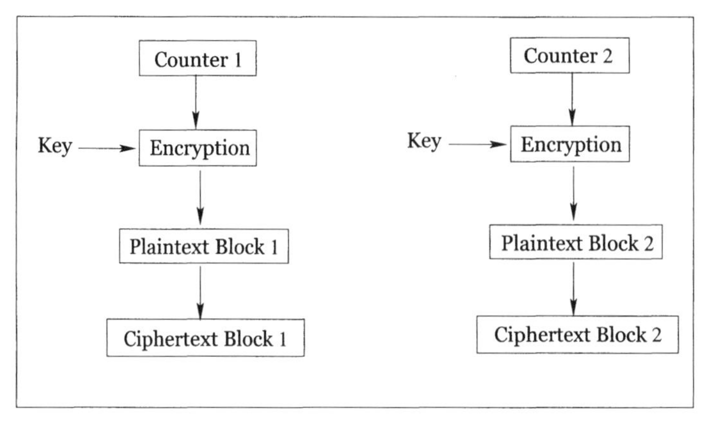

FIGURE 7.6

Counter (CTR) mode.

we use feedback in the sequence of blocks, similar blocks result in different ciphertext blocks, which is a desirable feature. The ciphertext for CFB is obtained by using $C_i = P_i \oplus E_K(C_{i-1}), C_0 = IV$, and the decryption is performed by $P_i = C_i \oplus E_K(C_{i-1})$. As opposed to the OFB mode, the CFB mode is self-synchronizing, but it requires $\lceil n/r \rceil$ ciphertext blocks to recover (cf. [322, Chapter 7]), where n is the length of the IV and r is the plaintext block length, $r \leq n$. In CBC and CFB, by predicting the messages' IVs, one may obtain information that allows the ciphertexts for certain messages to be distinguished (cf. [335,336]).

In addition to OFB, the *counter mode* (CTR) (see Figure 7.6), sometimes called *segmented integer counter* (SIC) mode, also transforms the cipher into a synchronous stream cipher, and uses the following encryption transformation $C_i = P_i \oplus O_i$, $O_i = E_K(IV \oplus i (\text{mod } 2^n))$ (the input of E_K is thought of as an n-bit string) (decryption is similar). In fact, the counter can be any function which produces a sequence whose period is sufficiently large. CTR has no error propagation, and the blocks can be decrypted independently.

7.3.2 Authentication modes

The *cipher block chaining-message authentication code* (CBC-MAC) (see [170,234,235]) is a message integrity method that uses block ciphers such as DES and AES. Each block of plaintext is encrypted with the cipher and then xor-ed with the next encrypted block.

The CBC-MAC algorithm [234] was improved to XCBC (*extended block cipher chaining*) algorithm by Gligor and Donescu [187] (see also the comments on XCBC-MAC by Black and Rogaway [40]), and further by Iwata and Kurosawa [236], who proposed OMAC to NIST. The CMAC authentication mode (specified in NIST's Special Publication 800-38B for use with any approved block cipher) stands for *cipher-based message authentication code*, analogous to HMAC, the hash-based MAC algorithm [342].

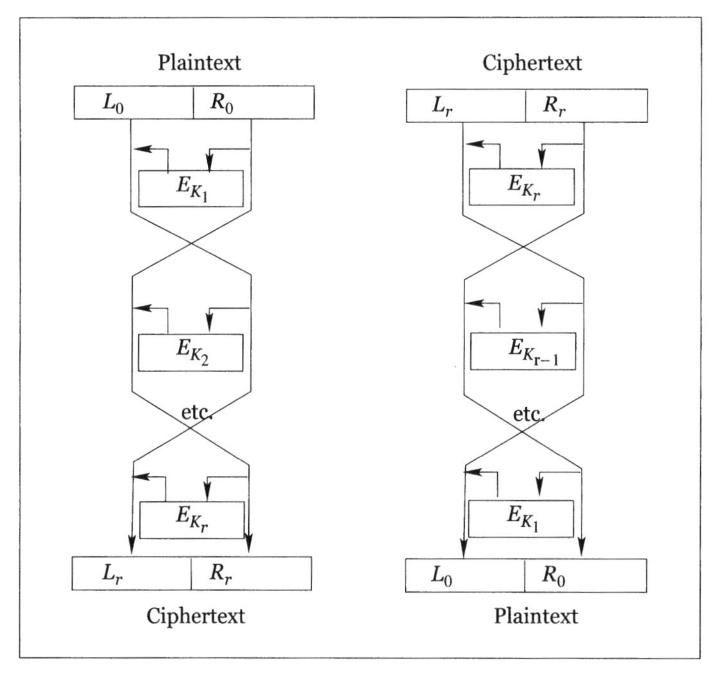

FIGURE 7.7

Encryption and decryption in a Feistel cipher with r rounds.

The materials from the two Modes of Operations workshops [335,336] are good references for further reading on various other confidentiality or authentication modes of operation for block ciphers.

7.4 DESIGN APPROACHES

General block ciphers combine simple operations, like transpositions, linear transformations, arithmetic operations, modular multiplication and substitutions, to construct a complex encryption transformation (see the excellent book of Menezes et al. [322, Chapter 7]). There are two main ways to achieve that: Feistel ciphers, and substitution-permutation networks (SPN).

7.4.1 Feistel ciphers

A *Feistel cipher* (named after the IBM cryptographer Horst Feistel, who helped design Lucifer and DES) is an iterated block cipher (depicted in Figure 7.7), which splits the input block into two parts L_{i-1} and R_{i-1}. In each round, R_{i-1} will form the left part of the next block L_i, and the left part L_{i-1} will be combined with the output of the encryption function of key K_i applied to R_{i-1}, precisely,

$$\begin{aligned} L_i &= R_{i-1} \\ R_i &= L_{i-1} \oplus E_{K_i}(R_{i-1}) = R_{i-2} \oplus E_{K_i}(R_{i-1}), \end{aligned} \tag{7.1}$$

where E_{K_i} is the round function. The subkeys K_i are all derived by the key scheduling from the cipher key K. Decryption is done using

$$\begin{aligned} R_{i-1} &= L_i \\ L_{i-1} &= R_i \oplus E_{K_i}(L_i) = L_{i+1} \oplus E_{K_i}(L_i), \end{aligned}$$

regardless of what the round function looks like (can be quite complex and not necessarily invertible).

Data Encryption Standard (DES) is an example of a Feistel cipher, which uses blocks of 64 bits (8 bytes), and outputs a ciphertext of the same length. The initial secret key has 56 bits; in fact, the bits $8, 16, \ldots, 64$ in a 64-bit key are used as parity bits [322, Chapter 7, Section 7.4.2]. We shall give full details of this important cipher in Section 7.5.1.

There are many more Feistel block ciphers based on the Feistel circuit: 3-DES, Blowfish, Camellia, CAST-128, RC5, Twofish, TEA, etc., some more secure than others. We will not discuss them here. There are some types of ciphers using the Feistel structure (called *unbalanced Feistel ciphers*), which split the input into two strings of nonequal size, but the general idea of a Feistel cipher is similar. Such a cipher is, for instance, the Skipjack encryption algorithm (developed by the NSA in 1992 and declassified in 1998) [49,36].

7.4.2 Substitution permutation networks

Another type of block cipher is the *substitution-permutation network* (SPN), which is a product cipher (involving two or more transformations) that builds a round function composed of a number of steps combining two invertible functions: substitutions (S-boxes) and permutations. For a general view on this type of cipher see Figure 7.8.

The permutation layers are linear transformations acting on the entire block, generating *diffusion* (cf. Shannon [418]). The substitution layers (S-boxes) satisfy various cryptographic properties (we referred to some of them previously: they have to be highly nonlinear, be correlation immune, have high avalanche features, etc.), and they act on small length (usually from 8 to 64) blocks, introducing *confusion* in the cipher (cf. [418]). The confusion must *mix* the plaintext bits and key bits, and the diffusion must spread the bits. The key mixing layers use xor or addition modulo some power of 2.

Precisely, an $m \times n$ S-box is a function $S : \mathbb{V}_m \to \mathbb{V}_n$, that is, an n-tuple of Boolean functions on $\mathbb{V}_m$.

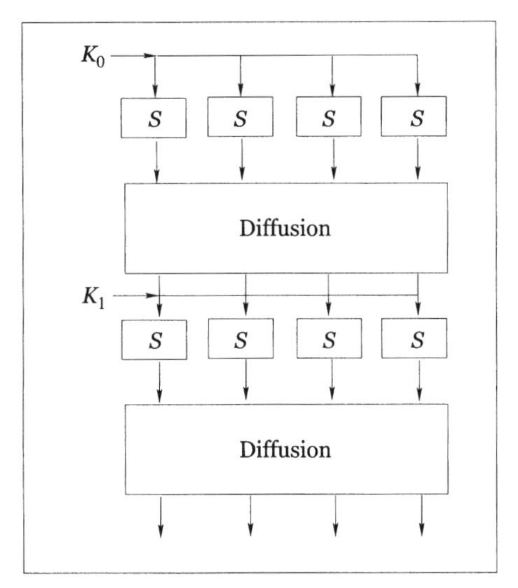

FIGURE 7.8

Substitution-permutation network (SPN).

7.5 NOTABLE SYMMETRIC CIPHERS

In any block cipher design, there are some universally accepted requirements. We shall refer to some of them here. Since the cipher must be computationally efficient, many ciphers employ a fixed key size. While this makes the cipher applicable and efficient in various commercial and other areas, it also makes the cipher vulnerable to at least one attack, *the exhaustive key search* (sometimes called *brute-force search*): for an n-bit cipher, the key K of size k can be recovered using a small number (say, $\lceil (k+4)/n \rceil$ [322, Chapter 7]) of plaintexts and the corresponding ciphertexts encrypted using K, in $O(2^{n-1})$ operations.

With the advent of improvement in computational power, one has to be careful in the choice of key size, perhaps designing a cipher that is flexible in accepting various key sizes, depending on the sensitivity of the encrypted data and how long it will need to be kept secret. One notable effort is the introduction of a *Variable Size Block Cipher* (VSBC), which is sufficiently flexible to encrypt blocks of arbitrary size, but produce good data diffusion across the entire block (cf. Ritter [377]). Various proposals support a variable key size, for example RC5 introduced by Rivest [378,379] (block size: 32, 64 or 128 bits; key size: 0 to 2040 bits; and number of rounds: 0 to 255).

If the key size is small, one may improve security by applying an encryption algorithm multiple times. One may use the same or a different encryption scheme, producing what is known as a *cascade cipher*: a concatenation of $s \geq 2$ block ciphers (stages), obviously, each using a different key, where the output of one cipher becomes the input of the next. We express this as

$$E(input) = E_{K_s} E_{K_{s-1}} \cdots E_{K_1}(input).$$

Multiple encryption is a type of cascade encryption, where the keys are not necessarily independent. Care should be taken, however, since iterating the same cipher may not improve security when the cipher's encryption transformations form a group. In that case, iterating a cipher amounts to composing these transformations, which (by the group structure) is simply another transformation, so the keyspace size will not increase.

7.5.1 Data Encryption Standard

Data Encryption Standard (DES) enciphers a 64-bit data block using: an initial permutation (IP); 16 rounds of a complex key dependent calculation f; a final permutation, the inverse of IP. It relies on the Feistel scheme (7.1) (for details on the algorithm, see [136]). Initially, there were about 12 criteria used in DES (we already described in the introduction of this chapter how DES was developed), resulting in about 1000 possible S-boxes, of which the implementers chose 8. Some conditions imposed on the S-boxes were: each row of an S-box is a permutation of 0 to 15; no S-box is a linear or affine function of the input; changing one input bit to an S-box results in changing at least two output bits; $S(\mathbf{x})$ and $S(\mathbf{x} \oplus 001100)$ must differ in at least 2 bits; the S-boxes were chosen to minimize the difference between the number of 1s and 0s in any S-box output when any single input is held constant. For more details on the design criteria for DES, see Coppersmith's report [118, pp. 247–250]. Coppersmith was on the DES design team. He chose to publish those details after differential cryptanalysis (which the DES team knew about in 1974) was finally described in the open literature by Biham and Shamir [37, 38]. DES required about 2^{55} operations in exhaustive key search attack. We shall describe DES below in more detail.

Biham and Shamir [37,38] published the first theoretical attack with smaller complexity than brute force, namely differential cryptanalysis. However, it requires an unrealistic 2^{47} chosen plaintexts. The first experimental cryptanalysis of DES was performed using linear cryptanalysis by Matsui [304]. With these attacks, and the work of Wiener [460, 461], the research community produced undoubtable evidence in the next few years that DES is insecure to a determined attacker and it must be replaced. Triple-DES followed DES and we expect this to be phased out by the next generation algorithm, namely AES.

DES uses a 56-bit key size K and a 64-bit plaintext P. In reality, the input key of DES is 64-bit, but the rightmost bit in each byte is a parity bit to ensure that the weight of each byte is odd. Ignoring these bits in each byte results in a 56-bit key. Each of the 16 rounds of DES uses a 48-bit subkey produced by a key scheduling algorithm, which will be described later. If each 64-bit plaintext block is encrypted individually, DES is in the ECB mode. One can use two other modes of DES encryption, namely CBC and CFB, by making each block dependent on all the previous blocks through an initial xor operation (see Section 7.3).

The DES algorithm (see [136,21]) starts by applying a permutation IP, displayed in Table 7.1, to the input plaintext P.

We read Table 7.1 (and subsequent ones) in the following way: for example, the bit 54 is at the intersection of the row labeled 17 and column labeled 1, and summing these labels we obtain 18, that is, bit 54 will be replaced by bit 18 in the new permuted key.

Table 7.1 Initial permutation, IP

Bit	0	1	2	3	4	5	6	7
1	58	50	42	34	26	18	10	2
9	60	52	44	36	28	20	12	4
17	62	54	46	38	30	22	14	6
25	64	56	48	40	32	24	16	8
33	57	49	41	33	25	17	9	1
41	59	51	43	35	27	19	11	3
49	61	53	45	37	29	21	13	5
57	63	55	47	39	31	23	15	7

Table 7.2 Inverse initial permutation, IP^{-1}

Bit	0	1	2	3	4	5	6	7
1	40	8	48	16	56	24	64	32
9	39	7	47	15	55	23	63	31
17	38	6	46	14	54	22	62	30
25	37	5	45	13	53	21	61	29
33	36	4	44	12	52	20	60	28
41	35	3	43	11	51	19	59	27
49	34	2	42	10	50	18	58	26
57	33	1	41	9	49	17	57	25

The permuted plaintext is split into two halves, left (L) and right (R). The Feistel round function (depicted in Figure 7.9) does the same thing each of the 16 rounds: takes as input $L_{r-1} \,||\, R_{r-1}$ and outputs $R_{r-1} \,||\, f(K_r, R_{r-1}) \oplus L_{r-1}$, via the function f that depends on the subkey K_r of round r. At the end of the 16th round the inverse permutation IP^{-1} of Table 7.2 is applied producing the ciphertext C.

Recall that in Shannon's paper [418], confusion refers to a complex relationship between the key and the ciphertext, while diffusion refers to the property that the statistical structure of the plaintext is 'dissipated' in the long range statistics of the ciphertext. Both confusion and diffusion occur in the Feistel round function f of DES.

We detail next the action of the 'core' function f on the 32 bit half-block. It consists of four operations:

(i) *Expansion*: the 32-bit half-block is expanded to 48 bits using the expansion permutation E, by duplicating some of the bits (see Table 7.3).

(ii) *Key mixing*: the output of (i) is combined with a subkey using an xor operation.

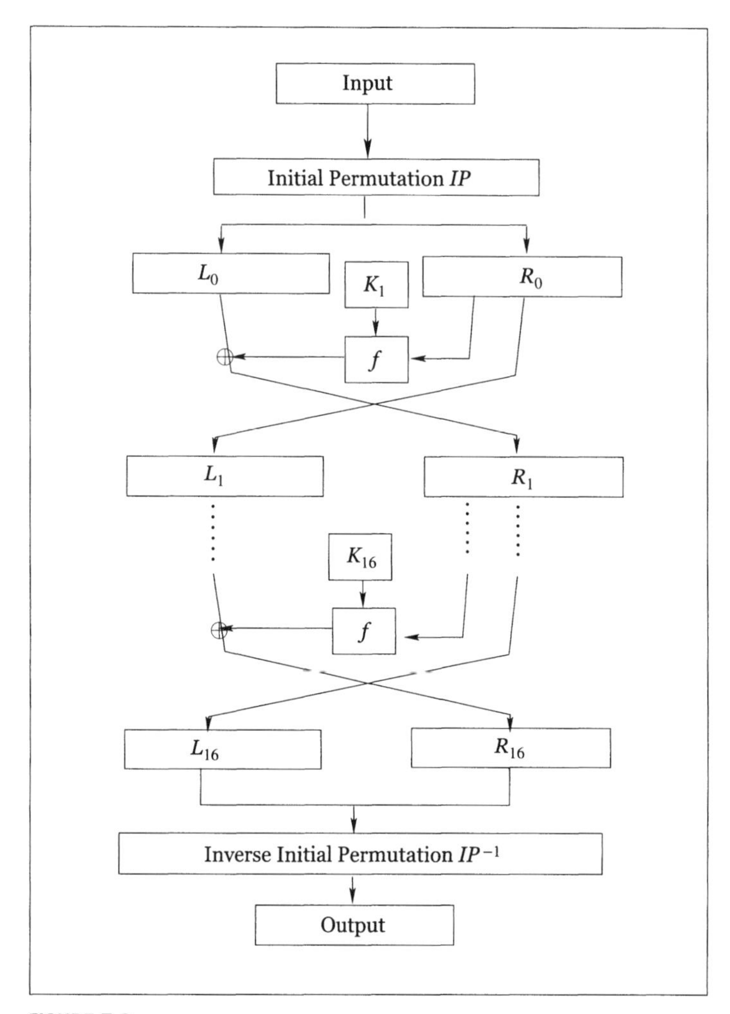

FIGURE 7.9

Data encryption standard (DES).

(iii) *Substitution*: the output of (ii) is divided into eight 6-bit words before processing by the S-boxes (substitution boxes). Each of the eight S-boxes (see Table 7.5) maps 6-bit inputs to 4-bit outputs according to a nonlinear transformation (lookup table). This operation stands at the core of the security of DES, and generates confusion.

(iv) *Permutation*: the 32 outputs from the S-boxes are rearranged according to a fixed permutation, the P-box (see Table 7.4). This permutation generates diffusion.

Here is a toy example to show how the S-boxes work. Take the following 48-bit binary number 011100100101010101000111101000110101101100011000. We split it into eight 6-

Table 7.3 The expansion function E

Bit	0	1	2	3	4	5
1	32	1	2	3	4	5
7	4	5	6	7	8	9
13	8	9	10	11	12	13
19	12	13	14	15	16	17
25	16	17	18	19	20	21
31	20	21	22	23	24	25
37	24	25	26	27	28	29
43	28	29	30	31	32	1

Table 7.4 The permutation P

Bit	0	1	2	3
1	16	7	20	21
5	29	12	28	17
9	1	15	23	26
13	5	18	31	10
17	2	8	24	14
21	32	27	3	9
25	19	13	30	6
29	22	11	4	25

bit blocks, labeled B_1 to B_8 from left to right, say 011100 100101 010101 000111 101000 110101 101100 011000. Next, eight numbers (one from each box) are extracted from the S-boxes S_i, $1 \le i \le 8$:

$$B_1 = S_1(00, 1110) = S_1(0, 14) = 0 = 0000$$
$$B_2 = S_2(11, 0010) = S_2(3, 2) = 10 = 1010$$
$$B_3 = S_3(01, 1010) = S_3(1, 10) = 5 = 0101$$
$$B_4 = S_4(01, 0011) = S_4(1, 3) = 5 = 0101$$
$$B_5 = S_5(10, 0100) = S_5(2, 4) = 10 = 1010$$
$$B_6 = S_6(11, 1010) = S_6(3, 10) = 1 = 0001$$
$$B_7 = S_7(10, 0110) = S_7(2, 6) = 7 = 0111$$
$$B_8 = S_8(00, 1100) = S_8(0, 12) = 5 = 0101$$

Table 7.5 Substitution boxes (S-boxes): $S_i, 1 \leq i \leq 8$

Row/Col	0	1	2	3	4	5	6	7	8	9	10	11	12	13	14	15
0	14	4	13	1	2	15	11	8	3	10	6	12	5	9	0	7
1	0	15	7	4	14	2	13	1	10	6	12	11	9	5	3	8
2	4	1	14	8	13	6	2	11	15	12	9	7	3	10	5	0
3	15	12	8	2	4	9	1	7	5	11	3	14	10	0	6	13
Row/Col	**0**	**1**	**2**	**3**	**4**	**5**	**6**	**7**	**8**	**9**	**10**	**11**	**12**	**13**	**14**	**15**
0	15	1	8	14	6	11	3	4	9	7	2	13	12	0	5	10
1	3	13	4	7	15	2	8	14	12	0	1	10	6	9	11	5
2	0	14	7	11	10	4	13	1	5	8	12	6	9	3	2	15
3	13	8	10	1	3	15	4	2	11	6	7	12	0	5	14	9
Row/Col	**0**	**1**	**2**	**3**	**4**	**5**	**6**	**7**	**8**	**9**	**10**	**11**	**12**	**13**	**14**	**15**
0	10	0	9	14	6	3	15	5	1	13	12	7	11	4	2	8
1	13	7	0	9	3	4	6	10	2	8	5	14	12	11	15	1
2	13	6	4	9	8	15	3	0	11	1	2	12	5	10	14	7
3	1	10	13	0	6	9	8	7	4	15	14	3	11	5	2	12
Row/Col	**0**	**1**	**2**	**3**	**4**	**5**	**6**	**7**	**8**	**9**	**10**	**11**	**12**	**13**	**14**	**15**
0	7	13	14	3	0	6	9	10	1	2	8	5	11	12	4	15
1	13	8	11	5	6	15	0	3	4	7	2	12	1	10	14	9
2	10	6	9	0	12	11	7	13	15	1	3	14	5	2	8	4
3	3	15	0	6	10	1	13	8	9	4	5	11	12	7	2	14
Row/Col	**0**	**1**	**2**	**3**	**4**	**5**	**6**	**7**	**8**	**9**	**10**	**11**	**12**	**13**	**14**	**15**
0	2	12	4	1	7	10	11	6	8	5	3	15	13	0	14	9
1	14	11	2	12	4	7	13	1	5	0	15	10	3	9	8	6
2	4	2	1	11	10	13	7	8	15	9	12	5	6	3	0	14
3	11	8	12	7	1	14	2	13	6	15	0	9	10	4	5	3
Row/Col	**0**	**1**	**2**	**3**	**4**	**5**	**6**	**7**	**8**	**9**	**10**	**11**	**12**	**13**	**14**	**15**
0	12	1	10	15	9	2	6	8	0	13	3	4	14	7	5	11
1	10	15	4	2	7	12	9	5	6	1	13	14	0	11	3	8
2	9	14	15	5	2	8	12	3	7	0	4	10	1	13	11	6
3	4	3	2	12	9	5	15	10	11	14	1	7	6	0	8	13

(continued on next page)

Table 7.5 *(continued)*

Row/Col	0	1	2	3	4	5	6	7	8	9	10	11	12	13	14	15
0	4	11	2	14	15	0	8	13	3	12	9	7	5	10	6	1
1	13	0	11	7	4	9	1	10	14	3	5	12	2	15	8	6
2	1	4	11	13	12	3	7	14	10	15	6	8	0	5	9	2
3	6	11	13	8	1	4	10	7	9	5	0	15	14	2	3	12
Row/Col	0	1	2	3	4	5	6	7	8	9	10	11	12	13	14	15
0	13	2	8	4	6	15	11	1	10	9	3	14	5	0	12	7
1	1	15	13	8	10	3	7	4	12	5	6	11	0	14	9	2
2	7	11	4	1	9	12	14	2	0	6	10	13	15	3	5	8
3	2	1	14	7	4	10	8	13	15	12	9	0	3	5	6	11

Table 7.6 Rotation table - key scheduling

Round number	1	2	3	4	5	6	7	8	9	10	11	12	13	14	15	16
# bits to rotate	1	1	2	2	2	2	2	2	1	2	2	2	2	2	2	1

In $S_n(\text{row}, \text{column})$, the first and last bits of the current B_n are used as the row index, and the middle four bits as the column index. The results are concatenated to form a 32-bit number 00001010010101011010000101110101, to which one applies the P-box.

The key scheduling is fairly simple, and follows the following steps:

1. Apply a permutation PC-1 (PC stands for 'permuted choice') (see Table 7.7) to the initial 64-bit key, which removes the parity bits;
2. Set round number $r = 1$;
3. Split the current 56-bit key into two 28-bit blocks ($L \,||\, R$);
4. Rotate L left by the number of bits specified in Table 7.6, and rotate R left by the same number of bits;
5. Set the new subkey $K = L \,||\, R$;
6. Apply a new permutation PC-2 (see Table 7.8) to K to get the subkey K_r corresponding to round r;
7. Set $r = r + 1$, and go to 3.

Initially, some believed that trying out all $2^{56} = 72,057,594,037,927,936$ possible keys would be impossible because computers could not possibly ever become fast enough. In 1998 the Electronic Frontier Foundation (EFF) built a special-purpose machine (with a cost of less than $250,000) that could decrypt a message by trying out all possible keys in less than three days (searching over 88 billion keys per second). To increase the security, until an acceptable replacement for DES was found, an iterated DES was being used. This

Table 7.7 Permuted choice PC-1

Bit	0	1	2	3	4	5	6
1	57	49	41	33	25	17	9
8	1	58	50	42	34	26	18
15	10	2	59	51	43	35	27
22	19	11	3	60	52	44	36
29	63	55	47	39	31	23	15
36	7	62	54	46	38	30	22
43	14	6	61	53	45	37	29
50	21	13	5	28	20	12	4

Table 7.8 Permuted choice PC-2

Bit	0	1	2	3	4	5
1	14	17	11	24	1	5
7	3	28	15	6	21	10
13	23	19	12	4	26	8
19	16	7	27	20	13	2
25	41	52	31	37	47	55
31	30	40	51	45	33	48
37	44	49	39	56	34	53
43	46	42	50	36	29	32

version of DES, which does provide adequate security, namely triple-DES, is an iterated DES with two (or three) 56-bit keys applied: the first key is used to encrypt the initial message using DES and the second key is used to 'decrypt' the encrypted message (in reality, that will further encrypt the message since there is no connection between the first and second key). We apply again the encryption procedure using the first key (or a different third key) to obtain the final triple-DES encrypted ciphertext. The resultant key space is of size $\approx 2^{112}$, if two different keys are used. This was possible since DES encryption does not form a group [54], so multiple encryption is not equivalent to a single encryption.

Regardless of the controversy on how DES came into existence, it has been a great cipher, and it marked a new era for cryptology, since it was the first modern cipher of widespread use that was not classified.

7.5.2 Advanced Encryption Standard

The Advanced Encryption Standard (AES), published by the National Institute of Standards and Technology (NIST) as FIPS Publication 197 [4], is a SPN product cipher and specifies a cryptographic algorithm for use by the US government to protect sensitive, unclassified information, possibly for the next 20 years. Of course, as one would expect, AES will be certainly used also by private organizations in or outside of the United States.

AES is based on the algorithm Rijndael developed by Joan Daemen and Vincent Rijmen (see their book [133]) (the following pronunciation alternatives have been suggested: 'Reign Dahl', 'Rain Doll', 'Rhine Dahl').

AES is Rijndael with a fixed block size, so we see them in the literature used interchangeably. Rijndael supports various blocks sizes in multiples of 32, with key sizes between 128 and 256 bits, while AES assumes a fixed block size of 128 bits and a key size of 128, 192, or 256 bits.

We will follow [4,133,186,114] in the description of Rijndael/AES. The operations in AES are defined at byte level, with bytes representing elements in the finite field $\mathbb{F}_{2^8}$, that is, the byte $\mathbf{b}$ formed by the bits $b_7, b_6, \ldots, b_1, b_0$ is thought of as a polynomial in $\mathbb{F}_2[x]$, namely

$$b_7x^7 + b_6x^6 + \cdots + b_1x + b_0.$$

With this convention, the sum of two bytes is the byte corresponding to the polynomial sum modulo 2 of the two terms (that is, the xor of the two bytes), and the multiplication is done by taking the product of the two polynomials modulo an irreducible binary polynomial of degree 8. One such polynomial (used in Rijndael) is $m(x) = x^8 + x^4 + x^3 + x + 1$ ('11B' in hexadecimal, sometimes written as $1\{1B\}$). Using the extended Euclidean algorithm one can compute the inverse of any nonzero polynomial, which together with the previous observations helps to show that the set of 2^8 byte values with the two previously defined operations is isomorphic to the finite field $\mathbb{F}_{2^8}$.

In every round, Rijndael acts on a $4 \times N_c$ matrix (called a *state*) whose entries $s_{r,c}$ are bytes, where N_c is the input sequence length divided by 32. In AES, $N_c = 4$. The columns $s_{r,c}, 1 \leq r \leq 4$, of these state matrices are called *words* that can also be represented by cubic polynomials in $\mathbb{F}_{2^8}[x]$, say $a(x) = s_{4,c}x^3 + s_{3,c}x^2 + s_{2,c}x + s_{1,c}$. Multiplication of these cubic polynomials can be carried out modulo another relevant polynomial to Rijndael, namely $x^4 + 1$, to obtain another cubic polynomial. If a polynomial $a(x) = a_3x^3 + a_2x^2 + a_1x + a_0$ is fixed then the multiplication and reduction modulo $x^4 + 1$ can be simply done by matrix multiplication

$$\begin{pmatrix} d_3 \\ d_2 \\ d_1 \\ d_0 \end{pmatrix} = \begin{pmatrix} a_0 & a_1 & a_2 & a_3 \\ a_3 & a_0 & a_1 & a_2 \\ a_2 & a_3 & a_0 & a_1 \\ a_1 & a_2 & a_3 & a_0 \end{pmatrix} \begin{pmatrix} b_3 \\ b_2 \\ b_1 \\ b_0 \end{pmatrix}.$$

Since x^4+1 is not irreducible, not all of $a(x)$ are invertible. Rijndael uses the following polynomial which is invertible modulo x^4+1, namely

$$a(x)=\{03\}x^3+\{01\}x^2+\{01\}x+\{02\},$$

whose inverse (modulo x^4+1) is

$$a^{-1}(x)=\{0B\}x^3+\{0D\}x^2+\{09\}x+\{0E\}.$$

Further, it is worth noting that multiplication by x can be thought of as rotating the input word by one byte, and multiplications by higher powers of x are simply rotations by the appropriate number of bytes. We can express this also in terms of matrix multiplication, for $c(x)=x\cdot b(x) \pmod{x^4+1}=b_2x^3+b_1x^2+b_0x+b_3$, where $b(x)=b_3x^3+b_2x^2+b_1x+b_0$, namely

$$\begin{pmatrix} c_3 \\ c_2 \\ c_1 \\ c_0 \end{pmatrix} = \begin{pmatrix} 0 & 1 & 0 & 0 \\ 0 & 0 & 1 & 0 \\ 0 & 0 & 0 & 1 \\ 1 & 0 & 0 & 0 \end{pmatrix} \begin{pmatrix} b_3 \\ b_2 \\ b_1 \\ b_0 \end{pmatrix}.$$

The criteria that went into the design of Rijndael were [133]: resistance against all known attacks, speed and code compactness on a wide range of platforms, design simplicity.

We now describe the cipher. The cipher key (divided by 32) has the same length N_k chosen from the set $\{4, 6, 8\}$ in the AES standard. First, an initial round key *AddRoundKey* is added to the input and then a round function, consisting of four different transformations, is applied to the input every round, except in the final round. The number of rounds N_n varies with the key length and block size, including the additional round at the end: 10, if both the block and the key are 128 bits long; 12, if either the block or the key is 192 bits long, but neither is longer than that; 14, if either the block or the key is 256 bits long. Since it concerns security, the brute force attack on AES would need to check: 3.4×10^{38} possible 128-bit keys; 6.2×10^{57} possible 192-bit keys; 1.1×10^{77} possible 256-bit keys. If one could build a machine to check 2^{55} keys per second, then it would take that machine approximately 149 thousand-billion (149 trillion) years to crack a 128-bit AES key.

The four mentioned transformations are: *ByteSub, ShiftRow, MixColumn, AddRound-Key*, and they will be detailed below. The last step applies all, with the exception of the MixColumn transformation. We picture the round function in Figure 7.10 in Savard's vision [399].

The *ByteSub* transformation (S-box) replaces every byte of the block by its substitute in an S-box, which is constructed in the following way [133,186], using the composition $g \circ f$ of two functions $f, g : \mathbb{F}_{2^8} \to \mathbb{F}_{2^8}$:

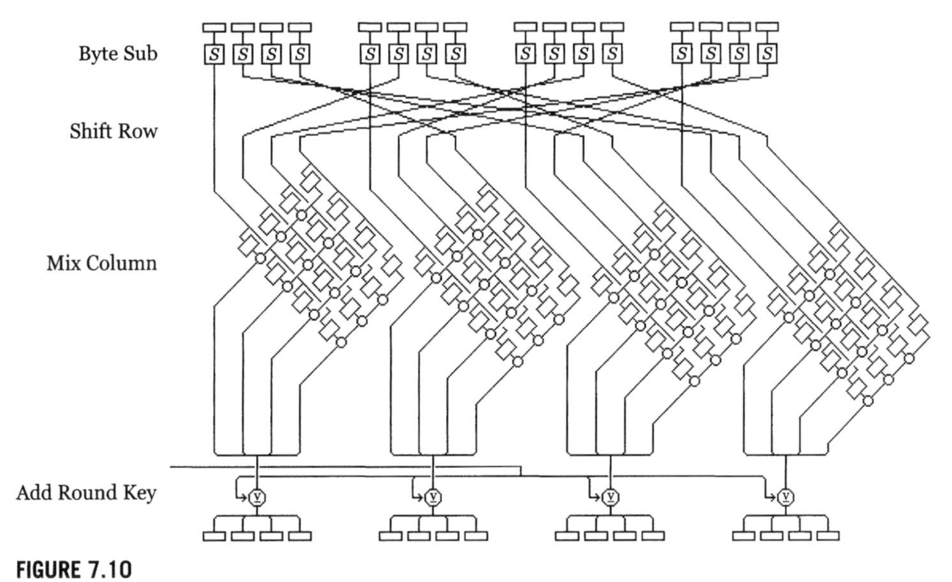

FIGURE 7.10

AES/Rijndael cipher.

1. Take the multiplicative inverse in $\mathbb{F}_{2^8}$ (except for 0, which obviously has no inverse, and it is mapped to itself), that is

$$f(x)=\begin{cases} x^{-1} & \text{if } x \neq 0 \\ 0 & \text{if } x = 0. \end{cases}$$

2. Apply an affine transformation (over $\mathbb{F}_2$), namely

$$b_i' = b_i \oplus b_{(i+4)(\text{mod } 8)} \oplus b_{(i+5)(\text{mod } 8)} \oplus b_{(i+6)(\text{mod } 8)} \oplus b_{(i+7)(\text{mod } 8)} \oplus c_i,$$

$0 \leq i < 8$, where b_i is the ith bit of the input byte, $c = \{63\}$, and b_i' is the ith bit of the output byte, or in matrix form $\mathbf{b}' = g(\mathbf{b}) = A\mathbf{b} \oplus \mathbf{c}$

$$\begin{pmatrix} b_0' \\ b_1' \\ b_2' \\ b_3' \\ b_4' \\ b_5' \\ b_6' \\ b_7' \end{pmatrix} = \begin{pmatrix} 1 & 0 & 0 & 0 & 1 & 1 & 1 & 1 \\ 1 & 1 & 0 & 0 & 0 & 1 & 1 & 1 \\ 1 & 1 & 1 & 0 & 0 & 0 & 1 & 1 \\ 1 & 1 & 1 & 1 & 0 & 0 & 0 & 1 \\ 1 & 1 & 1 & 1 & 1 & 0 & 0 & 0 \\ 0 & 1 & 1 & 1 & 1 & 1 & 0 & 0 \\ 0 & 0 & 1 & 1 & 1 & 1 & 1 & 0 \\ 0 & 0 & 0 & 1 & 1 & 1 & 1 & 1 \end{pmatrix} \begin{pmatrix} b_0 \\ b_1 \\ b_2 \\ b_3 \\ b_4 \\ b_5 \\ b_6 \\ b_7 \end{pmatrix} \oplus \begin{pmatrix} 1 \\ 1 \\ 0 \\ 0 \\ 0 \\ 1 \\ 1 \\ 0 \end{pmatrix}$$

Table 7.9 The action of the ByteSub transformation

$s_{0,0}$	$s_{0,1}$	$s_{0,2}$	$s_{0,3}$		$s'_{0,0}$	$s'_{0,1}$	$s'_{0,2}$	$s'_{0,3}$
$s_{1,0}$	$s_{1,1}$	$s_{1,2}$	$s_{1,3}$	*ByteSub* *S-box* →	$s'_{1,0}$	$s'_{1,1}$	$s'_{1,2}$	$s'_{1,3}$
$s_{2,0}$	$s_{2,1}$	$s_{2,2}$	$s_{2,3}$		$s'_{2,0}$	$s'_{2,1}$	$s'_{2,2}$	$s'_{2,3}$
$s_{3,0}$	$s_{3,1}$	$s_{3,2}$	$s_{3,3}$		$s'_{3,0}$	$s'_{3,1}$	$s'_{3,2}$	$s'_{3,3}$

Table 7.10 AES *S*-box

hex	0	1	2	3	4	5	6	7	8	9	A	B	C	D	E	F
0	63	7C	77	7B	F2	6B	6F	C5	30	01	67	2B	FE	D7	AB	76
1	CA	82	C9	7D	FA	59	47	F0	AD	D4	A2	AF	9C	A4	72	C0
2	B7	FD	93	26	36	3F	F7	CC	34	A5	E5	F1	71	D8	31	15
3	04	C7	23	C3	18	96	05	9A	07	12	80	E2	EB	27	B2	75
4	09	83	2C	1A	1B	6E	5A	A0	52	3B	D6	B3	29	E3	2F	84
5	53	D1	00	ED	20	FC	B1	5B	6A	CB	BE	39	4A	4C	58	CF
6	D0	EF	AA	FB	43	4D	33	85	45	F9	02	7F	50	3C	9F	A8
7	51	A3	40	8F	92	9D	38	F5	BC	B6	DA	21	10	FF	F3	D2
8	CD	0C	13	EC	5F	97	44	17	C4	A7	7E	3D	64	5D	19	73
9	60	81	4F	DC	22	2A	90	88	46	EE	B8	14	DE	5E	0B	DB
A	E0	32	3A	0A	49	06	24	5C	C2	D3	AC	62	91	95	E4	79
B	E7	C8	37	6D	8D	D5	4E	A9	6C	56	F4	EA	65	7A	AE	08
C	BA	78	25	2E	1C	A6	B4	C6	E8	DD	74	1F	4B	BD	8B	8A
D	70	3E	B5	66	48	03	F6	0E	61	35	57	B9	86	C1	1D	9E
E	E1	F8	98	11	69	D9	8E	94	9B	1E	87	E9	CE	55	28	DF
F	8C	A1	89	0D	BF	E6	42	68	41	99	2D	0F	B0	54	BB	16

The action of ByteSub is illustrated in Table 7.9. Although it is not explicitly mentioned in the initial submission, the final *S*-box transformation result is in fact (in hexadecimal) as in Table 7.10 [186].

The inverse ByteSub transformation, say *InvByteSub*, is similar to the direct ByteSub transformation, having two components, with the second component given by

$$b'_i = b_{(i+2)(\bmod 8)} \oplus b_{(i+5)(\bmod 8)} \oplus b_{(i+7)(\bmod 8)} \oplus d_i,$$

with $d = \{05\}$. The final inverse ByteSub transformation is (in hexadecimal) as in Table 7.11.

Table 7.11 Inverse AES S-box

hex	0	1	2	3	4	5	6	7	8	9	A	B	C	D	E	F
0	52	09	6A	D5	30	36	A5	38	BF	40	A3	9E	81	F3	D7	FB
1	7C	E3	39	82	9B	2F	FF	87	34	8E	43	44	C4	DE	E9	CB
2	54	7B	94	32	A6	C2	23	3D	EE	4C	95	0B	42	FA	C3	4E
3	08	2E	A1	66	28	D9	24	B2	76	5B	A2	49	6D	8B	D1	25
4	72	F8	F6	64	86	68	98	16	D4	A4	5C	CC	5D	65	B6	92
5	6C	70	48	50	FD	ED	B9	DA	5E	15	46	57	A7	8D	9D	84
6	90	D8	AB	00	8C	BC	D3	0A	F7	E4	58	05	B8	B3	45	06
7	D0	2C	1E	8F	CA	3F	0F	02	C1	AF	BD	03	01	13	8A	6B
8	3A	91	11	41	4F	67	DC	EA	97	F2	CF	CE	F0	B4	E6	73
9	96	AC	74	22	E7	AD	35	85	E2	F9	37	E8	1C	75	DF	6E
A	47	F1	1A	71	1D	29	C5	89	6F	B7	62	0E	AA	18	BE	1B
B	FC	56	3E	4B	C6	D2	79	20	9A	DB	C0	FE	78	CD	5A	F4
C	1F	DD	A8	33	88	07	C7	31	B1	12	10	59	27	80	EC	5F
D	60	51	7F	A9	19	B5	4A	0D	2D	E5	7A	9F	93	C9	9C	EF
E	A0	E0	3B	4D	AE	2A	F5	B0	C8	EB	BB	3C	83	53	99	61
F	17	2B	04	7E	BA	77	D6	26	E1	69	14	63	55	21	0C	7D

Table 7.12 The action of the ShiftRow transformation

$s_{0,0}$	$s_{0,1}$	$s_{0,2}$	$s_{0,3}$		$s_{0,0}$	$s_{0,1}$	$s_{0,2}$	$s_{0,3}$
$s_{1,0}$	$s_{1,1}$	$s_{1,2}$	$s_{1,3}$	*ShiftRow* $\longrightarrow$	$s_{1,1}$	$s_{1,2}$	$s_{1,3}$	$s_{1,0}$
$s_{2,0}$	$s_{2,1}$	$s_{2,2}$	$s_{2,3}$		$s_{2,2}$	$s_{2,3}$	$s_{2,0}$	$s_{2,1}$
$s_{3,0}$	$s_{3,1}$	$s_{3,2}$	$s_{3,3}$		$s_{3,3}$	$s_{3,0}$	$s_{3,1}$	$s_{3,2}$

The *ShiftRow* transformation cyclically shifts the byte entries of the last three rows of the state, that is,

$$s_{r,c} = s'_{r,[c+h(r,N_c)](\text{mod } N_c)}, \quad \text{for } 0 \leq c < N_c, 0 < r < 4,$$

where the shift $h(r, N_c)$ is 1, 2, 3 if the number of columns $N_c \in \{4, 5, 6\}$; 1, 2, 4 if $N_c = 7$; 1, 3, 4 if $N_c = 8$. In matrix form, for AES, ShiftRow acts as in Table 7.12.

Certainly, the inverse ShiftRow transformation, say *InvShiftRow*, also operates in a similar manner, by shifting the bytes in the last row according to

$$s'_{r,c} = s_{r,[c-h(r,N_c)](\text{mod } N_c)}, \quad \text{for } 0 \leq c < N_c, 0 < r < 4.$$

Table 7.13 The action of the MixColumn transformation

$s_{0,0}$	$s_{0,1}$	$s_{0,2}$	$s_{0,3}$		$s'_{0,0}$	$s'_{0,1}$	$s'_{0,2}$	$s'_{0,3}$
$s_{1,0}$	$s_{1,1}$	$s_{1,2}$	$s_{1,3}$	$\otimes c(x) \xrightarrow{MixColumn}$	$s'_{1,0}$	$s'_{1,1}$	$s'_{1,2}$	$s'_{1,3}$
$s_{2,0}$	$s_{2,1}$	$s_{2,2}$	$s_{2,3}$		$s'_{2,0}$	$s'_{2,1}$	$s'_{2,2}$	$s'_{2,3}$
$s_{3,0}$	$s_{3,1}$	$s_{3,2}$	$s_{3,3}$		$s'_{3,0}$	$s'_{3,1}$	$s'_{3,2}$	$s'_{3,3}$

The *MixColumn* transformation acts on every column independently. Its matrix form is (the entries of the 4×4 matrix are the finite fields elements we referred to, previously), for $0 \leq c \leq N_c$

$$\begin{pmatrix} s'_{3,c} \\ s'_{2,c} \\ s'_{1,c} \\ s'_{0,c} \end{pmatrix} = \begin{pmatrix} 02 & 01 & 01 & 03 \\ 03 & 02 & 01 & 01 \\ 01 & 03 & 02 & 01 \\ 01 & 01 & 03 & 02 \end{pmatrix} \begin{pmatrix} s_{3,c} \\ s_{2,c} \\ s_{1,c} \\ s_{0,c} \end{pmatrix}. \tag{7.2}$$

The MixColumn transformation is illustrated in Table 7.13.

The inverse MixColumn transformation, say *InvMixColumn*, acts also on every column, independently. In matrix form, InvMixColumn operates as follows

$$\begin{pmatrix} s'_{3,c} \\ s'_{2,c} \\ s'_{1,c} \\ s'_{0,c} \end{pmatrix} = \begin{pmatrix} 0E & 09 & 0D & 0B \\ 0B & 0E & 09 & 0D \\ 0D & 0B & 0E & 09 \\ 09 & 0D & 0B & 0E \end{pmatrix} \begin{pmatrix} s_{3,c} \\ s_{2,c} \\ s_{1,c} \\ s_{0,c} \end{pmatrix}, \tag{7.3}$$

where the used matrix is an MDS matrix. MDS stands for maximum distance separable, and in general it is a transformation matrix of a linear code that reaches the Singleton bound (see Vaudenay's paper [453], for further motivation on MDS matrices).

The transformation *AddRoundKey* acts (individually) on the word columns of the state by adding (xor-ing) the round key (see Table 7.14). We further note that the inverse AddRoundKey transformation is the AddRoundKey itself.

And now a few words about the key scheduling process. The cipher key is used to generate the round keys by using two operations, namely, *Key Expansion* and *Round Key selection*. Each round requires N_c words of key data, for a total of $N_c(N_n+1)$ words (recall that N_n is the number of rounds, N_c is the number of columns, which in AES is 4, and N_k is the key size, divided by 32). For instance, for a block of 128 bits and 10 rounds, we need 1408 round key bits. We can think of the key schedule either as a vector k of $N_c(N_n+1)$ 32-bit words, or as a two-dimensional $N_c \times (N_n+1)$ array, say $k(n,c)$.

The output of the Key Expansion operation is an *Expanded Key*, which is a linear array of 4-byte words. The Expanded Key is simply computed in the following way [133,458]:

Table 7.14 The action of the AddRoundKey transformation

$s_{0,0}$	$s_{0,1}$	$s_{0,2}$	$s_{0,3}$
$s_{1,0}$	$s_{1,1}$	$s_{1,2}$	$s_{1,3}$
$s_{2,0}$	$s_{2,1}$	$s_{2,2}$	$s_{2,3}$
$s_{3,0}$	$s_{3,1}$	$s_{3,2}$	$s_{3,3}$

$\longrightarrow \oplus \longleftarrow$

$k_{0,0}$	$k_{0,1}$	$k_{0,2}$	$k_{0,3}$
$k_{1,0}$	$k_{1,1}$	$k_{1,2}$	$k_{1,3}$
$k_{2,0}$	$k_{2,1}$	$k_{2,2}$	$k_{2,3}$
$k_{3,0}$	$k_{3,1}$	$k_{3,2}$	$k_{3,3}$

AddRoundKey

$\downarrow$

$s'_{0,0}$	$s'_{0,1}$	$s'_{0,2}$	$s'_{0,3}$
$s'_{1,0}$	$s'_{1,1}$	$s'_{1,2}$	$s'_{1,3}$
$s'_{2,0}$	$s'_{2,1}$	$s'_{2,2}$	$s'_{2,3}$
$s'_{3,0}$	$s'_{3,1}$	$s'_{3,2}$	$s'_{3,3}$

the first few columns, indexed $k_0, \ldots, k_{N_k-1}$, are equal to the cipher key, and the columns of index $i \geq N_k$ are computed as follows:

1. If $N_k \nmid i$, then the column i is the bitwise xor of column $i - N_k$ and column $i - 1$.
2. If $N_k | i$, then the columns i is the bitwise xor of column $i - N_k$ and the output of a nonlinear function applied to column $i - 1$ (the nonlinear function is formed of the SubWord and RotWord transformations explained below).

The Expanded Key proceeds according to the following pseudocode (cf. Daemen and Rijmen [133] and Gladman [186]):

```
KeyExpansion (byte key[4*N_k], word k[N_n+1,N_c], N_c, N_k, N_n)
begin
  i = 0
  while(i < N_k)
    k[i] = word[key[4*i+3],key[4*i+2],key[4*i+1],key[4*i]]
    i = i+1
  end while
  i = N_k
  while(i < N_c·(N_n+1))
    word temp = k[i-1]
    if (i(mod N_k) = 0)
      temp = SubWord(RotWord(temp)) ⊕ Rcon[i/N_k]
    else if ((N_k > 6) and (i(mod N_k) = 4))
      temp = SubWord(temp)
    end if
    k[i] = k[i-N_k] ⊕ temp
    i = i+1
  end while
end
```

Note that the function *SubWord*() applies the substitution transformation of the S-box on a 4-byte input word to produce a 4-byte output word, while the *RotWord*() function performs a cyclic permutation on a 4-byte word, that is, $RotWord[b_3, b_2, b_1, b_0] = [b_0, b_3, b_2, b_1]$. The word array $Rcon[i]$ contains the values $[0, 0, 0, x^{i+1}]$ where x^{i+1} is thought of in the finite field representation we discussed earlier.

The *Round Key selection* uses the Expanded Key in the following way: the first Round Key consists of the first N_c words, the second one of the following N_c words, and the process continues, that is, Round Key i is given by the words from $k[N_c \cdot i]$ to $k[N_c \cdot (i+1) - 1]$.

It should be noted that Rijndael/AES uses the CBC and ECB modes by default, but one can extend it to other modes, such as CFB, OFB, OCB, and CTR (see Section 7.3). We will not go into the software/hardware implementation issues, suffice it to say that much research has gone into this area: for instance, Canright [55,56] does an extensive analysis of AES when one uses different irreducible polynomials in $\mathbb{F}_{2^8}$ and found some improvements on the implementation side.

7.6 PERIODS OF RIJNDAEL TRANSFORMATIONS

It is rather a straightforward exercise to investigate cyclic properties of each Rijndael transformation and various authors did just that (see Song and Seberry [432] and references therein).

Theorem 7.1. *Every input byte x of the S-box returns to the initial state after t repeated applications of the substitutions, that is*

$$S^t(x) = x, \quad \textit{for some integer } t.$$

The period t could be either 87, 81, 59, 27, *or* 2. *Since the least common multiple of these values is* 277182, *we obtain*

$$ByteSub^{277182}(X) = X,$$

for any input block X. Further, the ShiftRow and MixColumn operation (for AES) will have period 4, *that is,*

$$ShiftRow^4(X) = X,$$
$$MixColumn^4(X) = X.$$

By analyzing various combinations of AES transformations, Song and Seberry concluded that the maximal period for the composition of the ByteSub and MixColumn transformation is about 2^{110}. Shorter periods are bound to reveal some information in the cryptanalysis of the cipher. For instance, Song and Seberry [432] observed that if the ByteSub and MixColumn transformations are applied to the bytes of an input block X that are all 'EF', then X is mapped to itself after 27 iterations. Also, if all bytes in X are the same and are either '73' or '8F' then X is mapped to itself after two iterations of the two transformations.

7.7 ALGEBRAIC REPRESENTATIONS OF RIJNDAEL/AES

We start this section with a result that basically describes every block cipher as a system of polynomial equations, which is the basis of the many attacks on Rijndael/AES.

Theorem 7.2. *If u, v are positive integers, $P_0, \ldots, P_u$ (respectively, $C_0, \ldots, C_u$) are variables representing the plaintext (respectively, the ciphertext), and $K_0, \ldots, K_v$ are variables representing the key, then every block cipher can be written as a system of polynomial equations over some extension of $\mathbb{F}_2$.*

Proof. Fairly straightforward (see [458, p. 16] for a sketch). ■

We mentioned previously that an exhaustive attack on AES is futile at this stage, as one cannot search through the keyspace efficiently. So, one chance of breaking AES would be to reduce the size of the keyspace. Since there are references dedicated entirely to this subject, and how they pertain to AES (see the book of Cid et al. [114] and the references therein), and we referred to them in Chapter 6, we shall refrain from discussing the algebraic attacks in detail.

If we apply consecutively in AES the four operations (ByteSub, ShiftRow, MixColumn, AddRoundKey) to a 4-byte column we obtain the following algebraic equation that depends on the S-box (call it S) and the round key k (see [186] for further details)

$$\begin{pmatrix} s'_{3,c} \\ s'_{2,c} \\ s'_{1,c} \\ s'_{0,c} \end{pmatrix} = S[s_{3,c(3)}] \cdot \begin{pmatrix} 02 \\ 03 \\ 01 \\ 01 \end{pmatrix} \oplus S[s_{2,c(2)}] \cdot \begin{pmatrix} 01 \\ 02 \\ 03 \\ 01 \end{pmatrix} \oplus S[s_{1,c(1)}] \cdot \begin{pmatrix} 01 \\ 01 \\ 02 \\ 03 \end{pmatrix}$$
$$\oplus S[s_{0,c(0)}] \cdot \begin{pmatrix} 03 \\ 01 \\ 01 \\ 02 \end{pmatrix} \oplus \begin{pmatrix} k_{3,c} \\ k_{2,c} \\ k_{1,c} \\ k_{0,c} \end{pmatrix},$$

where $c(0) = c$ and $c(r) = [c + h(r, N_c)] \pmod{N_c}$ (recall that $h(r, N_c)$ is 1, 2, 3 if the number of columns $N_c \in \{4,5,6\}$; 1, 2, 4 if $N_c = 7$; 1, 3, 4 if $N_c = 8$). Every round will use the previous equation except for the last round (recall that the last round is missing the MixColumn transformation), and the equation one obtains leaving out that transformation is easy to obtain. Each (complete) round of the inverse cipher can also be expressed as an equation that depends on the inverse S-box and the round key for the inverse cipher (call it ki)

$$\begin{pmatrix} s'_{3,c} \\ s'_{2,c} \\ s'_{1,c} \\ s'_{0,c} \end{pmatrix} = S^{-1}[s_{3,c(3)}] \cdot \begin{pmatrix} 0E \\ 0B \\ 0D \\ 09 \end{pmatrix} \oplus S^{-1}[s_{2,c(2)}] \cdot \begin{pmatrix} 09 \\ 0E \\ 0B \\ 0D \end{pmatrix}$$

$$\oplus S^{-1}[s_{1,c(1)}] \cdot \begin{pmatrix} 0D \\ 09 \\ 0E \\ 0B \end{pmatrix} \oplus S^{-1}[s_{0,c(0)}] \cdot \begin{pmatrix} 0B \\ 0D \\ 09 \\ 0E \end{pmatrix} \oplus \begin{pmatrix} ki_{3,c} \\ ki_{2,c} \\ ki_{1,c} \\ ki_{0,c} \end{pmatrix},$$

where $c_0 = c, c(r) = [c - h(r, N_c)] \pmod{N_c}$ and $ki_{r,c}$ is the round key for the inverse cipher. We see that the cipher relies heavily on the properties of the S-box, so is not surprising that the S-boxes of AES received quite a bit of attention.

The designers of Rijndael, Daemen and Rijmen, noted that the S-box can be written as an equation of the form [133,168]

$$S(x) = w_8 + \sum_{d=0}^{7} w_d x^{255-2^d},$$

for some constants $w_0, \ldots, w_8$. In 2001, Ferguson et al. [168] took advantage of this highly algebraic structure of the Rijndael/AES cipher and were able to show that Rijndael (with a key and blocks size of 128) can be written as one algebraically closed equation, a (generalized) continued fraction expression over $\mathbb{F}_{2^8}$. We shall (changing slightly the notations) summarize their result, as it is very instructive. Let $a_{i,j}^{(r)}$, respectively, $k_{i,j}^{(r)}$ be the input, respectively, the round key byte at position (i, j) in round r (all coordinates are assumed reduced modulo 4). Define $E = \{0, 1, 2, 3\}$ and $D = \{0, 1, \ldots, 7\}$. Applying the S-box to each input byte of the state we get the output of this step to be

$$s_{i,j}^{(r)} = S[a_{i,j}^{(r)}] = \sum_{d_r=0}^{7} w_{d_r} (a_{i,j}^{(r)})^{-2^{d_r}},$$

and applying the ShiftRow transformation we get easily the output of this state to be

$$t_{i,j}^{(r)} = s_{i,i+j}^{(r)} = \sum_{d_r=0}^{7} w_{d_r} (a_{i,i+j}^{(r)})^{-2^{d_r}}.$$

Continue with the MixColumn transformation to obtain the following output

$$m_{i,j}^{(r)} = \sum_{e_r=0}^{3} v_{i,e_r} t_{e_r,j}^{(r)},$$

where $v_{i,j}$ are the coefficients of the MDS matrix used in the MixColumn transformation (7.3). We easily obtain now the following theorem.

Theorem 7.3. *The one-round Rijndael satisfies*

$$a_{i,j}^{(r+1)} = k_{i,j}^{(r)} + \sum_{\substack{e_r \in E \\ d_r \in D}} \frac{w_{i,e_r,d_r}}{(a_{e_r,e_r+j}^{(r)})^{2^{d_r}}}. \tag{7.4}$$

By substituting in (7.4), we obtain that the two-round Rijndael satisfies

$$a_{i,j}^{(3)} = k_{i,j}^{(2)} + \sum_{\substack{e_2 \in E \\ d_2 \in D}} \frac{w_{i,e_2,d_2}}{k_{e_2,e_2+j}^{(1)} + \sum_{\substack{e_1 \in E \\ d_1 \in D}} \frac{w_{e_2,e_1,d_1}}{(a_{e_1,e_1+e_2+j}^{(1)})^{2^{d_1}}}}, \tag{7.5}$$

and, in general, any byte of the intermediate result after five rounds can be expressed as

$$x = K + \sum \frac{C_1}{K^* + \sum \frac{C_2}{K^* + \sum \frac{C_3}{K^* + \sum \frac{C_4}{K^* + \sum \frac{C_5}{K^* + p_*^*}}}}}, \tag{7.6}$$

where K is a byte depending on several bytes of the expanded key, C_i is a known constant and each $*$ is a known exponent or subscript, but these values depend on the summation variables that enclose the symbol.

The expanded version of (7.6) contains about 2^{25} terms [353]. For a 10 round AES-128, a cryptanalyst could use for each intermediate byte two equations of type (7.6), which would involve 2^{26} unknowns and about 2^{50} terms. If sufficient ciphertext/plaintext pairs are known, one should in theory be able to solve for the unknowns. Can this theoretical approach be put into practice? We will mention various attempts in that respect next.

The following proposition modifies Theorem 7.2 by bounding (by two) the degrees of the polynomials equations at the expense of increasing the number of unknowns.

Proposition 7.4. *Every system of polynomial equations over a finite field $\mathbb{F}$, $F \in \mathbb{F}[X]$ can be transformed into an equivalent system of quadratic equations $F' \in \mathbb{F}[X \cup X']$, $X' \cap X = \emptyset$, by replacing various polynomial expressions in F of degree at most 2 by new variables that form X'.*

We now give a toy example that illustrates the previous proposition.

Example 7.5. *Let $\mathbb{F}$ be a finite field and $X = \{x_1, x_2, x_3\}$. The following system over $\mathbb{F}[X]$,*

$$\begin{aligned} 2x_1^3 + x_1^2 x_2^2 + x_1 x_2^2 x_3 + 1 &= 0 \\ x_1^2 + 2x_1 x_2^3 + 3x_3^5 + 2 &= 0 \\ 3x_1 x_3^4 + x_2^3 + 1 &= 0 \\ x_1 x_2 + 2x_1^4 + x_3^3 &= 0, \end{aligned}$$

is equivalent to the quadratic system over $\mathbb{F}[X \cup X']$, *with* $X' = \{x_4, \ldots, x_{10}\}$,

$$\begin{aligned}
2x_1x_4 + x_4x_5 + x_3x_6 + 1 &= 0 \\
x_1^2 + 2x_2x_6 + 3x_3x_8 + 2 &= 0 \\
3x_1x_{10} + x_2x_5 + 1 &= 0 \\
x_2x_2 + 2x_4^2 + x_3x_7 &= 0 \\
x_4 - x_1^2 &= 0 \\
x_5 - x_2^2 &= 0 \\
x_6 - x_1x_5 &= 0 \\
x_7 - x_3^2 &= 0 \\
x_8 - x_7^2 &= 0 \\
x_9 - x_3^2 &= 0 \\
x_{10} - x_9^2 &= 0.
\end{aligned}$$

Following the approach [168] of Ferguson et al., Courtois and Pieprzyk [124] showed that Rijndael can be written as an overdefined system of multivariate quadratic (MQ) equations. If w, x represent the input, respectively the output of an AES 'inversion', then we have that the equations

$$w \cdot x = 1, \quad w^2 \cdot x = w, \quad w \cdot x^2 = x$$

will produce 24 quadratic equations, 23 of which are true with probability 1 and one is true with probability 255/256 (see [124,369,15] for all equations and further analysis). For example, the authors show that for 128-bit Rijndael, recovering the secret key from one single plaintext is equivalent to solving a system of 8000 quadratic equations with 1600 unknowns. Can one find an efficient algorithm for solving such a system? Uncertain, and we do not plan on getting into the debate, as it is not the goal of this book. However, the problem deserves attention as it displays a 'weakness' in Rijndael/AES that may prove useful or not in breaking the cipher. The resolution of the AES equivalent system that Courtois and Pierpzyk came up with makes use of an algorithm called Extended Sparse Linearization (XSL) which was defined by Shamir et al. [416]. The idea behind XSL is to: generate equations of higher degree from the original set of equations, regard the system of equations as linear combinations of formal terms, and solve this linear system. There are already some successful applications of XSL on reduced rounds (Baby) Rijndael cipher [260] (we are not sure, however, whether the approach is practical even for the four rounds cipher as the 1,807,740 equations obtained in this case were not solved).

It is uncertain at the time of writing this book if this method can work in practice for the full-round AES. However, we should mention that in a recent paper [122], Courtois and Bard effectively break (up to) 10 round DES by an algebraic attack using only one known plaintext. That is indeed an improvement since the algebraic attacks move from the theoretical into the practical realm. One advantage of this attack is that one needs very few known plaintext/ciphertext pairs (just one, in the mentioned case) to recover the key as opposed to other known attacks (differential, linear cryptanalysis, boomerang attack, cache-timing attacks, square attack) where one needs a large number of plaintext/ciphertext pairs (see de Cannière et al. [145]).

7.8 EMBEDDING AES IN BES

In 2002, Murphy and Robshaw [339] embedded AES-128 into a new 128-byte cipher called Big Encryption System (BES), which can be expressed as a sparser quadratic system (over $\mathbb{F}_{2^8}$) than the one of AES alone.

BES involves only operations in $\mathbb{F} := \mathbb{F}_{2^8} = \frac{\mathbb{F}_2[X]}{(X^8+X^4+X^3+X+1)} = \mathbb{F}_2(\theta)$ and uses bytes as state vectors (as well as AES). The state space of BES is $\mathbf{B} = \mathbb{F}^{128}$ as opposed to AES where we saw that the state space is $\mathbf{A} = \mathbb{F}^{16}$. The embedding is done by using the conjugation ϕ, which on $\mathbb{F}$, respectively, $\mathbb{F}^n$ acts as follows

$$\phi(a) = \tilde{a} = (a^{2^0}, a^{2^1}, \ldots, a^{2^7}), \quad a \in \mathbb{F}$$
$$\phi(\mathbf{a}) = \tilde{\mathbf{a}} = (\phi(a_0), \phi(a_1), \ldots, \phi(a_7)), \quad \mathbf{a} \in \mathbb{F}^{8n}.$$

With the convention that $0^{-1} = 0$, ϕ becomes a ring homomorphism, that is,

$$\phi(\mathbf{a} + \mathbf{b}) = \phi(\mathbf{a}) + \phi(\mathbf{b})$$
$$\phi(\mathbf{a}^{-1}) = \phi(\mathbf{a})^{-1}.$$

Let $\mathbf{B}_\mathrm{A} = \phi(\mathbf{A}) \subset \mathbf{B}$. The cipher BES (see [114,339,451]) is defined in such a way that the diagram (7.7) remains commutative.

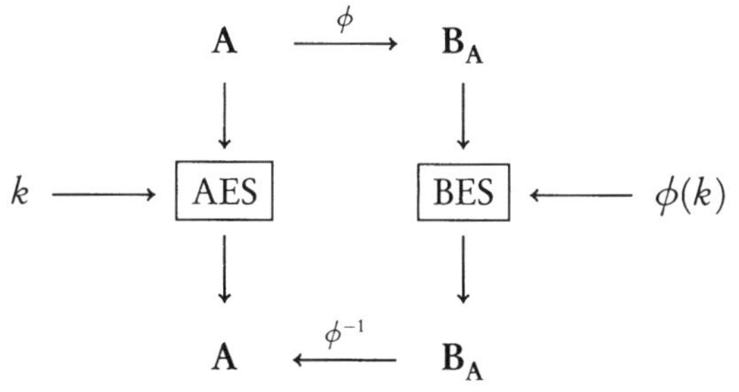

$$\begin{array}{ccccccc} & & \mathbf{A} & \xrightarrow{\ \phi\ } & \mathbf{B}_\mathrm{A} & & \\ & & \downarrow & & \downarrow & & \\ k & \longrightarrow & \boxed{\text{AES}} & & \boxed{\text{BES}} & \longleftarrow & \phi(k) \\ & & \downarrow & & \downarrow & & \\ & & \mathbf{A} & \xleftarrow{\ \phi^{-1}\ } & \mathbf{B}_\mathrm{A} & & \end{array} \qquad (7.7)$$

Let $\mathbf{b} \in \mathbf{B}$ be a state vector for BES (written as a vector, rather than as a matrix as in AES), $(k_\mathbf{B})_i \in \mathbf{B}$ be a subkey in round i. We shall describe next the involved transformations of BES as they are cleverly defined.

The ShiftRow operation in AES can be represented by a matrix $R_\mathrm{A} : \mathbb{F}^{16} \to \mathbb{F}^{16}$. Replacing each 1 in R_A by the identity matrix I_8, and each 0 by a zero 8×8 matrix, we obtain a matrix $R_\mathrm{B} : \mathbb{F}^{128} \to \mathbb{F}^{128}$.

The affine component $L_{\mathbf{A}} : \mathbb{F} \to \mathbb{F}$ (acting on each byte) of the S-box of AES can be represented by the polynomial $f : \mathbb{F} \to \mathbb{F}$

$$f(a) = \sum_{k=0}^{7} \lambda_k a^{2^k},$$

where $(\lambda_0, \lambda_1, \lambda_2, \lambda_3, \lambda_4, \lambda_5, \lambda_6, \lambda_7) = (05, 09, F9, 25, F4, 01, B5, 8F)$, or in polynomial form,

$$\lambda_0 = t^2 + 1; \quad \lambda_4 = t^7 + t^6 + t^5 + t^4 + t^2;$$
$$\lambda_1 = t^3 + 1; \quad \lambda_5 = 1;$$
$$\lambda_2 = t^7 + t^6 + t^5 + t^4 + t^3 + 1; \quad \lambda_6 = t^7 + t^5 + t^4 + t^2 + 1;$$
$$\lambda_3 = t^5 + t^2 + 1; \quad \lambda_7 = t^7 + t^3 + t^2 + t + 1.$$

$L_{\mathbf{A}}$ can be extended to $\mathbf{B}$ by

$$L_{\mathbf{B}}(a) = \phi(L_{\mathbf{A}}(a)) = (f(a)^{2^0}, \ldots, f(a)^{2^7}).$$

Next, define the global transformation $Lin_{\mathbf{B}} : \mathbb{F}^{128} \to \mathbb{F}^{128}$ as a block diagonal matrix with 16 blocks equal to $L_{\mathbf{B}} = [l_{ij}]_{i,j=0,\ldots,7}$, where $l_{ij} = \lambda^{2^i}_{(8-i+j)(\mathrm{mod}\ 8)}$.

Recall that the constant $c_{\mathbf{A}}$ of the S-box transformation is

$$c_{\mathbf{A}} = \theta^6 + \theta^5 + \theta + 1 \in \mathbb{F}.$$

The homomorphism ϕ applied to $c_{\mathbf{A}}$ renders

$$\phi(c_{\mathbf{A}}) = (63, C2, 35, 66, D3, 2F, 39, 36),$$

which, of course, can be written in polynomial form. Repeating this ϕ image 16 times, we get a constant

$$c_{\mathbf{B}} = (\phi(c_{\mathbf{A}}), \ldots, \phi(c_{\mathbf{A}})); \quad [c_{\mathbf{B}}]_i = [\phi(c_{\mathbf{A}})]_{i(\mathrm{mod}\ 8)}.$$

The MixColumn matrix (7.2) can be translated into the polynomial form as

$$C_{\mathbf{A}} = \begin{pmatrix} \theta & 1 & 1 & \theta+1 \\ \theta+1 & \theta & 1 & 1 \\ 1 & \theta+1 & \theta & 1 \\ 1 & 1 & \theta+1 & \theta \end{pmatrix}, \tag{7.8}$$

and the MixColumn transformation is given by $Mix_{\mathbf{A}} : \mathbb{F}^{16} \to \mathbb{F}^{16}$ which is a block diagonal matrix with four copies of $C_{\mathbf{A}}$.

We extend it to a block diagonal matrix $M_{\mathbf{B}} : \mathbb{F}^{128} \to \mathbb{F}^{128}$ (in an appropriate basis [451]), with four consecutive copies of the following matrix $C^{(k)}_{\mathbf{B}}$ for all possible k

$$C_{\mathbf{B}}^{(k)} = \begin{pmatrix} \theta^{2^k} & 1 & 1 & (\theta+1)^{2^k} \\ (\theta+1)^{2^k} & \theta^{2^k} & 1 & 1 \\ 1 & (\theta+1)^{2^k} & \theta^{2^k} & 1 \\ 1 & 1 & (\theta+1)^{2^k} & \theta^{2^k} \end{pmatrix}, \tag{7.9}$$

with the appropriate reductions

$$\theta^{2^0} = \theta; \quad \theta^{2^1} = \theta^2; \quad \theta^{2^2} = \theta^4;$$
$$\theta^{2^3} = \theta^4 + \theta^3 + \theta + 1; \quad \theta^{2^4} = \theta^6 + \theta^4 + \theta^3 + \theta^2 + \theta;$$
$$\theta^{2^5} = \theta^7 + \theta^6 + \theta^5 + \theta^2; \quad \theta^{2^6} = \theta^6 + \theta^3 + \theta^2 + 1;$$
$$\theta^{2^7} = \theta^7 + \theta^6 + \theta^5 + \theta^4 + \theta^3 + \theta.$$

It is not difficult to show that the minimal polynomial of $M_{\mathbf{B}}$ is $(X+1)^{15}$. Represent the transformation $Mix_{\mathbf{B}} : \mathbb{F}^{128} \rightarrow \mathbb{F}^{128}$ by $Mix_{\mathbf{B}} = Perm_{\mathbf{B}}^{-1} \cdot M_{\mathbf{B}} \cdot Perm_{\mathbf{B}}$, where the permutation matrix $Perm_{\mathbf{B}}$ is made up of 16×8 submatrices P_{hk} defined by $[P_{hk}]_{ij} = 1$ if $i = k$ and $j = h$, and $[P_{hk}]_{ij} = 0$, otherwise. The inverse permutation matrix $Perm_{\mathbf{B}}^{-1}$ is described similarly [451]. All 32 (4×4) submatrices in $Mix_{\mathbf{B}}$ are MDS matrices so they contribute to the diffusion property.

A round of BES accepts an input $\mathbf{b} \in \mathbf{B}$ and a subkey $(h_{\mathbf{B}})_i$ and acts as follows:

$$\begin{aligned} R_i(\mathbf{b}, k_{\mathbf{B}})_i &= Mix_{\mathbf{B}}(R_{\mathbf{B}}(Lin_{\mathbf{B}}(\mathbf{b}^{-1}) + c_{\mathbf{B}})) + (h_{\mathbf{B}})_i \\ &= M_{\mathbf{B}} \cdot (b^{-1}) + (C_{\mathbf{B}}(c_{\mathbf{B}}) + (h_{\mathbf{B}})_i) \\ &= M_{\mathbf{B}} \cdot \mathbf{b}^{-1} + (k_{\mathbf{B}})_i, \end{aligned}$$

where $M_{\mathbf{B}} = Mix_{\mathbf{B}} \cdot R_{\mathbf{B}} \cdot Lin_{\mathbf{B}}$ is the 128×128 matrix in $\mathbb{F}$ defined above, and performs linear diffusion within BES, $C_{\mathbf{B}} = Mix_{\mathbf{B}} \cdot R_{\mathbf{B}}$, $(h_{\mathbf{B}})_i = C_{\mathbf{B}}(c_{\mathbf{B}}) + (h_{\mathbf{B}})_i$.

The key schedule of AES is also extended to BES in the following way. The RotWord operation of Section 7.5 can be represented by the matrix $RW_{\mathbf{A}} : \mathbb{F}^4 \rightarrow \mathbb{F}^4$

$$RW_{\mathbf{A}} = \begin{pmatrix} 0 & 1 & 0 & 0 \\ 0 & 0 & 1 & 0 \\ 0 & 0 & 0 & 1 \\ 1 & 0 & 0 & 0 \end{pmatrix}.$$

It is now not surprising that the corresponding matrix $RW_{\mathbf{B}} : \mathbb{F}^{32} \rightarrow \mathbb{F}^{32}$ is obtained by replacing the 1s by the identity matrix I_8 and 0s by the 8×8 zero matrix. The S-box acts only on a subword, and we define $Lin_{\mathbf{B}}^k : \mathbb{F}^{32} \rightarrow \mathbb{F}^{32}$ as a diagonal matrix with four blocks equal to $L_{\mathbf{B}}$ [451]. The constant involved in the key scheduling S-box, say $c_{\mathbf{B}}^k$, is simply

$$c_{\mathbf{B}}^k = \phi(c_{\mathbf{A}}, c_{\mathbf{A}}, c_{\mathbf{A}}, c_{\mathbf{A}}) = (\phi(c_{\mathbf{A}}), \ldots, \phi(c_{\mathbf{A}})); \quad [c_{\mathbf{B}}^k]_i = [\phi(c_{\mathbf{A}})]_{i (\mathrm{mod}\ 8)}.$$

The word array $Rcon[i]$ of AES is mapped into

$$Rcon_{\mathbf{B}}[i] = \phi(Rcon[i]) = (\underbrace{0,\ldots,0}_{24}, \phi(\theta^{i-1})).$$

The round i mapping of BES is

$$\phi_{\mathbf{B}}^{i}(\mathbf{x}) = Lin_{\mathbf{B}}^{k}(RW_{\mathbf{B}}(\mathbf{x}))^{-1} + c_{\mathbf{B}}^{k} + Rcon_{\mathbf{B}}[i],$$

and the generic key round matrices of AES, respectively, BES are

$$MK_{\mathbf{A}}^{i} = \begin{pmatrix} 0 & 0 & 0 & \phi_{\mathbf{A}}^{i} \\ 0 & I_4 & 0 & \phi_{\mathbf{A}}^{i} \\ 0 & I_4 & I_4 & \phi_{\mathbf{A}}^{i} \\ 0 & I_4 & I_4 & I_4 + \phi_{\mathbf{A}}^{i} \end{pmatrix}, \quad MK_{\mathbf{B}}^{i} = \begin{pmatrix} 0 & 0 & 0 & \phi_{\mathbf{B}}^{i} \\ 0 & I_{32} & 0 & \phi_{\mathbf{B}}^{i} \\ 0 & I_{32} & I_{32} & \phi_{\mathbf{B}}^{i} \\ 0 & I_{32} & I_{32} & I_{32} + \phi_{\mathbf{B}}^{i} \end{pmatrix},$$

with the key schedule computation being given by $h_i = MK_{\mathbf{B}}^{i}(h_{i-1})$.

We denote the plaintext and ciphertext by $\mathbf{p} \in \mathbf{B}$ and $\mathbf{c} \in \mathbf{B}$, respectively, and the state vectors before and after the ith inversion $\mathbf{w}_i \in \mathbf{B}$ and $\mathbf{x}_i \in \mathbf{B}$ $(0 \leq i \leq 9)$, respectively. Recalling that the final round in the BES (AES also) does not use the MixColumn operation, denoting $M_{\mathbf{B}}^{*} = R_B \cdot Lin_{\mathbf{B}} = Mix_{\mathbf{B}}^{-1} \cdot M_{\mathbf{B}}$, the encryption of BES, is given by [339,451]:

$$\begin{aligned} &\mathbf{w}_0 = \mathbf{p} + k_0, \\ &\mathbf{x}_i = \mathbf{w}_i^{-1}, \quad \text{for } 0 \leq i \leq 9, \\ &\mathbf{w}_i = M_{\mathbf{B}}\mathbf{x}_{i-1} + k_i, \quad \text{for } 1 \leq i \leq 9, \\ &\mathbf{c} = M_{\mathbf{B}}^{*}\mathbf{x}_9 + k_{10}, \end{aligned} \tag{7.10}$$

including an initial addition key operation and a different matrix in the last round. The reader can certainly write explicitly the algebraic equations governing BES (and implicitly AES):

$$\begin{aligned} &0 = w_{0,(j,m)} + p_{(j,m)} + k_{0,(j,m)}, \\ &0 = x_{i,(j,m)} w_{i,(j,m)} + 1 \quad \text{for } 0 \leq i \leq 9, \\ &0 = w_{i,(j,m)} + (M_{\mathbf{B}}\mathbf{x}_{i-1})_{(j,m)} + k_{i,(j,m)} \quad \text{for } 1 \leq i \leq 9, \\ &0 = c_{(j,m)} + (M_{\mathbf{B}}^{*}\mathbf{x}_9)_{(j,m)} + k_{10,(j,m)}, \end{aligned} \tag{7.11}$$

where (j,m) $(0 \leq j \leq 15, 0 \leq m \leq 7)$ denotes the $(8j + m)$th component of all the vectors. By letting $\alpha_{(j,m)}, \beta_{(j,m)}$ be the entries of $M_{\mathbf{B}}$, respectively, $M_{\mathbf{B}}^{*}$, and assuming that no 0-inversion occurs (true for 53% of encryptions and 85% of 128-bit keys [339]) we get the system of MQ equations

$$\begin{aligned}
0 &= w_{0,(j,m)} + p_{(j,m)} + k_{0,(j,m)},\\
0 &= x_{i,(j,m)} w_{i,(j,m)} + 1 \quad \text{for } 0 \le i \le 9,\\
0 &= w_{i,(j,m)} + k_{i,(j,m)} + \sum_{(j',m')} \alpha_{(j,m),(j',m')} x_{i-1,(j',m')} \quad \text{for } 1 \le i \le 9,\\
0 &= c_{(j,m)} + k_{10,(j,m)} + \sum_{(j',m')} \beta_{(j,m),(j',m')} x_{9,(j',m')}.
\end{aligned} \tag{7.12}$$

The BES encryption is thus described as an overdetermined MQ system with 2688 equations (over $\mathbb{F}$), of which 1280 are quadratic (and very sparse, as we mentioned earlier) and the rest are linear. This MQ system contains 5248 terms (2560 state variables and 1408 key variables). Let δ_i be the components of $c\mathbf{R}_i = c\mathbf{B}^k + Rcon_{\mathbf{B}}[i]$ which is the vector in each round. The key schedule for BES can also be described by an MQ system (see [114,339,451] for further details):

$$\begin{aligned}
0 &= z_{i,(\tilde{j},m)} + h^{254}_{i-1,(12+[(\tilde{j}+1)(\text{mod } 4)],m)}\\
0 &= h_{i,(\tilde{j},m)} + \delta_{i,(\tilde{j},m)} + \sum_{(\tilde{j}',m')} \gamma_{(\tilde{j},m)(\tilde{j}',m')} z_{i,(\tilde{j}',m')}\\
0 &= h_{i,(4s+\tilde{j},m)} + h_{i,(4(s-1)+\tilde{j},m)} + h_{i-1,(4s+\tilde{j},m)} \quad 1 \le s \le 3\\
0 &= k_{i,(t,m)} + (C_{\mathbf{B}}(c_{\mathbf{B}}))_{(t,m)} + h_{i,(t,m)}\\
0 &= z^2_{i,(\tilde{j},m)} + z_{i,(\tilde{j},m+1)}\\
0 &= h_{i,(\tilde{j},m)} + h_{i,(\tilde{j},m+1)},
\end{aligned} \tag{7.13}$$

where $\gamma_{(j,m)}$ are the entries of $Lin^k_{\mathbf{B}}$, and $1 \le i \le 10$, $0 \le \tilde{j}, \tilde{j}' \le 3$, $0 \le m, m' \le 7$.

The (sub)cipher AES, as an embedded cipher in BES, generates more (artifact of the embedding nature) MQ equations (linear and quadratic), namely

$$\begin{aligned}
0 &= w_{0,(j,m)} + p_{(j,m)} + k_{0,(j,m)};\\
0 &= w_{i,(j,m)} + k_{i,(j,m)} + \sum_{(j',m')} \alpha_{(j,m),(j',m')} x_{i-1,(j',m')} \quad \text{for } 0 \le i \le 9,\\
0 &= c_{(j,m)} + k_{10,(j,m)} + \sum_{(j',m')} \beta_{(j,m),(j',m')} x_{9,(j',m')};\\
0 &= x_{i,(j,m)} w_{i,(j,m)} + 1 \quad \text{for } 0 \le i \le 9,\\
0 &= x^2_{i,(j,m)} + x_{i,(j,m+1)} \quad \text{for } 0 \le i \le 9,\\
0 &= w^2_{i,(j,m)} + w_{i,(j,m+1)} \quad \text{for } 0 \le i \le 9.
\end{aligned} \tag{7.14}$$

One can count now [114,339,451] the number of equations that the AES encryption satisfies. Thus we have 5248 equations, of which 3840 are sparse quadratic and 1408 are linear equations. The MQ system contains 7808 terms and 2560 state variables and 1408 key variables. The key schedule of AES can also be expressed as an MQ system which contains about 2560 equations (over $\mathbb{F}$), of which 960 are (very sparse) quadratic and 1600 are linear equations. The system contains 2368 terms with 2048 variables (1408 are basic key variables and 640 are auxiliary variables) [339].

Murphy and Robshaw in [339] pointed out that since BES's operations can be defined in a single field, the representation we also described has several nice properties that may make it easier to cryptanalyze, with 'potentially very few plaintext-ciphertext pairs'. The XSL method [416] may have a chance of success to reduce the complexity of the key recovery since the BES MQ system is extremely sparse. In a subsequent note [340] Murphy and Robshaw argue that if XSL is a valid method then it has better chance of success if applied to the MQ system of AES obtained as an embedding in BES (over $\mathbb{F} = \mathbb{F}_{2^8}$) rather than to the MQ system obtained over $\mathbb{F}_2$.

7.9 FURTHER EMBEDDINGS OF AES

Monnerat and Vaudenay introduced two new embeddings of AES into what they called CES and Big-BES [337].

We detail here the big-BES extension of AES, which was motivated by the p-adic extension of $\mathbb{F}_p$ (we follow [337,113]. Recall that $\mathbb{F} = \mathbb{F}_{2^8} = \frac{\mathbb{F}_2[X]}{(X^8+X^4+X^3+X+1)}$. Define on $\mathbf{R} = \mathbb{F} \times \mathbb{F}$ the following binary operations

$$(x_1, y_1) \oplus (x_2, y_2) = (x_1 + x_2, y_1 + y_2)$$
$$(x_1, y_1) \otimes (x_2, y_2) = (x_1 x_2, x_1 y_2 + x_2 y_1).$$

The state space for big-BES is the algebra $\mathbf{C} = \mathbf{R}^{16}$ over $\mathbb{F}$, with scalar multiplication by $\lambda \in \mathbb{F}$ defined using the ring multiplication by $(\lambda, 0, \lambda, 0, \ldots, \lambda, 0) \in \mathbf{C}$.

Thus $\mathbf{R}$ becomes a commutative ring with unit (1, 0). Since no element of the form $(0, y)$ is invertible because the inverse is obviously given by

$$(x, y)^{-1} = (x^{-1}, y x^{-2}),$$

one modifies the inversion map (as was the case for AES, as well) by mapping $(0, y)$ into $(0, y^{-1})$. BES is embedded into big-BES based on the injective algebra homomorphism $\phi : \mathbb{F}^{16} \to \mathbf{C}$

$$\phi(\mathbf{a}) = ((a_0, 0), \ldots, (a_{15}, 0)),$$

making the following diagram commutative

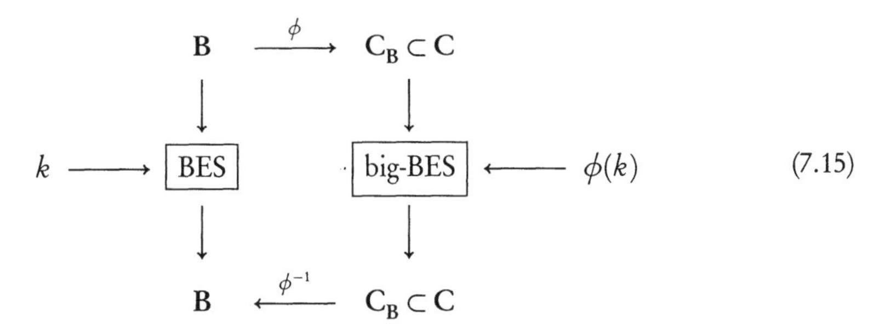

We will not go into details regarding the encryption process of big-BES as it is defined exactly as the one of BES where the entries of the matrices involved are moved into C by means of the embedding $\theta : \mathbb{F} \to \mathbf{R}$, $\theta(x) = (x, 0)$.

The motivation for this new embedding was to show that in these new ciphers some cryptanalytic attacks could be easily mounted [337]. It was unclear whether these attacks have any relevance to AES (or BES, for that matter) [337,113].

In spite of these doubts of the usefulness of such an extension of BES (and implicitly AES), the following question remains: can one find a framework where such an embedding of AES (or any other block cipher, for that matter) might provide the additional tools it needs for the cipher to be cryptanalyzed successfully? We mention here work by Cid et al. [113] where they propose such an algebraic framework (which is yet to prove practical). Only time and further research will tell.

CHAPTER

Boolean Cayley graphs

8

8.1 INTRODUCTION

One cannot escape the feeling that mathematical formulas have an independent existence and intelligence of their own, that they are wiser than we are.

Heinrich Hertz (1857–1894)

In this chapter we will concentrate on a new technique for dealing with Boolean functions, which will be used successfully to find a characterization of bent Boolean functions in terms of graph spectra.

Definition 8.1. *A* simple graph Γ *consists of a finite set V (of cardinality v) of* vertices *and a set E of 2-subsets of V called* edges. *The* adjacency matrix *of Γ (of course, dependent on the labeling of vertices) is a $v \times v$ matrix $A_\Gamma = (a_{ij})$ such that $a_{ij} = 1$ if $\{\mathbf{x}_i, \mathbf{x}_j\} \in E$ and $a_{ij} = 0$ otherwise.*

The *degree* (or *valency*) of a vertex $\mathbf{v}$ is the number of edges incident to $\mathbf{v}$, and it is labeled by $\deg(\mathbf{v})$.

Definition 8.2. *A graph is* regular of degree *r (or r-regular) if every vertex has degree r. An r-regular graph Γ is a* strongly regular graph *(srg) if there exist nonnegative integers e, d such that for all vertices $\mathbf{u}, \mathbf{v}$ the number of vertices adjacent to both $\mathbf{u}, \mathbf{v}$ is e, d, if $\mathbf{u}, \mathbf{v}$ are adjacent, respectively, nonadjacent (we say that Γ is an srg with parameters (v, r, e, d)).*

A graph $\Gamma = (V(\Gamma), E(\Gamma))$ is *bipartite* if the vertex set $V(\Gamma)$ can be partitioned into two sets S_1, S_2 in such a way that no two vertices from the same set are adjacent. We can also define the *diameter* (diam) of a graph as the maximum distance between any two vertices. Since the distance between two vertices is the length of the shortest path between them, then the diameter is the longest of the shortest paths. More about graphs in [14,35,132, 430] or your favorite book on graph theory.

Definition 8.3. *A subset C in a group G with identity element e is called a* Cayley set *if and only if whenever $g \in C$, then $g^{-1} \in C$ and $e \notin C$. We modify slightly the definition by allowing $e \in C$, if needed.*

Take $G = \mathbb{V}_n$. Since in characteristic 2 under addition, $-\mathbf{w} = \mathbf{w}$, for any $\mathbf{w} \in \mathbb{V}_n$, we see that any subset $C \subseteq \mathbb{V}_n$ is a Cayley set.

Definition 8.4. *Let f be a Boolean function on $\mathbb{V}_n$. We define the* Cayley graph *of f to be the graph $\Gamma_f = (\mathbb{V}_n, E_f)$ whose vertices set is $\mathbb{V}_n$ and the set of edges is defined by*

$$E_f = \{(\mathbf{w},\mathbf{u}) \in \mathbb{V}_n \times \mathbb{V}_n : f(\mathbf{w} \oplus \mathbf{u}) = 1\}.$$

The adjacency matrix A_f is the matrix whose entries are $a_{i,j} = f(\mathbf{b}(i) \oplus \mathbf{b}(j))$, where $\mathbf{b}(i) = \alpha_i$ (in the notations of Chapter 2) is the binary representation of the argument, written as a vector ($\mathbf{b}(i)$ is of length 2^n, if $i \le 2^n - 1$). It is simple to prove that A_f has the dyadic property: $a_{i,j} = a_{i+2^{n-1}, j+2^{n-1}}$. As before, let $\Omega_f = \{\mathbf{x} \in \mathbb{V}_n : f(\mathbf{x}) = 1\}$. From its definition we derive that Γ_f is a regular graph of degree $wt(f) = |\Omega_f|$ (see the book of Cvetkovic et al. [132, Chapter 3] for further definitions).

By $\langle \Omega_f \rangle$ we mean the space of the 0, 1 sequences generated by Ω_f. We denote its dimension by $\dim\langle \Omega_f \rangle$.

Given a graph f and its adjacency matrix A, the *spectrum* of Γ_f is the set of eigenvalues of A (called also the *eigenvalues* or *spectral coefficients* of Γ_f). All of our theorems will assume that Γ_f is connected, that is, there is a path between every two vertices. One can show easily that all connected components of Γ_f are isomorphic (we shall often point out what changes in our arguments in case Γ_f is not connected).

An easy counting argument shows that if a graph is strongly regular, then

$$r(r - d - 1) = e(v - r - 1). \tag{8.1}$$

The *complementary* graph $\bar{\Gamma}_f$ of the strongly regular graph Γ_f is also strongly regular with parameters $(v, v - r - 1, v - 2r + e - 2, v - 2r + d)$.

We do not want to get into details, but the knowledgeable reader may observe that a strongly regular graph is essentially the same as an association scheme of class 2 (see [117,243] and the references therein). In spite of their (apparently) strict arithmetic nature, strongly regular graphs are difficult to investigate. P.J. Cameron [52] mentions that '*Strongly regular graphs lie on the cusp between highly structured and unstructured. For example, there is a unique strongly regular graph with parameters* (36; 10; 4; 2), *but there are 32548 nonisomorphic graphs with parameters* (36; 15; 6; 6). *(The first assertion is a special case of a theorem of Shrikhande* (our reference [421]), *while the second is the result of a computer search by McKay and Spence* (our reference [314]).) *In the light of this, it will be difficult to develop a theory of random strongly regular graphs!*'

If G is a group of order v and D is an r-element subset of G, then D is called a (v, r, d, e)-*partial difference set* if for every nonidentity element g in G, the equation $d_1 d_2^{-1} = g$ has exactly d solutions (d_1, d_2) in D^2, if $g \in D$, respectively, e solutions, if $g \in G \backslash D$. It is known that a subset D of G (the identity $e \notin D$ and $\{d^{-1} : d \in D\} = D$) is a partial difference set if and only if the Cayley graph generated by D is strongly regular (see [152,153,311]).

As bent Boolean functions are as elusive as the strongly regular graphs, perhaps it is then not surprising that there should be some connections between graph theory and

Boolean functions, as we saw some connections between bent Boolean functions and difference sets in Chapter 5.

8.2 SPECTRA OF BOOLEAN CAYLEY GRAPHS

Here and throughout this chapter we assume that $n \geq 4$. The following theorem is a compilation of various results from [28] and [132] (we slightly changed the notations).

Theorem 8.5. *The following statements are true:*

(i) *Let $f : \mathbb{V}_n \to \mathbb{F}_2$, and let $\lambda_i, 0 \leq i \leq 2^n - 1$ be the eigenvalues of its associated graph Γ_f. Then $\lambda_i = W(f)(\mathbf{b}(i))$, for any i.*

(ii) *The multiplicity of the largest spectral coefficient of f, $W(f)(\mathbf{b}(0))$, is equal to $2^{n\text{-}\dim(\Omega_f)}$.*

(iii) *If Γ_f is connected, then f has a spectral coefficient equal to $-wt(f)$ if and only if its Walsh spectrum is symmetric with respect to zero.*

(iv) *The number of nonzero spectral coefficients is equal to $\mathrm{rank}(A_f)$, the rank of A_f, which satisfies $2^{d_2} \leq \mathrm{rank}(A_f) \leq \sum_{i=1}^{d} \binom{n}{i}$ (d_2, respectively, d is the degree of f over $\mathbb{F}_2$, respectively $\mathbb{R}$).*

Proof. (i) is quite simple, as the eigenvectors of the Cayley graph Γ_f are the characters $Q_{\mathbf{w}}(\mathbf{x}) = (-1)^{\mathbf{w}\cdot\mathbf{x}}$ of $\mathbb{V}_n$. Moreover, the ith eigenvalue of A_f, corresponding to the eigenvector $Q_{\mathbf{b}(i)}$ is given by $\lambda_i = \sum_{\mathbf{x}}(-1)^{\mathbf{b}(i)\cdot\mathbf{x}} f(\mathbf{x}) = W(f)(\mathbf{b}(i))$.

(ii) follows from the fact that the multiplicity of the largest eigenvalue of a regular graph is equal to the number of its connected components (see [132, Theorems 3.23 and 3.35]).

In [132, Theorem 3.37 and p. 109] it is proven that the smallest eigenvalue of a regular connected graph is equal to $-r$, where r is the degree, if and only if the graph is bipartite. It is also known that a graph is bipartite if and only if its spectrum is symmetric with respect to the origin. Thus (iii) is proven.

The first part of (iv) follows from the fact that A_f is symmetric. Since the degree of f as a real polynomial (in multiple variables) equals the maximum order of the nonzero spectral coefficients, and the order of the coefficient $W(f)(\mathbf{w})$ is the number of 1s in the word $\mathbf{w}$, the upper bound follows. Now, we observe that for any subfunction of f, say f', $A_{f'}$ is a submatrix of A_f, therefore $\mathrm{rank}(A_{f'}) \leq \mathrm{rank}(A_f)$. Obviously, all the spectral coefficients of a function that takes the value 1 on an odd number of words are nonzero, so, for any subfunction f' such that $|\Omega_{f'}|$ is an odd integer, $A_{f'}$ has maximum rank. But, then the degree of a Boolean function can be characterized as the size of the biggest subfunction that takes the value 1 on an odd number of strings, so $\mathrm{rank}(A_f) \geq 2^{d_2}$. ∎

8.3 FEW SPECTRAL COEFFICIENTS OF BOOLEAN FUNCTIONS

A natural question is whether one can characterize those functions with few spectral coefficients. In [28] the following theorem is proven. We provide here the simple proof.

Theorem 8.6. *If Γ_f has two distinct eigenvalues, then its connected components are complete graphs and $\Omega_f \cup \{\mathbf{b}(0)\}$ is a group.*

Proof. From [132, Theorem 3.13, p. 88] we know that if a graph has m distinct eigenvalues, then its diameter $diam \leq m-1$. Thus, since in this case $m=2$, it follows that $diam \leq 1$, which proves the first claim (see also [132, p. 162]).

Now, any $\mathbf{u}, \mathbf{v} \in \Omega_f$ will belong to the component with vertex set $\langle \Omega_f \rangle$. But any connected component is a complete graph, so $\mathbf{u}, \mathbf{v}$ must be adjacent, which means that $f(\mathbf{u} \oplus \mathbf{v}) = 1$, that is $\mathbf{u} \oplus \mathbf{v} \in \Omega_f$. Furthermore, $\mathbf{x} \oplus \mathbf{x} = 0$, so the second claim is proved. ∎

The following theorem is known [132, Theorem 3.32, p. 103].

Theorem 8.7. *A connected r-regular graph is strongly regular if and only if it has exactly three distinct eigenvalues $\lambda_0 = r, \lambda_1, \lambda_2$ (so $e = r + \lambda_1\lambda_2 + \lambda_1 + \lambda_2$, $d = r + \lambda_1\lambda_2$). Furthermore, the adjacency matrix satisfies*

$$A^2 = (d-e)A + (r-e)I + eJ,$$

where J is the all 1 matrix.

If Γ_f has three eigenvalues with at most one of them zero, one can completely describe Γ_f [132, pp. 162–163].

Theorem 8.8. *Γ_f has three distinct eigenvalues $\lambda_0 > \lambda_1 = 0 > \lambda_2 = -\lambda_0$ if and only if Γ_f is the complete bipartite graph between the vertices of Ω_f and $\mathbb{V}_n \setminus \Omega_f$. Γ_f has three distinct eigenvalues $\lambda_0 = |\Omega_f| > \lambda_1 = 0 > \lambda_2 \neq -\lambda_0$ if and only if the complement of Γ_f is the direct sum of $-(r/\lambda_2)+1$ complete graphs of order $-\lambda_2$ (that is, Γ_f is a complete multipartite graph).*

If none of the eigenvalues is zero then the following result holds [132, pp. 194–195].

Theorem 8.9. *If Γ_f has three distinct eigenvalues none of which is zero, then these eigenvalues are*

$$\lambda_0 = |\Omega_f| = wt(f), \quad \lambda_2 = -\lambda_1 = \sqrt{|\Omega_f| - e}, \tag{8.2}$$

of multiplicities

$$m_0 = 1, \quad m_1 = \frac{(2^n-1)\sqrt{|\Omega_f|-e} - |\Omega_f|}{2\sqrt{|\Omega_f|-e}}, \quad m_2 = \frac{(2^n-1)\sqrt{|\Omega_f|-e} + |\Omega_f|}{2\sqrt{|\Omega_f|-e}}. \tag{8.3}$$

It is assumed above that Γ_f is connected. If it is not connected, then the multiplicities are to be multiplied by $2^{n-\dim(\Omega_f)}$ (since the connected components of Γ_f are isomorphic).

8.4 BENT BOOLEAN CAYLEY GRAPHS

Bernasconi and Codenotti [28], and Bernasconi, Codenotti and VanderKam [31] proved some results which we have gathered in the following theorem that describes completely the class of bent Boolean functions in terms of a subset of the class of strongly regular graphs.

Theorem 8.10. *The Boolean function f defined on $\mathbb{V}_n$ (n even) is bent if and only if its associated Cayley graph Γ_f is a strongly regular graph with the additional property $e = d$. Moreover, under this assumption, the adjacency matrix satisfies*

$$A^2 = \left(2^{n-1} \pm 2^{n/2-1} - e\right) I + eJ,$$

for some choice of the $\pm$ sign.

Proof. Denote $r = |\Omega_f|$ and assume that f is bent. Using Theorem 8.7 and the fact that $\lambda_1 = -\lambda_2$, it follows that Γ_f is strongly regular and further $e = d$.

Conversely, assume the Γ_f is strongly regular and $e = d$. Rewriting Parseval's identity $W(\hat{f})(\mathbf{b}(0)) = \sum_{i=0}^{2^n-1} W(\hat{f})(\mathbf{b}(i))^2$ in terms of eigenvalues and using Theorem 8.9, we get the equation $(r - \lambda_1)(r - \lambda_2) = 2^n(r + \lambda_1\lambda_2)$. Since f is bent, one gets further that the previous equation is equivalent to (labeling $\lambda := \lambda_1 = -\lambda_2$)

$$r^2 - 2^n r + \lambda^2(2^n - 1) = 0. \tag{8.4}$$

Two integer solutions of (8.4) are found by taking $\lambda = 0, \lambda^2 = 1$, but this will correspond to graphs with only two distinct eigenvalues. A proper solution for the three eigenvalues is found by taking $\lambda^2 = 2^{n-2}$, which corresponds to the class of bent functions. It will be sufficient to prove that these are the only possibilities for integer solutions of Equation (8.4).

Assume that $n > 2$, since $n = 2$ follows easily. The discriminant of the equation is

$$\Delta = 2^{2n} - 4\lambda^2(2^n - 1).$$

To have integer solutions, Δ must be an (even) integer square, say $\Delta = (2a)^2$. Further, (8.4) will become

$$2^{2n-2} - \lambda^2(2^n - 1) = a^2. \tag{8.5}$$

Assuming that λ, a are positive or zero, we get that

$$\lambda^2 < 2^{2n-2}/(2^n - 1) < 2^{n-2} + 1, \tag{8.6}$$

$$a \le 2^{n-1}. \tag{8.7}$$

Since for $n > 2$, we have $2^{2n-2} > 2^n$, Equation (8.5) leads to $\lambda^2 \equiv a^2 (\mathrm{mod}\ 2^n)$. Assume that $2^p \,||\, \lambda$ (that is, $2^p | \lambda$ and $2^{p+1} \nmid \lambda$). Thus $\lambda = 2^p \lambda'$ and $a = 2^p a'$, for some odd λ', a'. Therefore,

$$2^{2n-2p-2} - \lambda'^2(2^n - 1) = a'^2. \tag{8.8}$$

From (8.6), we get $2p \leq n-2$, so $2n-2p-2 \geq n$. Then, (8.8) implies $\lambda'^2 \equiv a'^2 \pmod{2^n}$, that is, $\lambda' \equiv \pm a' \pmod{2^{n-1}}$ (because both λ', a' are odd). Therefore, since $\lambda', a' \leq 2^{n-1}$, we have $a' = \lambda'$ or $a' = 2^{n-1} - \lambda'$. Substituting these values into (8.8), we get $\lambda^2 = 2^{n-2}$, respectively, $2^n(\lambda' - \lambda'^2) = 2^{2n-2p-2}(2^{2p}-1)$. If nonzero, the right side is positive and the left side is negative. So both sides must be zero, that is, $p = 0$, and $\lambda' = 0, 1$, which proves the claim. ■

A more detailed analysis of Theorem 8.9 easily enabled Huang and You [231] to find further algebraic properties of a strongly regular graph Γ of parameters (v, r, e, e) $(e > 1)$, even though Γ may not be a Cayley graph: using the notations of Theorem 8.9, then λ_2 divides e, and $(v, r) = \left(\frac{(\lambda_2^2+\lambda_2+e)(\lambda_2^2-\lambda_2+e)}{e}, \lambda_2^2 + e\right)$. Further, if $e = pq$, where $p > q$ are two distinct primes, then:

1. If $q \geq 3$, then $(v, \lambda_2) = \left(\frac{p(p+q-1)(p+q+1)}{q}, p\right)$, and $p = 2cq \pm 1$, for some integer c.
2. If $q = 2$, then $(v, \lambda_2) = \left(\frac{p(p+1)(p+3)}{2}, p\right)$, or (16, 2).

The following theorem will be used to prove some results on the Cayley graph of a bent Boolean function (see the book of Asratian et al. [14, Chapter 2]).

Theorem 8.11. *The following are equivalent for a graph* Γ*:*

1. Γ *is bipartite.*
2. Γ *has no cycles of odd length.*
3. *Every subgraph* H *of* Γ *has at least* $|V(H)|/2$ *mutually nonadjacent vertices.*
4. *The spectrum of* Γ *is symmetric with respect with 0, that is, if* λ *is an eigenvalue, then* $-\lambda$ *is also an eigenvalue.*

Stănică [436] found the following necessary conditions on the Cayley graph associated to bent functions.

Theorem 8.12. *The Cayley graph associated to a bent function is not bipartite.*

Proof. Using Theorem 8.11, we gather that for Γ_f to be bipartite it is necessary to have its spectrum symmetric about the origin. But according to Theorem 8.10, that is impossible since $-\lambda_1 = -wt(f)$ it is not an eigenvalue of Γ_f. ■

From Theorem 8.11, a graph is bipartite if and only if it does not contain cycles of odd length. Thus, if f is bent then the associated Cayley graph contains a cycle of odd length. One can get more precise results. Recall that Γ_f is $|\Omega_f|$-regular.

Theorem 8.13. *Assume* $n > 4$. *If* Γ_f *is triangle-free (that is, there are no paths of the form* **xyzx**, *where the vertices* **x**, **y**, **z** *are distinct), then* f *is not bent.*

Proof. Assume on the contrary that f is bent. Erdös and Sós proved in 1974 (cf. [14, Chapter 2]) that a triangle-free graph G on p vertices with minimum degree $\delta(G) > 2p/5$ is bipartite. Recall that Γ_f is a regular graph of degree $|\Omega_f|$ of order $p = 2^n$. Since $n > 4$, then $2^{n/2} > 5$ is equivalent to $5(2^{n-1} - 2^{n/2-1}) > 2^{n+1}$, which implies $|\Omega_f| = wt(f) > 2^{n+1}/5$. Thus, G is bipartite, which is false by Theorem 8.12, contradicting our assumption that f is bent. ■

In the previous proof it is sufficient to assume that Γ_f is regular of degree greater than $2^{n+1}/5$ (if the degree is $<2^{n+1}/5$, then the function is certainly not bent).

A more constructive argument would be the following: assuming that f is bent (one may replace f by its complement, so we assume that $f(0) = 0$), we prove that we can always find triangles in Γ_f. Since in this case Γ_f is strongly regular, Lemma 8 of [28] shows that $e = |(\mathbf{x} \oplus \Omega_f) \cap (\mathbf{y} \oplus \Omega_f)| \geq 1$. Applying this for $\mathbf{x} = 0$ and an arbitrary vector $\mathbf{y} \in \Omega_f$, implies that $e = |\Omega_f \cap (\mathbf{y} \oplus \Omega_f)| \geq 1$. That is, there exists $\mathbf{z} \in \Omega_f$ such that $\mathbf{y} \oplus \mathbf{z} \in \Omega_f$. Thus, $f(\mathbf{y} \oplus \mathbf{z}) = f(\mathbf{z}) = f(\mathbf{y}) = 1$. It follows that $0, \mathbf{y}, \mathbf{z}$ is a triangle in Γ_f.

The converse of Theorem 8.13 is not true, as it can be seen by considering on $\mathbb{V}_6$ the function $f(x_1, x_2, x_3, x_4, x_5, x_6) = x_1x_2x_3 \oplus x_2x_3x_4 \oplus x_3x_4x_5 \oplus x_4x_5x_6 \oplus x_5x_6x_1 \oplus x_6x_1x_2$ and the associated Cayley graph which has plenty of triangles, but f is not bent.

The number of triangles sitting on any two (fixed) adjacent vertices is equal to e. We know that $e = |(\Omega_f \oplus \mathbf{v}) \cap (\Omega_f \oplus \mathbf{w})|$ (Lemma 8 of [28]) for any pair of vertices $\mathbf{v} \neq \mathbf{w}$. We note that $e \neq |\Omega_f|$, since the equality forces two eigenvalues to become 0. That is not possible since in that case (see [132]) the graph Γ_f cannot be strongly connected. Thus $e < |\Omega_f|$. There are other restrictions on e. A simple corollary of Theorem 8.10 is that e must differ from $|\Omega_f|$ by a perfect square.

Let $\mathcal{B}_0(n)$, respectively, $\mathcal{F}_0(n)$, denote the set of bipartite respectively, triangle-free graphs on n vertices. Let $\mathcal{B}_1(n)$ be the set of all triangle-free graphs that can be made bipartite by removing a single vertex (*quasi-bipartite*), and $\mathcal{F}_1(n) = \mathcal{F}_0(n) \setminus \mathcal{B}_0(n)$. It is known [372] that $\mathcal{B}_0(n) \asymp 2^{\frac{1}{4}n^2 - \frac{1}{2}\log n}$. What is interesting is that almost all triangle-free graphs are bipartite, that is,

$$|\mathcal{F}_0(n)| = |\mathcal{B}_0(n)|(1 + o(1)|.$$

This was shown by Erdös, Kleitman and Rothchild in [162]. Their work was followed by Prömel et al. [372], who proved, by using similar techniques, the following result.

Theorem 8.14. *Almost all triangle-free graphs which are not bipartite are quasi-bipartite, that is*

$$|\mathcal{F}_1(n)| = |\mathcal{B}_1(n)|(1 + o(1)|.$$

An interesting question (looking at bent functions) would be: how many vertices should we remove from the Cayley graph associated to a bent function, in order to transform the graph into a bipartite graph? We do not have an answer to this question.

Further, is it possible to detect that a strongly regular graph is not representing a bent function by showing that it is triangle-free? Certainly, but unfortunately there are not many triangle-free strongly regular graphs that are known. In fact, there are only seven such graphs that were known when this book was being written (see [188]).

8.5 COLORING THE BOOLEAN CAYLEY GRAPH

A *k-coloring* of a graph Γ is a function $f : V(\Gamma) \to \{1,2,\ldots,k\}$, such that no edge $e = (u,v)$ has $f(u) = f(v)$. If a graph Γ contains a loop, then Γ has no k-coloring for any k. A graph which has a k-coloring is called k-colorable. The *chromatic number* $\chi(\Gamma)$ of a graph Γ is the minimum integer k such that Γ has a k-coloring (Skiena [430, p. 210]). For example, Γ is bipartite if and only if Γ is 2-colorable.

If e is an edge of a graph Γ, we write $\Gamma - e$ for the subgraph of Γ with vertex set $V(\Gamma)$ and edge set $E(\Gamma)\setminus\{e\}$. Then, $\Gamma - e$ is the subgraph that results from deleting the edge e from Γ. We write $G\cdot e$ for the graph which results from contracting the edge e (the graph $\Gamma\cdot e$ is not necessarily a subgraph of Γ). If Γ is a simple graph (no loops, no multiple edges), then $\Gamma - e$ must be simple, but $\Gamma \cdot e$ may have loops or multiple edges.

Let $\pi_\Gamma(k)$ denote the number of k-colorings of the graph Γ, where k is a positive integer. Let e be an edge of Γ. It is an easy exercise to show that $\pi_\Gamma(k) = \pi_{\Gamma-e}(k) - \pi_{\Gamma\cdot e}(k)$. A simple induction proof will show that $\pi_\Gamma(k)$ is a polynomial in k with sign alternating integer coefficients, called *chromatic polynomial*. The chromatic number of a graph Γ can also be interpreted as the smallest positive integer k such that the chromatic polynomial $\pi_\Gamma(k) > 0$. The computation of the chromatic number of a graph is an NP-complete problem [430, pp. 211–212].

Assume that the eigenvalues of Γ_f are ordered as $\lambda_1 \geq \lambda_2 \geq \cdots \geq \lambda_v$.

Theorem 8.15. *Let f be a Boolean function, and let Γ_f be the associated Cayley graph with g being the multiplicity of its lowest eigenvalue $\lambda_v(\Gamma_f)$. Then,* $\min\left\{g+1, 1-\frac{\lambda_v(\Gamma_f)}{\lambda_2(\Gamma_f)}\right\} \leq \chi(\Gamma_f) \leq |\Omega_f|$ *(provided $\lambda_2(\Gamma_f)\neq 0$).*

Proof. The first inequality $\min\left\{g+1, 1-\frac{\lambda_v(G)}{\lambda_2(\Gamma)}\right\} \leq \chi(\Gamma)$ can be found in Haemers [217], holding for arbitrary graphs Γ. Cao proved in [62] that the chromatic number satisfies $\chi(\Gamma) \leq \sqrt{T(\Gamma)}+1$, for any graph Γ, where $T(\Gamma)$ is the maximum sum of degrees of vertices adjacent to any vertex v (that is, the maximum number of 2-walks in Γ). When $\Gamma = \Gamma_f$, since Γ_f is Ω_f-regular, then $T(\Gamma_f) = |\Omega|^2$, so we get $\chi(\Gamma_f) \leq |\Omega|+1$. By Wilf's theorem [464], the equality $\chi(\Gamma_f) = |\Omega|+1$ holds if and only if Γ_f is a complete graph or an odd cycle. Since Γ_f is neither, we obtain $\chi(\Gamma_f)\leq|\Omega|$. ■

Corollary 8.16. *With the notations of the previous theorem, assuming that Γ_f is a strongly regular (connected) graph, with $e = d$, then* $\max\left\{2, 1+\frac{|\Omega_f|}{\sqrt{|\Omega_f|-e}}\right\} \leq \chi(\Gamma_f) \leq |\Omega_f|$.

Proof. The corollary follows easily observing that under the imposed conditions $v = 3$, $\lambda_3 = -\lambda_2$. Using Theorem 8.15 (with $g \geq 1$), Hoffman's famous bound on the chromatic number $\chi(\Gamma_f) \geq 1 - \frac{\lambda_1(\Gamma_f)}{\lambda_v(\Gamma_f)}$ (cf. Hoffman [226]), and Theorem 8.10, we get the result. ■

Recall that a graph is Ramanujan if it is r-regular and all eigenvalues $\neq r$ are $\leq 2\sqrt{r-1}$. One can see that Γ_f, for a bent function f, is a Ramanujan graph, since $r = |\Omega_f|$ and the eigenvalues in absolute value are $\sqrt{|\Omega_f| - e} \leq 2\sqrt{|\Omega_f| - 1}$. One cannot get better bounds on the chromatic number by using the fact that Γ_f is always a Ramanujan graph. If Γ_f is connected and nonbipartite, r-regular then $\chi(\Gamma_f) \geq \frac{r}{2\sqrt{r-1}} \sim \frac{\sqrt{r}}{2}$ (see the book of Davidoff et al. [138]). But this bound is not better than the one obtained by Corollary 8.16.

8.6 AVALANCHE FEATURES OF THE CAYLEY GRAPHS

Recall that if a Boolean function f can be written as

$$f(x_1, \ldots, x_i, \ldots, x_n) = g(x_1, \ldots, \hat{x}_i, \ldots, x_n) \oplus x_i$$

($\hat{x}_i$ means that the variable x_i is missing), we say that f *depends linearly* on x_i.

It is easy to see that a Boolean function f depends on the variable x_i linearly if and only if the Walsh transform $W(\hat{f})(\mathbf{u}) = 0$ for all $\mathbf{u}$ with the ith component $u_i = 0$ (see also [448,450]). Using the result of Lemma 2.9 that shows the relationship between the Walsh transform of f and $\hat{f}$,

$$W(\hat{f})(\mathbf{u}) = -2W(f)(\mathbf{u}) + 2^n \delta(\mathbf{u}), \quad \text{on } \mathbb{V}_n, \tag{8.9}$$

it is rather easy to deduce the following result.

Proposition 8.17. *A Boolean function f depends on a variable x_i linearly if and only if the eigenvalues for the Cayley graph Γ_f satisfy $\lambda_0 = 2^{n-1}$ and $\lambda_{j \neq 0} = 0$, if $\mathbf{b}(j)$ has its ith component equal to 0.*

We recall from Chapter 4 that a function f on $\mathbb{V}_n$ is correlation immune of order ℓ if its Walsh transform satisfies $W(\hat{f})(\mathbf{v}) = 0$ for all $1 \leq wt(\mathbf{v}) \leq \ell$. If, in addition, $W(\hat{f})(0) = 0$, then f is called resilient of order ℓ.

It is not hard to derive the following characterization of these properties in terms of graph spectra.

Proposition 8.18. *A function f on $\mathbb{V}_n$ is correlation immune of order ℓ if and only if the eigenvalues of the associated Cayley graph Γ_f satisfy $\lambda_i = 0$ for all i with $1 \leq wt(\mathbf{b}(i)) \leq \ell$. Further, f is resilient of order ℓ if and only if $\lambda_i = 0$ for all $1 \leq wt(\mathbf{b}(i)) \leq \ell$ and $\lambda_0 = 2^{n-1}$.*

Proof. We know that $\lambda_i = W(f)(\mathbf{b}(i))$, for any $0 \le i \le 2^n - 1$. By the definition of correlation immune of order ℓ functions, we derive the result. ■

Corollary 8.19. *For an unbalanced correlation immune of order ℓ function f on $\mathbb{V}_n$, there are $\sum_{s=1}^{\ell} \binom{2^n}{s}$ zero eigenvalues of Γ_f.*

One can compute the Walsh spectrum by using $f = H_n W(f)$, and $W(f) = 2^{-n} H_n f$. Recall from Section 2.9 that the Sylvester-Hadamard matrix H_n is defined by $H_1 = \begin{pmatrix} 1 & 1 \\ 1 & -1 \end{pmatrix}$ and $H_n = \begin{pmatrix} H_{n-1} & H_{n-1} \\ H_{n-1} & -H_{n-1} \end{pmatrix}$, that is, H_n is the Kronecker product $H_n = H_1 \otimes H_{n-1}$. The following result (proved by McFarland [310], cf. Dillon [151]) is Theorem 5.5 from Chapter 5. We give here yet another proof to put in evidence the eigenvalues approach.

Theorem 8.20. *If $H = H_n$ is the Sylvester-Hadamard matrix with (i, j) entries given by $(-1)^{\mathbf{v}_i \cdot \mathbf{v}_j}$, where $\mathbf{v}_i, \mathbf{v}_j$ are the vectors of $\mathbb{V}_n$, then*

$$H \cdot A_f \cdot H^t = 2^n D,$$

where D is the diagonal matrix formed by the eigenvalues of A_f.

Proof. Since $H \cdot H^t = 2^n I_{2^n}$, it suffices to show that $HA_f = DH$. Now, for $H = (h_{i,j})$ and $A_f = (a_{i,j})$, the left-hand side is

$$\begin{aligned}
(HA_f)_{i,j} &= \sum_{l=1}^{2^n} h_{i,l} a_{l,j} = \sum_{l=1}^{2^n} (-1)^{\mathbf{v}_i \cdot \mathbf{v}_l} f(\mathbf{v}_l \oplus \mathbf{v}_j) \\
&= \sum_{l=1}^{2^n} (-1)^{\mathbf{v}_i \cdot (\mathbf{v}_l \oplus \mathbf{v}_j) + \mathbf{v}_i \cdot \mathbf{v}_j} f(\mathbf{v}_l \oplus \mathbf{v}_j) \\
&= (-1)^{\mathbf{v}_i \cdot \mathbf{v}_j} \sum_{\mathbf{x} \in \mathbb{V}_n} (-1)^{\mathbf{v}_i \cdot \mathbf{x}} f(\mathbf{x}) \\
&= (-1)^{\mathbf{v}_i \cdot \mathbf{v}_j} W(f)(\mathbf{v}_i) = (-1)^{\mathbf{v}_i \cdot \mathbf{v}_j} \lambda_i.
\end{aligned}$$

■

Let f be a Boolean function on $\mathbb{V}_n$ and assume that $f(\mathbf{0}) = 0$. Moreover, assume that Γ_f is connected. Bernasconi and Codenotti [29] proved the next result.

Theorem 8.21. *The graph Γ_f is bipartite if and only if $\mathbb{V}_n \setminus \Omega_f$ contains a subspace of dimension $n - 1$.*

Let now S_0 be a subspace of dimension $n-1$ with basis $\{\alpha^{(1)}, \alpha^{(2)}, \ldots, \alpha^{(n-1)}\}$. Complete the previous basis with $\alpha^{(n)}$ to get a basis for $\mathbb{V}_n$. Let $\mathbf{b}$ be the unique solution in $\mathbb{V}_n$ of the system:

$$\begin{pmatrix} \alpha_1^{(1)} & \alpha_2^{(1)} & \dots & \alpha_n^{(1)} \\ \alpha_1^{(2)} & \alpha_2^{(2)} & \dots & \alpha_n^{(2)} \\ \vdots & \vdots & \dots & \vdots \\ \alpha_1^{(n)} & \alpha_2^{(n)} & \dots & \alpha_n^{(n)} \end{pmatrix} \begin{pmatrix} b_1 \\ b_2 \\ \vdots \\ b_n \end{pmatrix} = \begin{pmatrix} 0 \\ 0 \\ \vdots \\ 1 \end{pmatrix}. \tag{8.10}$$

Let $\mathbf{w} \in \mathbb{V}_n$. Since f is 0 on S_0, we have

$$W(f)(\mathbf{w} \oplus \mathbf{b}) = \sum_{\mathbf{x} \in \mathbb{V}_n} (-1)^{\mathbf{x} \cdot (\mathbf{w} \oplus \mathbf{b})} f(\mathbf{x}) = \sum_{\mathbf{x} \in \mathbb{V}_n \setminus S_0} (-1)^{\mathbf{x} \cdot \mathbf{w}} (-1)^{\mathbf{x} \cdot \mathbf{b}} f(\mathbf{x}).$$

Furthermore, since $\mathbf{b}$ is the solution to (8.10) and $\mathbf{x} \notin S_0$ is a linear combination of the vectors $\alpha^{(1)}, \dots, \alpha^{(n)}$ (with $\alpha^{(n)}$ always present), we get $(-1)^{\mathbf{x} \cdot \mathbf{b}} = -1$, and so, the following result is shown [29].

Lemma 8.22. *If* $\mathbf{b}$ *is given by* (8.10)), *then* $W(f)(\mathbf{w}) = -W(f)(\mathbf{w} \oplus \mathbf{b})$, *for any* $\mathbf{w} \in \mathbb{V}_n$.

Further, Bernasconi and Codenotti proved the following theorem [29] that describes the propagation features of f for all vectors with a specific property.

Theorem 8.23. *Let* $f : \mathbb{V}_n \to \mathbb{F}_2$ *be a Boolean function whose associated graph is bipartite, and let* $\mathbf{b} \in \mathbb{V}_n$ *be given by* (8.10). *If* $|\Omega_f| = 2^{n-2}$, *then* f *satisfies the PC with respect to all strings* $\mathbf{w}$ *such that* $\mathbf{w} \cdot \mathbf{b}$ *is an odd integer. If* $|\mathbf{b}| = n$, *then* f *satisfies the SAC.*

The previous theorem seems to be quite restrictive. Next, we state a result proved by Stănică [436], which connects the PC property with the symmetric difference in counting vertices of Γ_f. Let $\mathcal{N}(\mathbf{x})$ be the set of vertices adjacent to a vertex $\mathbf{x}$ in the graph Γ_f.

Theorem 8.24. *Let* $f : \mathbb{V}_n \to \mathbb{F}_2$ *be a Boolean function. Then the following statements are equivalent:*

1. f *satisfies the PC with respect to* $\mathbf{w}$;
2. $|\mathcal{N}(0) \setminus \mathcal{N}(\mathbf{w})| + |\mathcal{N}(\mathbf{w}) \setminus \mathcal{N}(0)| = 2^{n-1}$;
3. $\sum_{\mathbf{x} \in \mathbb{V}_n} (-1)^{\mathbf{x} \cdot \mathbf{w}} \lambda_{\mathbf{x}}^2 = 2^n \lambda_0 - 2^{2n-2} = 2^n wt(f) - 2^{2n-2}$.

Proof. It is easy to see that f is PC with respect to $\mathbf{w}$ if and only if the autocorrelation function

$$\begin{aligned} \hat{r}_{\hat{f}}(\mathbf{w}) &= \sum_{\mathbf{v} \in \mathbb{V}_n} (-1)^{f(\mathbf{v}) \oplus f(\mathbf{v} \oplus \mathbf{w})} \\ &= \sum_{\mathbf{v} \in \Omega_f} (-1)^{f(\mathbf{v}) \oplus f(\mathbf{v} \oplus \mathbf{w})} + \sum_{\mathbf{v} \notin \Omega_f} (-1)^{f(\mathbf{v}) \oplus f(\mathbf{v} \oplus \mathbf{w})} \\ &= \sum_{\mathbf{v} \in \Omega_f} (-1)^{1 \oplus f(\mathbf{v} \oplus \mathbf{w})} + \sum_{\mathbf{v} \notin \Omega_f} (-1)^{f(\mathbf{v} \oplus \mathbf{w})} \end{aligned}$$

$$= \sum_{\mathbf{v}\in\Omega_f\cap\mathcal{N}(\mathbf{w})} 1 + \sum_{\mathbf{v}\in\Omega_f\cap\overline{\mathcal{N}(\mathbf{w})}} (-1)$$
$$+ \sum_{\mathbf{v}\in\overline{\Omega_f}\cap\mathcal{N}(\mathbf{w})} (-1) + \sum_{\mathbf{v}\in\overline{\Omega_f}\cap\overline{\mathcal{N}(\mathbf{w})}} 1 = 0.$$

Thus, $|(\mathcal{N}(0)\cap\mathcal{N}(\mathbf{w}))\cup(\overline{\mathcal{N}(0)}\cap\overline{\mathcal{N}(\mathbf{w})})| = |(\mathcal{N}(0)\cap\overline{\mathcal{N}(\mathbf{w})})\cup(\overline{\mathcal{N}(0)}\cap\mathcal{N}(\mathbf{w}))|$. Further, using the inclusion–exclusion principle, the previous identity is equivalent to

$$|\mathcal{N}(0)\cap\mathcal{N}(\mathbf{w})| + |\overline{\mathcal{N}(0)\cup\mathcal{N}(\mathbf{w})}| = |\mathcal{N}(0)\cap\overline{\mathcal{N}(\mathbf{w})}| + |\overline{\mathcal{N}(0)\cup\overline{\mathcal{N}(\mathbf{w})}}| \iff$$
$$|\mathcal{N}(0)\cap\mathcal{N}(\mathbf{w})| + 2^n - |\mathcal{N}(0)\cup\mathcal{N}(\mathbf{w})| = |\mathcal{N}(0)\cap\overline{\mathcal{N}(\mathbf{w})}| + 2^n - |\mathcal{N}(0)\cup\overline{\mathcal{N}(\mathbf{w})}| \iff$$
$$|\mathcal{N}(0)\cup\mathcal{N}(\mathbf{w})| - |\mathcal{N}(0)\cap\mathcal{N}(\mathbf{w})| = |\mathcal{N}(0)\cup\overline{\mathcal{N}(\mathbf{w})}| - |\mathcal{N}(0)\cap\overline{\mathcal{N}(\mathbf{w})}| \iff$$
$$|\mathcal{N}(0)\setminus\mathcal{N}(\mathbf{w})| + |\mathcal{N}(\mathbf{w})\setminus\mathcal{N}(0)| = 2^n - |\mathcal{N}(0)\setminus\mathcal{N}(\mathbf{w})| - |\mathcal{N}(\mathbf{w})\setminus\mathcal{N}(0)|,$$

which proves the first claim.

Now, using Theorem 2.8, namely $W(\hat{r})(\mathbf{w}) = W(\hat{f})^2(\mathbf{w})$, Equation (8.9) and the autocorrelation definition one can deduce (see also [172]) that f satisfies the PC with respect to $\mathbf{w}$ if and only if

$$\sum_{\mathbf{w}\in\mathbb{V}_n}(-1)^{\mathbf{w}\cdot\mathbf{x}}W(\hat{f})^2(\mathbf{w}) = 0 \iff$$
$$\sum_{\mathbf{w}\in\mathbb{V}_n}(-1)^{\mathbf{w}\cdot\mathbf{x}}W(f)^2(\mathbf{w}) = 2^n W(f)(0) - 2^{2n-2}.$$

Since $W(f)(0)$ is equal to the number of ones in the truth table of f, that is, the weight of f, which is the eigenvalue corresponding to $(0,0,\ldots,0)$, we get the last claim. ■

8.7 SENSITIVITY OF HAMMING WEIGHT OF f TO $SPEC(\Gamma_f)$

We know that a strongly regular Cayley graph Γ_f with the extra condition $e = d$ corresponds to a bent Boolean function f. Is there any influence of arbitrary Cayley graph spectra on the weight (or nonlinearity) of f? We can only prove the following theorem and its corollary in this direction.

Theorem 8.25. *If Γ_f is connected and the spectrum of Γ_f, $Spec(\Gamma_f)$ contains exactly $m+1$ distinct eigenvalues ($m \le n/2$), then*

$$n \le \log_2\left(r + \binom{r}{m}\right),$$

where $r = wt(f)$.

Proof. We follow [436]. We know that if $|Spec(\Gamma_f)| = m+1$, then the diameter of Γ_f, $diam(\Gamma_f) \le m$ (cf. [132, Theorem 3.13, p. 88]). Thus, for any $\mathbf{w} \in \mathbb{V}_n \setminus \Omega_f$, there is a constant number of strings $\mathbf{w}^{(i)} \in \Omega_f$ such that $\mathbf{w} = \sum_i \mathbf{w}^{(i)}$. The number of such strings is less than or equal to m, say p. It follows that writing $\mathbf{w} = \sum_{j=1}^{r} c_j \mathbf{w}^{(j)}$, $c_j \in \mathbb{F}_2$, exactly p coefficients are nonzero. Thus, the number of elements of $\mathbb{V}_n \setminus \Omega_f$ is less than or equal to the number of ways of choosing p nonzero coefficients out of r. Thus, $2^n - r \leq \binom{r}{p} \leq \binom{r}{m}$ (since $m \leq n/2$). The result follows easily. ■

Corollary 8.26. *If the Cayley graph associated to a Boolean function f is connected and strongly regular, then* $wt(f) \geq \frac{-1+\sqrt{2^{n+3}+1}}{2}$.

Proof. If Γ_f is connected and strongly regular, then the number of distinct eigenvalues is $m = 3$. Therefore, $diam(\Gamma_f) \leq 2$. If $diam(\Gamma_f) = 1$, then Γ_f is complete, but then we would have only two distinct eigenvalues. So $diam(\Gamma_f) = 2$. Therefore, any $\mathbf{w} \in \mathbb{V}_n \setminus \Omega_f$ can be written as a sum of two elements in Ω_f. Writing, as before, $\mathbf{w} = \sum_{j=1}^{r} \mathbf{w}^{(j)}$, it follows that exactly two coefficients are nonzero. Therefore,

$$2^n - r \leq \binom{r}{2} \Longleftrightarrow r(r+1) \geq 2^{n+1},$$

which is equivalent to

$$r \geq \frac{-1+\sqrt{2^{n+3}+1}}{2},$$

thus proving the corollary. ■

8.8 BOOLEAN CAYLEY GRAPHS UNDER AFFINE TRANSFORMATIONS

We already referred to equivalent Boolean functions in Chapter 5, that is, functions that are equivalent under a set of affine transformations.

One could ask how the Cayley graph compares (or distinguishes) among Boolean functions in the same equivalence class. Bernasconi and Codenotti started that investigation [28] by displaying the Cayley graphs associated to each equivalence class representative of Boolean functions in 4 variables: obviously, there are $2^{2^4} = 65,536$ different Boolean functions in 4 variables, and the number of equivalence classes in four variables under affine transformations is only 8 (eight). We display the truth table and the Walsh spectrum of a representative of each class in Table 8.1 [28].

We note the structures of the Cayley graphs associated to the Boolean function representatives of the eight equivalence classes (under affine transformation) (we preserve the same configuration for the Cayley graphs as in [28]) from Table 8.1. The Cayley graph associated to the representative of the first equivalence class has only one eigenvalue, and is a totally disconnected graph (see Figure 8.1). The Cayley graph associated to the

Table 8.1 Truth table and Walsh spectrum of equivalence class representatives for Boolean functions in 4 variables under affine transformations

No.	Boolean representative	Walsh spectrum															
1	0000000000000000	0	0	0	0	0	0	0	0	0	0	0	0	0	0	0	0
2	00000000 00000001	1	−1	−1	1	−1	1	1	−1	−1	1	1	−1	1	−1	−1	1
3	0000000000000011	2	0	−2	0	−2	0	2	0	−2	0	2	0	2	0	−2	0
4	0000000000000111	3	−1	−1	−1	−3	1	1	1	−3	1	1	1	3	−1	−1	−1
5	0000000000001111	4	0	0	0	−4	0	0	0	−4	0	0	0	4	0	0	0
6	0000000000010111	4	−2	−2	0	−2	0	0	2	−4	2	2	0	2	0	0	−2
7	0000000100010111	5	−3	−3	1	−3	1	1	1	−3	1	1	1	1	1	1	−3
8	0000001101011001	6	−2	−2	2	−2	−2	2	−2	−2	2	−2	−2	−2	2	2	2

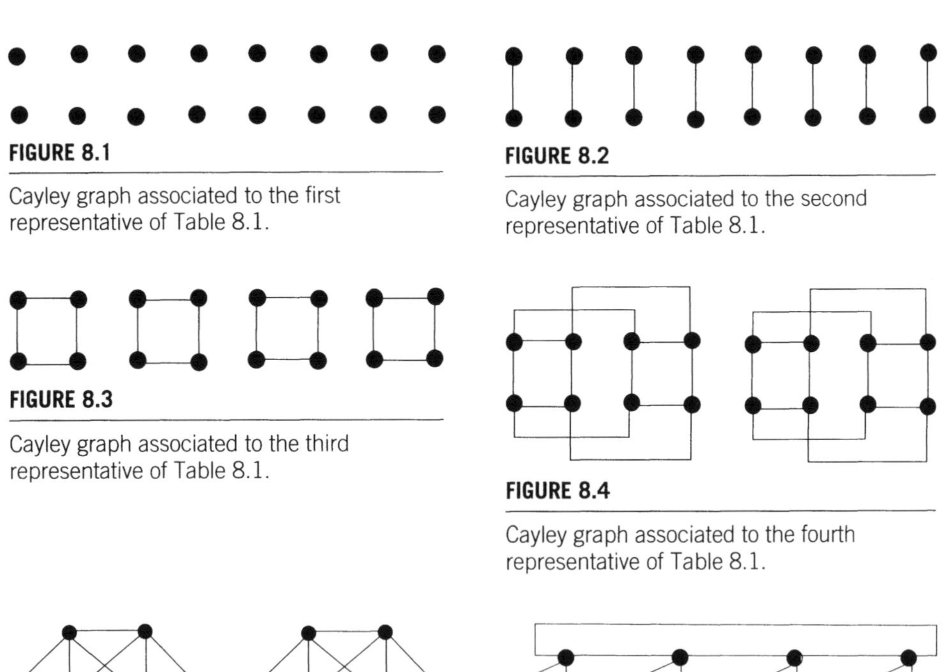

FIGURE 8.1

Cayley graph associated to the first representative of Table 8.1.

FIGURE 8.2

Cayley graph associated to the second representative of Table 8.1.

FIGURE 8.3

Cayley graph associated to the third representative of Table 8.1.

FIGURE 8.4

Cayley graph associated to the fourth representative of Table 8.1.

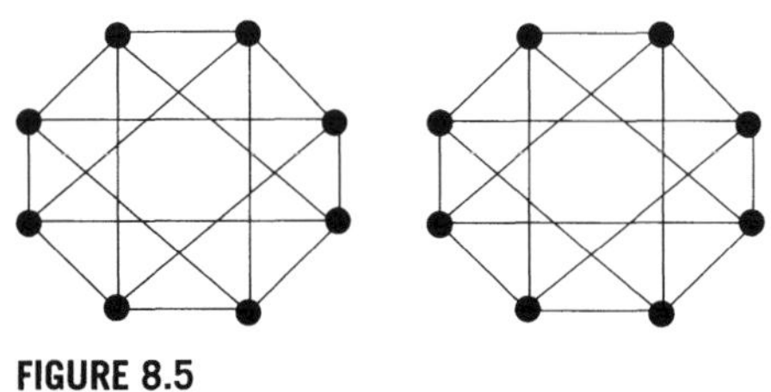

FIGURE 8.5

Cayley graph associated to the fifth representative of Table 8.1.

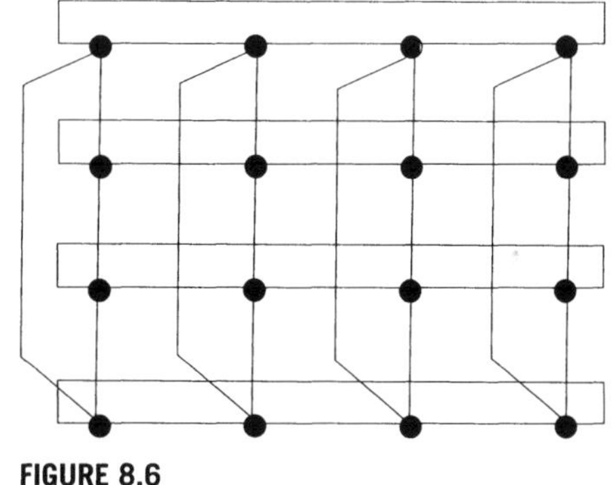

FIGURE 8.6

Cayley graph associated to the sixth representative of Table 8.1.

representative of the second equivalence class has two distinct spectral coefficients and its associated graph is a pairing, that is, a set of edges without common vertices (see Figure 8.2). The Cayley graph associated to the representative of the third equivalence class has four connected components and three distinct eigenvalues, one equal to 0 and two symmetric with respect to 0. Theorem 8.8 implies that each connected component is a complete bipartite graph (see Figure 8.3). The Cayley graph associated to the representative of the fourth equivalence class has two connected components, each corresponding to a three-dimensional cube (see Figure 8.4). The Cayley graph associated to the representative of the fifth equivalence class has two connected components and three distinct eigenvalues as for the third equivalence class, and so each connected component is a complete bipartite graph (see Figure 8.5). The Cayley graph associated

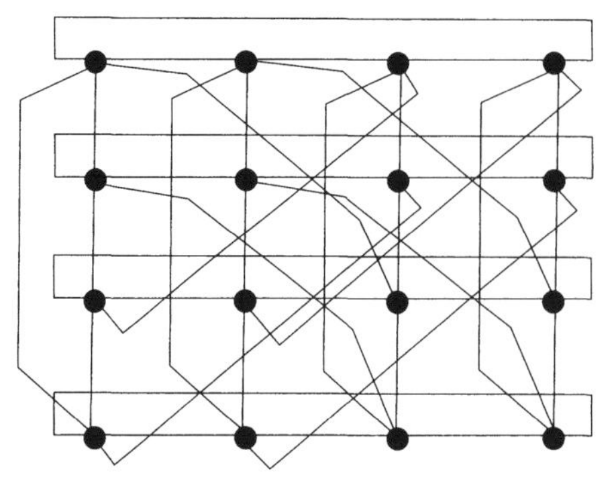

FIGURE 8.7

Cayley graph associated to the seventh representative of Table 8.1.

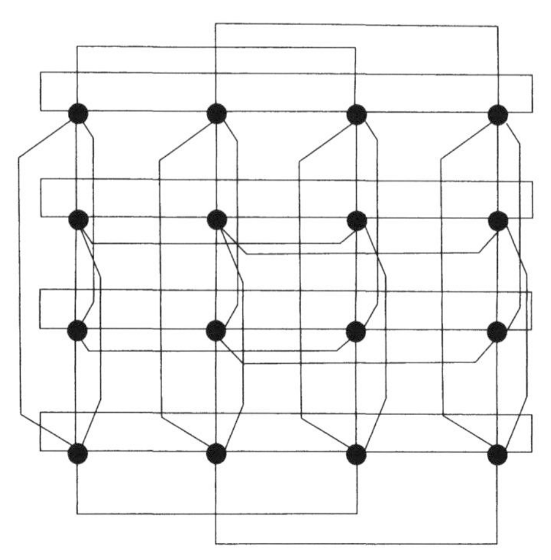

FIGURE 8.8

Cayley graph associated to the eighth representative of Table 8.1.

to the representative of the sixth equivalence class is a connected graph, with five distinct eigenvalues (see Figure 8.6). The Cayley graph associated to the representative of the seventh equivalence class has only three distinct eigenvalues and, therefore, is strongly regular (see Figure 8.7). Interestingly enough, the Cayley graph associated to the representative (which is a bent function) of the eighth equivalence class is strongly regular, with parameters $e = d = 2$ (see Figure 8.8).

Much remains to be done in this area. One could ask for indicators of a Boolean function f that are more sensitive to $Spec(\Gamma_f)$. Perhaps a collaboration between experts in the areas of cryptographic Boolean functions and graph theory might shed further light on these questions.

References

[1] Adams CM, Tavares SE. The structured design of cryptographically good *S*-boxes. J Cryptology 1990;3:27–41.

[2] Adams CM, Tavares SE. Generation and counting of bent sequences. IEEE Trans Inform Theory 1990;5:1170–3.

[3] Adams CM, Tavares SE. Generating bent sequences. Discrete Appl Math 1992;39:155–9.

[4] Advanced Encryption Standard. Federal Information Standard 197, http://csrc.nist.gov/publications/fips/fips197/fips-197.pdf; 2001.

[5] Advanced Encryption Standard. Third conference, attendee feedback form, http://csrc.nist.gov/encryption/aes/round2/conf3/AES3FeedbackForm-summary.pdf

[6] Agievich S. On the representation of bent functions by bent rectangles. In: Probabilistic methods in discrete mathematics. Proc. of the fifth international Petrozavodsk conference. p. 121–35; 2002.

[7] Alexi W, Chor B, Goldreich O, Schnorr C. RSA and Rabin functions: certain parts are as hard as the whole. SIAM J Comput 1988;17:194–209.

[8] Anderson RJ. Solving a class of stream ciphers. Cryptologia 1990;14:285–8.

[9] Anderson RJ. Faster attack on certain stream ciphers. Electr Lett 1993;29:1322–3.

[10] Anderson R. Searching for the optimum correlation attack. In: Fast software encryption–Leuven 1994. LNCS, vol. 1008. Berlin: Springer; 1995. p. 137–43.

[11] Anderson R. On Fibonacci keystream generators. In: Fast software encryption–Leuven 1994. LNCS, vol. 1008. Berlin: Springer; 1995. p. 346–52.

[12] Armknecht F. Improving fast algebraic attacks. In: Fast software encryption–Delhi 2004. LNCS, vol. 3017. Berlin: Springer; 2004. p. 65–82.

[13] Armknecht F, Carlet C, Gaborit P, Kuenzli S, Meier W, Ruatta O. Efficient computation of algebraic immunity for algebraic and fast algebraic attacks. In: Adv. in crypt.–Eurocrypt 2006. LNCS, vol. 4004. Berlin: Springer; 2006. p. 147–64.

[14] Asratian AS, Denley TMJ, Häggkvist R. Bipartite graphs and their applications. Cambridge University Press; 1998.

[15] Augot D, Biryukov A, Canteaut A, Cid C, Courtois N, De Cannière C et al. D.STVL.2-AES security report. ECRYPT report, http://www.ecrypt.eu.org/documents/D.STVL.2-1.0.pdf; January 2006.

[16] Babbage S. On the relevance of the strict avalanche criterion. Electr Lett 1990;26(7):461–2.

[17] Bakhtiari S, Safavi-Naini R, Pieprzyk JP. Cryptographic hash functions: a survey, Preprint 95-9. Department of Computer Science, The University of Wollongong; 1995.

[18] Bauer C, Gottlöb EJ. Results of an automated attack on the running key cipher. Cryptologia 2005;29:248–54.

[19] Beale M, Monaghan MF. Encryption using random Boolean functions. In: Proc. IMA conference on crypt. and coding. Cirencester. Oxford University Press; 1986. p. 219–30.

[20] Beauchamp KG. Walsh functions and their applications. New York: Academic Press; 1975.

[21] Bellare M, Rogaway P. Introduction to modern cryptography (lecture notes), http://www-cse.ucsd.edu/users/mihir/cse207/classnotes.html; 2006.

[22] Bending TD. Bent functions, SDP designs and their automorphism groups. Ph.D. thesis. University of London; 1993.

[23] Bennett CH, Brassard G, Robert J-M. How to reduce your enemy's information. In: Adv. in crypt.–Crypto '85. LNCS, vol. 218. Berlin: Springer; 1986. p. 468–76.

[24] Bennett CH, Brassard G, Robert J-M. Privacy amplification by public discussion. SIAM J Comput 1988;17:210–29.

[25] Berlekamp ER. Factoring polynomials over finite fields. Bell System Tech J 1967;1853–9.
[26] Berlekamp ER. Algebraic coding theory. New York: McGraw-Hill; 1968.
[27] Berlekamp ER. Factoring polynomials over large finite fields. Math Comput 1970;24:713–35.
[28] Bernasconi A, Codenotti B. Spectral analysis of Boolean functions as a graph eigenvalue problem. IEEE Trans Comput 1999;48:345–51.
[29] Bernasconi A, Codenotti B. On Boolean functions associated to bipartite Cayley graphs. In: Workshop on Boolean problems, Freiberg; 2000, p. 167–74.
[30] Bernasconi A, Codenotti B, Simon J. On the Fourier analysis of Boolean functions, preprint, 1-24, http://www.imc.pi.cnr.it/∼codenotti/ps_files/fourier.ps; 1996.
[31] Bernasconi A, Codenotti B, VanderKam JM. A characterization of bent functions in terms of strongly regular graphs. IEEE Trans Comput 2001;50:984–5.
[32] Bernstein DJ. Which eSTREAM ciphers have been broken? manuscript; 2008.
[33] Beth T, Piper F. The stop-and-go generator. In: Adv. in crypt.–Eurocrypt '84. LNCS, vol. 209. Berlin: Springer; 1985. p. 88–92.
[34] Betke U, Wills JM. Untere Schranken für zwei diophantische Approximations-Funktionen. Monatsh Math 1972;76:214–7.
[35] Biggs N. Algebraic graph theory. 2nd ed. Cambridge University Press; 1993.
[36] Biham E, Biryukov A, Shamir A. Cryptanalysis of Skipjack reduced to 31 rounds using impossible differentials. In: Adv. in crypt.–Eurocrypt '99. LNCS, vol. 1592. Berlin: Springer; 1999. p. 12–23.
[37] Biham E, Shamir A. Differential cryptanalysis of DES-like cryptosystems. In: Adv. in crypt.–Crypto '90. LNCS, vol. 537. Berlin: Springer; 1991. p. 2–21.
[38] Biham E, Shamir A. Differential cryptanalysis of the Data Encryption Standard. Springer; 1993.
[39] Biss DK. A lower bound on the number of functions satisfying the strict avalanche criterion. Discrete Math 1998;185:29–39.
[40] Black J, Rogaway P. A suggestion for handling arbitrary-length messages with the CBC MAC, Comments to NIST concerning AES modes of operation, http://csrc.nist.gov/groups/ST/toolkit/BCM/workshops.html
[41] Blackburn SR. The linear complexity of the self-shrinking generator. IEEE Trans Inform Theory 1999;45:2073–7.
[42] Blöcher U, Dichtl M. Fish: a fast software stream cipher. In: Fast software encryption–Cambridge 1993. LNCS, vol. 809. Berlin: Springer; 1994. p. 41–4.
[43] Blum L, Blum M, Shub M. A simple unpredictable pseudo-random number generator. SIAM J Comput 1986;15:364–83.
[44] Blum M, Micali S. How to generate cryptographically strong sequences of pseudo-random bits. In: Proc. 23rd IEEE symposium on foundations of computer science; 1982, p. 112–17.
[45] Blum M, Micali S. How to generate cryptographically strong sequences of pseudo-random bits. SIAM J Comput 1984;13:850–64.
[46] Boole G. The laws of thought. Reprint of the 1854 original. With an introduction by John Corcoran. Great Books in Philosophy. Amherst, NY: Prometheus Books; 2003.
[47] Braeken A, Preneel B. On the algebraic immunity of symmetric Boolean functions. In: Indocrypt 2005. LNCS, vol. 3797. Berlin: Springer; 2005. p. 35–48.
[48] Brands S, Gill R. Cryptography, statistics and pseudo-randomness (part I). Prob Math Stat 1995; 15:101–14.
[49] Brickell EF, Denning DE, Kent ST, Mahler DP, Tuchman W. SKIPJACK review. Interim report, July 28; 1993. 8 p.
[50] Brynielsson L. Below the unicity distance. In: Workshop on stream ciphers, Karlsruhe; 1989.
[51] Buchberger B. Introduction to Groebner bases, logic of computation (Marktoberdorf, 1995). NATO Adv. Sci. Inst. Ser. F Comput. Systems Sci., vol. 157. Berlin: Springer; 1997. p. 35–66.
[52] Cameron P. Random strongly regular graphs? In: Nesetril J, Noy M, Serra O, editors. Electronic Notes in Discrete Mathematics, vol. 10. Amsterdam: Elsevier; 2001.
[53] Camion P, Carlet C, Charpin P, Sendrier N. On correlation-immune functions. In: Adv. in crypt.–Crypto '91. LNCS, vol. 576. Berlin: Springer; 1992. p. 86–100.

[54] Campbell KW, Wiener MJ. DES is not a group. In: Adv. in crypt.–Crypto '92. LNCS, vol. 740. Berlin: Springer; 1992. p. 512–20.
[55] Canright D. A very compact S-box for AES. In: Workshop on cryptographic hardware and embedded systems (CHES2005). LNCS, vol. 3659. Berlin: Springer; 2005. p. 441–55.
[56] Canright D. A very compact Rijndael S-box, Naval Postgraduate School Technical Report: NPS-MA-05-001; 2005.
[57] Canteaut A. Open problems related to algebraic attacks on stream ciphers. In: Coding and cryptography–WCC 2005. LNCS, vol. 3969. Berlin: Springer; 2006. p. 120–34.
[58] Canteaut A, Carlet C, Charpin P, Fontaine C. Propagation characteristics and correlation-immunity of highly nonlinear Boolean functions. In: Adv. in crypt.–Eurocrypt 2000. LNCS, vol. 1807. Berlin: Springer; 2000. p. 507–22.
[59] Canteaut A, Daum H, Dobbertin H, Leander G. Normal and nonnormal bent functions. In: Coding and cryptography–WCC 2003, INRIA. p. 91–100.
[60] Canteaut A, Daum H, Dobbertin H, Leander G. Finding nonnormal bent functions. Discrete Appl Math 2006;154:202–18.
[61] Canteaut A, Trabbia M. Improved fast correlation attacks using parity check equations of weight 4 and 5. In: Adv. in crypt.–Eurocrypt 2000. LNCS, vol. 1807. Berlin: Springer; 2000. p. 573–88.
[62] Cao D. Bounds on eigenvalues and chromatic numbers. Linear Alg Appl 1998;270:1–13.
[63] Carlet C. A transformation on Boolean functions, its consequences on some problems related to Reed-Muller codes. In: Adv. in crypt.–Eurocrypt '90. LNCS, vol. 473. Berlin: Springer; 1991. p. 42–50.
[64] Carlet C. Partially-bent functions. Des Codes Cryptogr 1993;3:135–45.
[65] Carlet C. Two new classes of bent functions. In: Adv. in crypt.–Eurocrypt '93. LNCS, vol. 765. Berlin: Springer; 1994. p. 77–101.
[66] Carlet C. Generalized partial spreads. IEEE Trans Inform Theory 1995;41:1482–7.
[67] Carlet C. A construction of bent functions. Finite Fields Appl 1996;233:47–58.
[68] Carlet C. Recent results on binary bent functions. J Combin Inform System Sci 2000;25:133–49.
[69] Carlet C. Private communication; 1997.
[70] Carlet C. More correlation-immune and resilient functions over Galois fields and Galois rings. In: Adv. in crypt.–Eurocrypt '97. LNCS, vol. 1233. Berlin: Springer; 1997. p. 422–33.
[71] Carlet C. Sur le critère de propagation de degré l et d'ordre k. C R Acad Sci Paris Sr I Math 1998; 326:891–6.
[72] Carlet C. On the propagation criterion of degree l and order k. In: Adv. in crypt.–Eurocrypt '98. LNCS, vol. 1403. Berlin: Springer; 1998. p. 462–74.
[73] Carlet C. On cryptographic propagation criteria for Boolean functions. Inform Comput 1999; 151:32–56.
[74] Carlet C. On the divisibility properties and nonlinearity of resilient functions. C R Acad Sci Paris Sr I Math 2000;331:917–22.
[75] Carlet C. A larger class of cryptographic Boolean functions via a study of the Maiorana–McFarland construction. In: Adv. in crypt.–Crypto 2002. LNCS, vol. 2442. Berlin: Springer; 2002. p. 549–64.
[76] Carlet C. On the coset weight divisibility and nonlinearity of resilient and correlation-immune functions. In: Sequences and their applications (Bergen, 2001). Discrete math. theor. comput. sci. London: Springer; 2002. p. 131–44.
[77] Carlet C. On the complexity of cryptographic Boolean functions. In: Finite fields with applications to coding theory, cryptography and related areas, Oaxaca, 2001. Berlin: Springer; 2002. p. 53–69.
[78] Carlet C. On the degree, nonlinearity, algebraic thickness, and nonnormality of Boolean functions, with developments on symmetric functions. IEEE Trans Inform Theory 2004;50:2178–85.
[79] Carlet C. On the secondary constructions of resilient and bent functions. In: Coding, cryptography and combinatorics. Progr. Comput. Sci. Appl. Logic, vol. 2. Basel: Birkhäuser; 2004. p. 3–28.

[80] Carlet C. On the confusion and diffusion properties of Maiorana–McFarland's and extended Maiorana–McFarland's functions. J Complexity 2004;20:182–204.
[81] Carlet C. Improving the algebraic immunity of resilient and nonlinear functions and constructing bent functions, http://eprint.iacr.org/2004/276
[82] Carlet C. Boolean functions for cryptography and error correcting codes. Available at www.rocg.inria.fr/secret/Claude.Carlet/pubs.html.
[83] Carlet C, Charpin P, Zinoviev V. Codes bent functions and permutations suitable for DES-like cryptosystems. Des Codes Cryptogr 1998;15:125–56.
[84] Carlet C, Dalai DK, Gupta KC, Maitra S. Algebraic immunity for cryptographically significant Boolean functions: analysis and construction. IEEE Trans Inform Theory 2006;52:3105–21.
[85] Carlet C, Ding C. Highly nonlinear mappings. J Complexity 2004;20:205–44.
[86] Carlet C, Dobbertin H, Leander G. Normal extensions of bent functions. IEEE Trans Inform Theory 2004;50:2880–5.
[87] Carlet C, Dubuc S. On generalized bent and q-ary perfect nonlinear functions. In: Finite fields and applications, Augsburg, 1999. Berlin: Springer; 2001. p. 81–94.
[88] Carlet C, Gouget A. An upper bound on the number of m-resilient Boolean functions. In: Adv. in crypt.-Asiacrypt 2002. LNCS, vol. 2501. Berlin: Springer; 2002. p. 484–96.
[89] Carlet C, Guillot P. A characterization of binary bent functions. J Combin Theory A 1996;76: 328–35.
[90] Carlet C, Guillot P. An alternate characterization of the bentness of binary functions with uniqueness. Des Codes Cryptogr 1998;14:133–40.
[91] Carlet C, Guillot P. Bent, resilient functions and the numerical normal form. In: Codes and association schemes (Piscataway, NJ, 1999). DIMACS ser. discrete math. theoret. comput. sci., vol. 56. Providence, RI: Amer. Math. Soc.; 2001. p. 87–96.
[92] Carlet C, Klapper A. Upper bounds on the number of resilient functions and of bent functions. In: Proc. 23rd symposium on inform. theory in the Benelux; 2002. p. 307–14.
[93] Carlet C, Prouff E. On a new notion of nonlinearity relevant to multioutput pseudo-random generators. In: Selected areas crypt. 2003. LNCS, vol. 3006. Berlin: Springer; 2004. p. 291–305.
[94] Carlet C, Prouff E. Vectorial functions and covering sequences. In: Finite fields and applications. LNCS, vol. 2948. Berlin: Springer; 2004. p. 215–48.
[95] Carlet C, Sarkar P. Spectral domain analysis of correlation immune and resilient Boolean functions. Finite Fields Appl 2002;8:120–30.
[96] Carlet C, Tarannikov Y. Covering sequences of Boolean functions and their cryptographic significance. Des Codes Cryptogr 2002;25:263–79.
[97] Chabaud F, Vaudenay S. Links between differential and linear cryptanalysis. In: Adv. in crypt.-Eurocrypt '94. LNCS, vol. 950. Berlin: Springer; 1995. p. 363–74.
[98] Chambers WG. Clock-controlled shift registers in binary sequence generators. IEE Proc E 1988; 135:17–24.
[99] Chambers WG, Gollmann D. Lock-in effect in cascades of clock controlled shift-registers. In: Adv. in crypt.-Eurocrypt '88. LNCS, vol. 330. Berlin: Springer; 1988. p. 331–43.
[100] Chambers WG, Jennings SM. Linear equivalence of certain BRM shift-register sequences. Electr Lett 1984;20:1018–9.
[101] Charnes C, Pieprzyk JP. Linear nonequivalence versus nonlinearity. In: Adv. in crypt.-Auscrypt '92. LNCS, vol. 718. Berlin: Springer; 1993. p. 156–64.
[102] Charnes C, Rötteler M, Beth T. On homogeneous bent functions. In: Proceedings of applied algebra, algebraic algorithms and error correcting codes (AAECC-14). LNCS, vol. 2227. Berlin: Springer; 2001. p. 249–59.
[103] Charnes C, Rötteler M, Beth T. Homogeneous bent functions, invariants and designs. Des Codes Cryptogr 2002;26:139–54.
[104] Charpin P. Normal Boolean functions. J Complexity 2004;20:245–65.
[105] Charpin P, Pasalic E. On propagation characteristics of resilient functions. In: Selected areas crypt. 2002. LNCS, vol. 2595. Berlin: Springer; 2003. p. 175–95.

[106] Chee S, Lee S, Kim K. Semi-bent functions. In: Adv. in crypt.–Asiacrypt '94. LNCS, vol. 917. Berlin: Springer; 1995. p. 107–18.
[107] Chee S, Lee S, Kim K, Kim D. Correlation immune functions with controllable nonlinearity. ETRI J 1997;19:389–401.
[108] Chee S, Lee S, Lee D, Sung SH. On the correlation immune functions and their nonlinearity. In: Adv. in crypt.–Asiacrypt '96. LNCS, vol. 1123. Berlin: Springer; 1996. p. 232–43.
[109] Chepyzhov V, Johansson T, Smeets B. A simple algorithm for fast correlation attacks on stream ciphers. In: Fast software encryption–New York 2000. LNCS, vol. 1978. Berlin: Springer; 2001. p. 181–95.
[110] Chepyzhov V, Smeets B. On a fast correlation attack on certain stream ciphers. In: Adv. in crypt.–Eurocrypt '91. LNCS, vol. 547. Berlin: Springer; 1992. p. 176–85.
[111] Chor B, Goldreich O, Håstad J, Friedman J, Rudich S, Smolensky R. *The bit extraction problem or t-resilient functions*. In: Proc. 26th IEEE symposium on foundations of computer science; 1985, p. 396–407.
[112] Chowla S, Herstein IN, Moore K. On recursions connected with symmetric groups I. Canadian J Math 1951;3:328–34.
[113] Cid C, Murphy S, Robshaw M. An algebraic framework for cipher embeddings. In: Proc. of 10th IMA international conference on coding and cryptography. LNCS, vol. 3796. Berlin: Springer; 2005. p. 278–89.
[114] Cid C, Murphy S, Robshaw M. Algebraic aspects of the Advanced Encryption Standard. Berlin: Springer; 2006.
[115] Clark A, Golić JD, Dawson E. A comparison of fast correlation attacks. In: Fast software encryption - Cambridge 1996. LNCS, vol. 1039. Berlin: Springer; 1997. p. 145–57.
[116] Clark JA, Jacob JL, Maitra S, Stănică P. Almost Boolean functions: the design of Boolean functions by spectral inversion. Comput Intelligence 2004;20:450–62.
[117] Colbourn CJ, Dinitz J, editors. CRC handbook of combinatorial design. Boca Raton: CRC Press; 1996.
[118] Coppersmith D. The Data Encryption Standard (DES) and its strength against attacks. IBM J Res Develop 1994;38:243–50.
[119] Coppersmith D, Krawczyk H, Mansour Y. The shrinking generator. In: Adv. in crypt.–Crypto '93. LNCS, vol. 773. Berlin: Springer; 1994. p. 22–39.
[120] Coppersmith D, Winograd S. Matrix multiplication via arithmetic progressions. J Symbolic Comput 1990;9:251–80.
[121] Courtois N. Higher order correlation attacks, XL algorithm and cryptanalysis of Toyocrypt, http://eprint.iacr.org/2002/087
[122] Courtois N, Bard G. Algebraic cryptanalysis of the Data Encryption Standard, http://eprint.iacr.org/2006/402
[123] Courtois N, Meier W. Algebraic attacks on stream ciphers with linear feedback. In: Adv. in crypt.–Eurocrypt 2003. LNCS, vol. 2656. Berlin: Springer; 2003. p. 345–59.
[124] Courtois N, Pieprzyk JP. Cryptanalysis of block ciphers with overdefined systems of equations. In: Adv. in crypt.–Asiacrypt 2002. LNCS, vol. 2501. Berlin: Springer; 2002. p. 267–87.
[125] Courtois N, Meier W. Algebraic attacks on stream ciphers with linear feedback (extended version of [123]), http://www.cryptosystem.net/stream/
[126] Cusick TW. Boolean functions satisfying a higher order strict avalanche criterion. In: Adv. in crypt.–Eurocrypt '93. LNCS, vol. 765. Berlin: Springer; 1994. p. 102–17.
[127] Cusick TW. Bounds on the number of functions satisfying the strict avalanche criterion. Inform Process Lett 1996;57:261–3.
[128] Cusick TW. On constructing balanced correlation immune functions. In: Sequences and their applications (Singapore, 1998). Springer ser. discrete math. theor. comput. sci. London: Springer; 1999. p. 184–90.
[129] Cusick TW, Ding C, Renvall A. Stream ciphers and number theory. North-Holland mathematical library, vol. 55. Amsterdam: North-Holland/Elsevier; 1998.

[130] Cusick TW, Stănică P. Bounds on the number of functions satisfying the strict avalanche criterion. Inform Process Lett 1996;60:215–9.
[131] Cusick TW, Stănică P. Fast evaluation, weights and nonlinearity of rotation-symmetric functions. Discrete Math 2002;258:289–301.
[132] Cvetkovic DM, Doob M, Sachs H. Spectra of graphs. Academic Press; 1979.
[133] Daemen J, Rijmen V. The Design of Rijndael. AES–the Advanced Encryption Standard. In: Information security and cryptography. Berlin: Springer; 2002.
[134] Dalai DK, Maitra S, Sarkar P. Basic theory in construction of Boolean functions with maximum possible annihilator immunity. Des Codes Cryptogr 2006;40:41–58.
[135] Dalai DK, Maitra S. Reducing the number of homogeneous linear equations in finding annihilators, http://eprint.iacr.org/2006/032
[136] Data Encryption Standard. National Bureau of Standards, Federal Information Standard 46, USA; 1977.
[137] Daum M, Dobbertin H, Leander G. An algorithm for checking normality of Boolean functions. Coding and cryptography–WCC 2003, INRIA, p. 133–42.
[138] Davidoff G, Sarnak P, Valette A. Elementary number theory, group theory and Ramanujan graphs. Cambridge University Press; 2003.
[139] Davio M, Desmedt Y, Quisquater J-J. Propagation characteristics of the DES. In: Adv. in crypt.-Eurocrypt '84. LNCS, vol. 209. Berlin: Springer; 1985. p. 62–73.
[140] Davis JA, Jedwab J. A survey of Hadamard difference sets. In: Groups, difference sets and the monster. Berlin–New York: de Gruyter; 1996. p. 145–56.
[141] Dawson E. Cryptanalysis of summation generator. In: Adv. in crypt.–Auscrypt '92. LNCS, vol. 718. Berlin: Springer; 1993. p. 209–15.
[142] Dawson E, Clark A. Divide and conquer attacks on certain classes of stream ciphers. Cryptologia 1994;18:25–40.
[143] Dawson E, Nielsen L. Automated cryptanalysis of XOR plaintext strings. Cryptologia 1996;20: 165–81.
[144] de Bruijn NG. A combinatorial problem. Nederl Akad Wetensch Proc 1946;49:758–64; or Indagationes Math 1946;8:461–7.
[145] de Cannière C, Biryukov A, Preneel B. An introduction to block cipher cryptanalysis. Proc IEEE 2006;94:346–56.
[146] de Cannière C, Preneel B. Trivium specifications. http://www.ecrypt.eu.org/stream/p3ciphers/trivium/trivium_p3.pdf
[147] Delsarte P. Four fundamental parameters of a code and their combinatorial significance. Inform Control 1973;23:407–38.
[148] Denisov OV. An asymptotic formula for the number of correlation immune of order k Boolean functions. Discrete Math Appl 1992;2:407–26 [translation of Russian article in Diskretnaya Matematika 3 (1991), 25–46].
[149] Denisov OV. A local limit theorem for the distribution of a part of the spectrum of a random binary function. Discrete Math Appl 2000;12:87–101 [translation of Russian article in Diskretnaya Matematika 12 (2000), 82–95].
[150] Diffie W, Hellman M. New directions in cryptography. IEEE Trans Inform Theory 1976;22: 644–54.
[151] Dillon JF. A survey of bent functions. NSA Technical Journal – unclassified 1972;191–215.
[152] Dillon JF. Elementary Hadamard difference sets. Ph.D. thesis. University of Maryland; 1974.
[153] Dillon JF. Elementary Hadamard difference sets. In: 6th Southeastern conf. on combinatorics, graph theory & computing. 1975. p. 237–49.
[154] Dillon JF. Private communication; 1997.
[155] Dillon JF, Dobbertin H. New cyclic difference sets with Singer parameters. Finite Fields Appl 2004;10:342–89.
[156] Ding C, Xiao G, Shan W. The stability theory of stream ciphers. LNCS, vol. 561. Berlin: Springer; 1991.

[157] Dobbertin H. Construction of bent functions and balanced Boolean functions with high nonlinearity. In: Fast software encryption–Leuven 1994. LNCS, vol. 1008. Berlin: Springer; 1995. p. 61–74.

[158] Dobbertin H, Leander G. A survey of some recent results on bent functions. In: Sequences and their applications–SETA 2004. LNCS, vol. 3486. Berlin: Springer; 2005. p. 1–29.

[159] Dobbertin H, Leander G. Cryptographer's toolkit for construction of 8-bit bent functions, http://eprint.iacr.org/2005/089

[160] Dubuc S. Etude des propriétés de dégénérescence et de normalité des fonctions booléennes et construction de fonctions q-aires parfaitement nonlinéaires. Ph.D. thesis. Université de Caen; 2001.

[161] eCRYPT – European Network of Excellence for Cryptology, http://www.ecrypt.eu.org/index.html

[162] Erdös P, Kleitman DJ, Rothschild BL. Asymptotic enumeration of K_n-free graphs. In: Colloquio Internazionale sulle Teorie Combinatorie (Rome, 1976), Tomo II, 19–27. Atti dei Convegni Lincei, No. 17, Accad. Naz. Lincei.

[163] Evertse J-H. Linear structures in block ciphers. In: Adv. in crypt.–Eurocrypt '87. LNCS, vol. 304. Berlin: Springer; 1988. p. 249–66.

[164] Faugère J-C. A new efficient algorithm for computing Gröbner bases ($\mathbb{F}_4$). J Pure Appl Alg 1999; 139:61–88.

[165] Fedorova M, Tarannikov Y. On the constructing of highly nonlinear resilient Boolean functions by means of special matrices. In: Indocrypt 2001. LNCS, vol. 2247. Berlin: Springer; 2001. p. 254–66.

[166] Feistel H. Cryptography and computer privacy. Scientific American 1973;228:15–23.

[167] Ferguson N, Kelsey J, Lucks S, Schneier B, Stay M, Wagner D et al. Improved cryptanalysis of Rijndael. In: Fast software encryption–New York 2000. LNCS, vol. 1978. Berlin: Springer; 2001. p. 213–30.

[168] Ferguson N, Schroeppel R, Whiting D. A simple algebraic representation of Rijndael. In: Selected areas crypt. 2001. LNCS, vol. 2259. Berlin: Springer; 2001. p. 149–65.

[169] Filiol E, Fontaine C. Highly nonlinear balanced Boolean functions with a good correlation immunity. In: Adv. in crypt.–Eurocrypt '98. LNCS, vol. 1403. Berlin: Springer; 1998. p. 475–88.

[170] FIPS 113. Computer data authentication. Federal Information Processing Standards Publication 113, National Bureau of Standards. Springfield, Virginia; 1994.

[171] Fischer J-B, Stern J. An efficient pseudo-random generator provably as secure as syndrome decoding. In: Adv. in crypt.–Eurocrypt '96. LNCS, vol. 1070. Berlin: Springer; 1996. p. 245–55.

[172] Forré R. The strict avalanche criterion: spectral properties of Boolean functions and an extended definition. In: Adv. in crypt.–Crypto '88. LNCS, vol. 403. Berlin: Springer; 1990. p. 450–68.

[173] Forré R. A fast correlation attack on nonlinearly feedforward filtered shift-register sequences. In: Adv. in crypt.–Eurocrypt '89. LNCS, vol. 434. Berlin: Springer; 1990. p. 586–95.

[174] Forré R. Methods and instruments for designing S-boxes. J Cryptology 1990;2:115–30.

[175] Fredricksen H. The lexicographically least de Bruijn cycle. J Combin Theory A 1970;9:1–5.

[176] Fredricksen H. A class of nonlinear de Bruijn cycles. J Combin Theory A 1975;19:192–9.

[177] Fredricksen H. A survey of full length nonlinear shift register cycle algorithms. SIAM Review 1982;24:195–221.

[178] Fredricksen H, Kessler IJ. An algorithm for generating necklaces of beads in two colors. Discrete Math 1986;61:181–8.

[179] Fredricksen H, Maiorana J. Necklaces of beads in k colors and k-ary de Bruijn sequences. Discrete Math 1978;23:207–10.

[180] Friedman J. On the bit extraction problem. In: Proc. 33rd IEEE symposium on foundations of computer science; 1992. p. 314–19.

[181] Gallager RG. Low-density parity-check codes. Cambridge, MA: MIT Press; 1963.

[182] Garey MR, Johnson DS. Computers and intractability: a guide to the theory of NP-completeness. San Francisco: W. H. Freeman; 1979.

[183] Geffe PR. How to protect data with ciphers that are really hard to break. Electronics 1973;46: 99–101.
[184] Gennaro R. An improved pseduo-random generator based on discrete log. In: Adv. in crypt.–Crypto 2000. LNCS, vol. 1880. Berlin: Springer; 2000. p. 469–81.
[185] Gilbert EN, Riordan J. Symmetry types of periodic sequences. Illinois J Math 1961;5:657–65.
[186] Gladman B. A specification for Rijndael, the AES algorithm, www.techheap.com/cryptography/encryption/spec.v36.pdf
[187] Gligor VD, Donescu P. Fast encryption and authentication: XCBC Encryption and XECB Authentication Modes, Submission to NIST, http://csrc.nist.gov/CryptoToolkit/modes/proposedmodes/index.html
[188] Godsil CD. Problems in algebraic combinatorics. Elec J Combin 1995; # F1.
[189] Golić JD. On the security of shift register based keystream generators. In: Fast software encryption–Cambridge 1993. LNCS, vol. 809. Berlin: Springer; 1994. p. 90–100.
[190] Golić JD. Intrinsic statistical weakness of keystream generators. In: Adv. in crypt.–Asiacrypt '94. LNCS, vol. 917. Berlin: Springer; 1995. p. 91–103.
[191] Golić JD. Linear cryptanalysis of stream ciphers. In: Fast software encryption - Leuven 1994. LNCS, vol. 1008. Berlin: Springer; 1995. p. 154–69.
[192] Golić JD. On the security of nonlinear filter generators. In: Fast software encryption–Cambridge 1996. LNCS, vol. 1039. Berlin: Springer; 1997. p. 173–88.
[193] Golić JD. Correlation analysis of the shrinking generator. In: Adv. in crypt.–Crypto 2001. LNCS, vol. 2139. Berlin: Springer; 2001. p. 440–57.
[194] Golić JD. Vectorial Boolean functions and induced algebraic equations. IEEE Trans Inform Theory 2006;52:528–37.
[195] Golić JD. Embedding probabilities for the alternating step generator. IEEE Trans Inform Theory 2005;51:2543–53.
[196] Golić JD, Menicocci R. Edit distance correlation attack on the alternating step generator. In: Adv. in crypt.–Crypto '97. LNCS, vol. 1294. Berlin: Springer; 1997. p. 499–512.
[197] Golić JD, Menicocci R. Edit probability correlation attacks on stop/go clocked keystream generators. J Cryptology 2003;16:41–68.
[198] Golić JD, Menicocci R. Correlation analysis of the alternating step generator. Des Codes Cryptogr 2004;31:51–74.
[199] Golić JD, Mihaljević MJ. A generalized correlation attack on a class of stream ciphers based on the Levenshtein distance. J Cryptology 1991;3:201–12.
[200] Golić JD, Morgari G. Vectorial fast correlation attacks, http://eprint.iacr.org/2004/247
[201] Golić JD, O'Connor L. Embedding and probabilistic correlation attacks on clock-controlled shift registers. In: Adv. in crypt.–Eurocrypt '94. LNCS, vol. 950. Berlin: Springer; 1995. p. 230–43.
[202] Golić JD, Petrović SV. A generalized correlation attack with a probabilistic constrained edit distance. In: Adv. in crypt.–Eurocrypt '92. LNCS, vol. 658. Berlin: Springer; 1993. p. 472–6.
[203] Golić JD, Petrović SV. Correlation attacks on clock-controlled shift registers in keystream generators. IEEE Trans Comput 1996;45:482–6.
[204] Golić JD, Salmasizadeh M, Dawson E. Fast correlation attacks on the summation generator. J Cryptology 2000;13:245–62.
[205] Golić JD, Salmasizadeh M, Simpson L, Dawson E. Fast correlation attacks on nonlinear filter generators. Inform Process Lett 1997;64:37–42.
[206] Gollmann D. Pseudo random properties of cascade connections of clock controlled shift registers. In: Adv. in crypt.–Eurocrypt '84. LNCS, vol. 209. Berlin: Springer; 1985. p. 93–8.
[207] Gollmann D. Correlation analysis of cascaded sequences. In: Cryptography and coding. 1986 IMA conference, Oxford: Clarendon Press; 1989. p. 289–97.
[208] Gollmann D. Transition matrices of clock-controlled shift registers. In: Cryptography and coding III. 1991 IMA conference. Oxford: Clarendon Press; 1993. p. 197–210.
[209] Gollmann D, Chambers WG. Clock-controlled shift registers: a review. IEEE J Selected Areas Communications 1989;7:525–33.

[210] Gollmann D, Chambers WG. A cryptanalysis of $step_{k,m}$ cascades. In: Adv. in crypt. Eurocrypt '89. LNCS, vol. 434. Berlin: Springer; 1990. p. 680–7.
[211] Golomb SW. Shift register sequences. San Francisco: Holden-Day; 1967; revised edition Laguna Hills, CA: Aegean Park Press; 1982.
[212] Gopalakrishnan K, Stinson DR. A short proof of the non-existence of certain cryptographic functions. J Combin Math Combin Comput 1996;20:129–37.
[213] Gordon B, Mills WH, Welch LR. Some new difference sets. Canadian J Math 1962;14:614–25.
[214] Göttfert R, Niederreiter H. On the linear complexity of products of shift-register sequences. In: Adv. in crypt.–Eurocrypt '93. LNCS, vol. 765. Berlin: Springer; 1994. p. 151–8.
[215] Günther CG. Alternating step generators controlled by de Bruijn sequences. In: Adv. in crypt.–Eurocrypt '87. LNCS, vol. 304. Berlin: Springer; 1988. p. 5–14.
[216] Gupta KC, Sarkar P. Computing partial Walsh transform from the algebraic normal form of a Boolean function, preprint, http://www.cacr.math.uwaterloo.ca/techreports/2006/cacr2006-25.ps; 2006.
[217] Haemers WH. Eigenvalue techniques in design and graph theory. Ph.D. thesis. Eindhoven University of Technology; 1979 (also found in Math. Centre Tract 121, Mathematical Centre, Amsterdam, 1980).
[218] Hall M. Combinatorial theory. Waltham, MA: Blaisdell; 1967.
[219] Hall P, Dowling G. Approximate string matching. Computing Surveys 1980;12:381–402.
[220] Hardy GH, Wright EM. An introduction to the theory of numbers. 4th ed. Oxford: Clarendon Press; 1960.
[221] Håstad J, Schrift A, Shamir A. The discrete logarithm modulo a composite hides $O(n)$ bits. J Computer System Sci 1993;47:376–404.
[222] Hauge ER, Mykkeltveit J. The analysis of de Bruijn sequences of non-extremal weight. Discrete Math 1998;189:133–47.
[223] Hedayat AS, Sloane NJA, Stufken J. Orthogonal arrays: theory and applications. Springer series in statistics. New York: Springer; 1999.
[224] Hell M, Johansson T. Two new attacks on the self-shrinking generator. IEEE Trans Inform Theory 2006;52:3837–43.
[225] Herlestam T. On functions of linear shift register sequences. In: Adv. in crypt.–Eurocrypt '85. LNCS, vol. 219. Berlin: Springer; 1986. p. 119–29.
[226] Hoffman AJ. On eigenvalues and colorings of graphs. In: Harris B, editor. Graph theory and its applications. Academic Press; 1970. p. 78–91.
[227] Hou X-D. New constructions of bent functions. J Combin Inform System Sci 2000;25:173–89.
[228] Hou X-D. On the norm and covering radius of the first-order Reed-Muller codes. IEEE Trans Inform Theory 1997;43:1025–7.
[229] Hou X-D. Cubic bent functions. Discrete Math 1998;189:149–61.
[230] Hou X-D, Langevin P. Results on bent functions. J Combin Theory A 1997;80:232–46.
[231] Huang T, You K-H. Strongly regular graphs associated with bent functions. In: Proc. 7th internat. symp. parallel architect., algo. and networks; 2004. p. 380–4.
[232] Huynh DT. A superexponential lower bound for Gröbner bases and Church-Rosser commutative Thue systems. Inform Control 1986;68:196–206.
[233] Impagliazzo R, Naor M. Efficient cryptographic schemes provably as secure as subset sum. In: Proc. 30th IEEE symposium on foundations of computer science; 1989. p. 236–41.
[234] ISO/IEC 9797. Data cryptographic techniques – data integrity mechanism using a cryptographic check function employing a block cipher algorithm; 1989.
[235] ISO/IEC 9797-1. Information technology – security techniques – data integrity, mechanism using a cryptographic check function employing a block cipher algorithm. International Organization for Standards, Geneva, Switzerland; 1999.
[236] Iwata T, Kurosawa K. OMAC: One-key CBC MAC, submitted for consideration to NIST (available at: http://csrc.nist.gov/CryptoToolkit/modes/proposedmodes/ProposedModesPage.html); 2002.

[237] Jakobsen T, Knudsen L. The interpolation attack on block ciphers. In: Fast software encryption–Haifa 1997. LNCS, vol. 1267. Berlin: Springer; 1997. p. 28–40.
[238] Jennings SM. Multiplexed sequences: some properties of the minimum polynomial. In: Proc. workshop on cryptography, Burg Feuerstein 1982. LNCS, vol. 149. Berlin: Springer; 1983. p. 189–206.
[239] Johansson T. Reduced complexity correlation attacks on two clock controlled generators. In: Adv. in crypt.–Asiacrypt '98. LNCS, vol. 1514. Berlin: Springer; 1998. p. 342–56.
[240] Johansson T, Jönsson F. Improved fast correlation attacks on stream ciphers via convolutional codes. In: Adv. in crypt.–Eurocrypt '99. LNCS, vol. 1592. Berlin: Springer; 1999. p. 347–62.
[241] Johansson T, Jönsson F. Fast correlation attacks based on turbo code techniques. In: Adv. in crypt.–Crypto '99. LNCS, vol. 1666. Berlin: Springer; 1999. p. 181–97.
[242] Johansson T, Jönsson F. Fast correlation attacks through reconstruction of linear polynomials. In: Adv. in crypt.–Crypto 2000. LNCS, vol. 1880. Berlin: Springer; 2000. p. 300–15.
[243] Jørgensen LK, Klin M. Switching of edges in strongly regular graphs. I. A family of partial difference sets on 100 vertices. Electronic J Combin 2003;10: # R17.
[244] Jungnickel D, Pott A. Difference sets: an introduction. In: Difference sets, sequences and their correlation properties (Bad Windsheim, 1998). NATO Adv. Sci. Inst. Ser. C Math. Phys. Sci., vol. 542. Dordrecht: Kluwer Acad. Publ.; 1999. p. 259–95.
[245] Junod P. On the complexity of Matsui's attack. In: Selected areas crypt. 2001. LNCS, vol. 2259. Berlin: Springer; 2001. p. 199–211.
[246] Junod P. Statistical cryptanalysis of block ciphers. Ph.D. thesis 3179. École Polytechnique Fédérale de Lausanne, Switzerland; 2004.
[247] Kam JB, Davida GI. Structured design of substitution-permutation encryption networks. IEEE Trans Comput 1979;28(10):747–53.
[248] Kahn D. The codebreakers. New York: Macmillan; 1967.
[249] Kavut S, Maitra S, Yücel MD. Enumeration of 9-variable rotation symmetric Boolean functions having nonlinearity >240. In: Indocrypt 2006. LNCS, vol. 4329. Berlin: Springer; 2006. p. 266–79.
[250] Kavut S, Maitra S, Yücel MD. Search for Boolean functions with excellent profiles in the rotation symmetric class. IEEE Trans Inform Theory 2007;53:1743–51.
[251] Kavut S, Yücel D. Generalized rotation symmetric and dihedral symmetric boolean functions - 9 variable Boolean functions with nonlinearity 242. In: Proceedings of applied algebra, algebraic algorithms and error correcting codes. LNCS, vol. 4851. Berlin: Springer; 2007. p. 321–9.
[252] Kesava Menon P. Difference sets in Abelian groups. Proc AMS 1960;11:368–77.
[253] Key EL. An analysis of the structure and complexity of nonlinear binary sequence generators. IEEE Trans Inform Theory 1976;22:732–6.
[254] Kharaghani H, Tayfeh-Rezaie B. A Hadamard matrix of order 428. J Combin Designs 2005;13: 435–40.
[255] Khoo K, Gong G, Stinson DR. A new characterization of semi-bent and bent functions on finite fields. Des Codes Cryptogr 2006;38:279–95.
[256] Kim K, Matsumoto T, Imai H. A recursive construction method of S-boxes satisfying strict avalanche criteria. In: Adv. in crypt.–Crypto '90. LNCS, vol. 537. Berlin: Springer; 1991. p. 564–74.
[257] Kim K. Construction of DES-like S-boxes based on Boolean functions satisfying the SAC. In: Adv. in crypt.–Asiacrypt '91. LNCS, vol. 739. Berlin: Springer; 1993. p. 59–72.
[258] Kipnis A, Shamir A. Cryptanalysis of the HFE public key cryptosystem by relinearization. In: Adv. in crypt.–Crypto '99. LNCS, vol. 1666. Berlin: Springer; 1999. p. 19–30.
[259] Klapper A, Goresky M. Cryptanalysis based on 2-adic rational approximation. In: Adv. in crypt.–Crypto '95. LNCS, vol. 963. Berlin: Springer; 1995. p. 262–73.
[260] Kleiman E. The XL and XSL attacks on Baby Rijndael. Master of Science Thesis. Iowa State University, http://orion.math.iastate.edu/dept/thesisarchive/MS/EKleimanMSSS05.pdf; 2005.
[261] Knuth DE. The art of computer programming, vol. 2: Seminumerical algorithms. 2nd ed. Reading, MA: Addison-Wesley; 1981.

[262] Krause M. BDD-based cryptanalysis of keystream generators. In: Adv. in crypt.–Eurocrypt 2002. LNCS, vol. 2332. Berlin: Springer; 2002. p. 222–37.
[263] Krawczyk H. The shrinking generator: some practical considerations. In: Fast software encryption–Cambridge 1993. LNCS, vol. 809. Berlin: Springer; 1994. p. 45–6.
[264] Kumar PV, Scholtz RA, Welch LR. Generalized bent functions and their properties. J Combin Theory A 1985;40:90–107.
[265] Kurosawa K, Satoh T. Design of SAC/PC(l) of order k Boolean functions and three other cryptographic criteria. In: Adv. in crypt.–Eurocrypt '97. LNCS, vol. 1233. Berlin: Springer; 1997. p. 434–49.
[266] Lachaud G, Wolfmann J. Sommes de Kloosterman Courbes elliptique et codes cyclique en characteristique 2. C R Acad Sci Paris, Série 1 1987;305:881–3.
[267] Lai X. On the design and security of block ciphers. In: Massey JL, editor. ETH Series in Information Processing, vol. 1. Kostanz: Hartung-Gorre Verlag; 1992.
[268] Lai X. Additive and linear structures of cryptographic functions. In: Fast software encryption–Leuven 1994. LNCS, vol. 1008. Berlin: Springer; 1995. p. 75–85.
[269] Langevin P, Leander G, Rabizzoni P, Véron P, Zanotti J-P. The number of bent functions with 8 variables, http://langevin.univ-tln.fr/project/quartics
[270] Langevin P, Rabizzoni P, Véron P, Zanotti J-P. On the number of bent functions with 8 variables. In: Boolean functions: cryptography and applications, proceedings–BFCA'06. 2006. p. 125–35.
[271] Leander NG. Normality of bent functions, monomial and binomial bent functions. Ph.D. thesis. Ruhr Universität Bochum; 2004.
[272] Lechner RJ. Harmonic analysis of switching functions. In: Mukhopadhyay A, editor. Recent developments in switching theory. New York: Academic Press; 1971.
[273] Lee DH, Kim J, Hong J, Han JW, Moon D. Algebraic attacks on summation generators. In: Fast software encryption–Delhi 2004. LNCS, vol. 3017. Berlin: Springer; 2004. p. 34–48.
[274] Lee S, Chee S, Park S, Park S. Conditional correlation attack on nonlinear filter generators. In: Adv. in crypt.–Asiacrypt '96. LNCS, vol. 1163. Berlin: Springer; 1996. p. 360–7.
[275] Lempel A, Cohn M. Maximal families of bent sequences. IEEE Trans Inform Theory 1982;28: 865–8.
[276] Lempel A, Weinberger MJ. Self-complementary normal bases in finite fields. SIAM J Discrete Math 1988;1:193–8.
[277] Leonard B, editor. Defending secrets, sharing data: new locks and keys for electronic information. DIANE Publishing Company; 1993.
[278] Lidl R, Niederreiter H. Introduction to finite fields and their applications. Cambridge: Cambridge University Press; 1986.
[279] Lloyd S. Counting functions satisfying a higher order strict avalanche criterion. In: Adv. in crypt.–Eurocrypt '89. LNCS, vol. 434. Berlin: Springer; 1990. p. 63–74.
[280] Lloyd S. Properties of binary functions. In: Adv. in crypt.–Eurocrypt '90. LNCS, vol. 473. Berlin: Springer; 1991. p. 124–39.
[281] Lloyd S. Characterizing and counting functions satisfying the strict avalanche criterion of order $(n-3)$. In: Mitchell C, editor. Cryptography and coding II. Oxford: Clarendon Press; 1992. p. 165–72.
[282] Lloyd S. Counting binary functions with certain cryptographic properties. J Cryptology 1992;5: 107–31.
[283] Lloyd S. Balance uncorrelatedness and the strict avalanche criterion. Discrete Appl Math 1993;41: 223–33.
[284] Löhlein B. Attacks based on conditional correlations against the nonlinear filter generator, http://eprint.iacr.org/2003/020
[285] Long DL, Wigderson A. The discrete logarithm hides $O(\log n)$ bits. SIAM J Comput 1988;17: 363–72.
[286] Machale D. George Boole: his life and work (profiles of genius series). Boole Press; 1985.

[287] MacWilliams FJ, Sloane NJA. The theory of error-correcting codes. Amsterdam: North-Holland Publishing Company; 1978.

[288] Maiorana JA. A class of bent functions. R41 Technical paper. August 1970.

[289] Maitra S. Highly nonlinear balanced Boolean functions with very good autocorrelation property. In: Coding and cryptography–WCCC 2001, Paris. Electron. Notes Discrete Math., vol. 6. Amsterdam: Elsevier; 2001. 10 pages.

[290] Maitra S. Highly nonlinear balanced Boolean functions with good local and global avalanche characteristics. Inform Process Lett 2002;83:281–6.

[291] Maitra S. On nonlinearity and autocorrelation properties of correlation immune Boolean functions. JISE J Inf Sci Eng 2004;20:305–23.

[292] Maitra S, Pasalic E. Further constructions of resilient Boolean functions with very high nonlinearity. In: Sequences and their applications (Bergen, 2001). Discrete Math. Theor. Comput. Sci. London: Springer; 2002. p. 265–80.

[293] Maitra S, Pasalic E. Further constructions of resilient Boolean functions with very high nonlinearity. IEEE Trans Inform Theory 2002;48:1825–34.

[294] Maitra S, Sarkar P. Highly nonlinear resilient functions optimizing Siegenthaler's inequality. In: Adv. in crypt.–Crypto '99. LNCS, vol. 1666. Berlin: Springer; 1999. p. 198–215.

[295] Maitra S, Sarkar P. Hamming weights of correlation immune Boolean functions. Inform Process Lett 1999;71:149–53.

[296] Maitra S, Sarkar P. Characterization of symmetric bent functions – an elementary proof. J Combin Math Combin Comput 2002;43:227–30.

[297] Maitra S, Sarkar P. Maximum nonlinearity of symmetric Boolean functions on odd number of variables. IEEE Trans Inform Theory 2002;48:2626–30.

[298] Maitra S, Sarkar P. Cryptographically significant Boolean functions with five valued Walsh spectra. Theoret Comput Sci 2002;276:133–46.

[299] Maitra S, Sarkar P. Modifications of Patterson–Wiedemann functions for cryptographic applications. IEEE Trans Inform Theory 2002;48:278–84.

[300] Massey JL. Shift-register synthesis and BCH decoding. IEEE Trans Inform Theory 1969;15:122–7.

[301] Massey JL, Serconek S. A Fourier transform approach to the linear complexity of nonlinearly filtered sequences. In: Adv. in crypt.–Crypto '94. LNCS, vol. 839. Berlin: Springer; 1994. p. 332–40.

[302] Massey JL, Serconek S. Linear complexity of periodic sequences: a general theory. In: Adv. in crypt.–Crypto '96. LNCS, vol. 1109. Berlin: Springer; 1996. p. 358–71.

[303] Matsui M, Yamagishi A. A new method for known plaintext attack of FEAL cipher. In: Adv. in crypt.–Eurocrypt '92. LNCS, vol. 658. Berlin: Springer; 1993. p. 81–91.

[304] Matsui M. Linear cryptanalysis method for DES cipher. In: Adv. in crypt.–Eurocrypt '93. LNCS, vol. 765. Berlin: Springer; 1994. p. 386–97.

[305] Matsui M. On correlation between the order of S-boxes and the strength of DES. In: Adv. in crypt.–Eurocrypt '94. LNCS, vol. 950. Berlin: Springer; 1995. p. 366–75.

[306] Maximov A, Hell M, Maitra S. Plateaued rotation symmetric Boolean functions on odd number of variables. In: First Workshop on Boolean Functions: cryptography and applications (BFCA05), 2005, Rouen, France (also available at eprint.iacr.org/2004/144.ps).

[307] Mayhew GL. Weight class distributions of de Bruijn sequences. Discrete Math 1994;126:425–9.

[308] Mayhew GL. Further results on de Bruijn weight classes. Discrete Math 2001;232:171–3.

[309] Mayr EW, Meyer A. The complexity of the word problem for commutative semigroups and polynomial ideals. Adv Math 1982;46:305–29.

[310] McFarland R. A discrete Fourier theory for binary functions. R41 Technical paper; June 1971.

[311] McFarland R. A family of difference sets in non-cyclic groups. J Combin Theory A 1973;15:1–10.

[312] McGrew D, Viega J. The Galois/counter mode of operation (GCM), submission to NIST modes of operation process, January 15; 2004.

[313] McGrew D, Viega J. The security and performance of the Galois/counter mode (GCM) of operation. In: Indocrypt 2004. LNCS, vol. 3348. Berlin: Springer; 2004. p. 343–55.

[314] McKay BD, Spence E. Classification of regular two-graphs on 36 and 38 vertices. Australas J Combin 2001;24:293–300.
[315] Meier W, Pasalic E, Carlet C. Algebraic attacks and decomposition of Boolean functions. In: Adv. in crypt.–Eurocrypt 2004. LNCS, vol. 3027. Berlin: Springer; 2004. p. 474–91.
[316] Meier W, Staffelbach O. Fast correlation attacks on certain stream ciphers. In: Adv. in crypt.–Eurocrypt '88. LNCS, vol. 330. Berlin: Springer; 1989. p. 301–14.
[317] Meier W, Staffelbach O. Fast correlation attacks on certain stream ciphers. J Cryptology 1988–89; 1:159–76.
[318] Meier W, Staffelbach O. Nonlinearity criteria for cryptographic functions. In: Adv. in crypt.–Eurocrypt '89. LNCS, vol. 434. Berlin: Springer; 1990. p. 549–62.
[319] Meier W, Staffelbach O. Correlation properties of combiners with memory in stream ciphers. In: Adv. in crypt.–Eurocrypt '90. LNCS, vol. 473. Berlin: Springer; 1991. p. 204–13.
[320] Meier W, Staffelbach O. Correlation properties of combiners with memory in stream ciphers. J Cryptology 1992;5:67–86.
[321] Meier W, Staffelbach O. The self-shrinking generator. In: Adv. in crypt.–Eurocrypt '94. LNCS, vol. 950. Berlin: Springer; 1995. p. 205–14.
[322] Menezes AJ, Van Oorschot PC, Vanstone SA. Handbook of applied cryptography. Boca Raton: CRC Press; 1997.
[323] Meng Q, Yang M, Zhang H, Cui J. Discrete Math 2008;308:5576–84.
[324] Meng Q, Zhang H, Yang M, Cui J. On the degree of homogeneous bent functions. Discrete Appl Math 2007;155:665–9.
[325] Menicocci R. Short Gollmann generators may be insecure. In: Codes and cyphers, cryptography and coding IV. 1993 IMA conference. Southend-on-Sea: Formara; 1995. p. 281–97.
[326] Mihaljević MJ. An approach to the initial state reconstruction of a clock-controlled shift register based on a novel distance measure. In: Adv. in crypt.–Auscrypt '92. LNCS, vol. 718. Berlin: Springer; 1993. p. 349–56.
[327] Mihaljević MJ. A faster cryptanalysis of the self-shrinking generator. In: Information security and privacy 1996. LNCS, vol. 1172. Berlin: Springer; 1996. p. 182–9.
[328] Mihaljević MJ, Fossorier M, Imai H. A low complexity and high performance algorithm for the fast correlation attack. In: Fast software encryption–New York 2000. LNCS, vol. 1978. Berlin: Springer; 2001. p. 196–212.
[329] Mihaljević MJ, Fossorier M, Imai H. Fast correlation attack algorithm with list decoding and an application. In: Fast software encryption–Yokohama 2001. LNCS, vol. 2355. Berlin: Springer; 2002. p. 196–210.
[330] Mihaljević MJ, Golić JD. A fast iterative algorithm for a shift register initial state reconstruction given the noisy output sequence. In: Adv. in crypt.–Auscrypt '90. LNCS, vol. 453. Berlin: Springer; 1990. p. 165–75.
[331] Mihaljević MJ, Golić JD. A comparison of cryptanalytic principles based on iterative error-correction. In: Adv. in crypt.–Eurocrypt '91. LNCS, vol. 547. Berlin: Springer; 1992. p. 527–31.
[332] Mihaljević MJ, Golić JD. Convergence of a Bayesian iterative error-correction procedure on a noisy shift register sequence. In: Adv. in crypt.–Eurocrypt '92. LNCS, vol. 658. Berlin: Springer; 1993. p. 124–37.
[333] Miranovich K. Spectral analysis of Boolean functions under nonuniformity of arguments, http://eprint.iacr.org/2002/021
[334] Mitchell C. Enumerating Boolean functions of cryptographic significance. J Cryptology 1990;2: 155–70.
[335] Modes of Operation. Workshop 1 (October 20, 2000), Baltimore, Maryland, USA, http://csrc.nist.gov/CryptoToolkit/modes/
[336] Modes of Operation. Workshop 2 (August 24, 2001), Goleta, California, USA, http://csrc.nist.gov/CryptoToolkit/modes/

[337] Monnerat J, Vaudenay S. On some weak extensions of AES and BES. In: The 6th international conference on information and communications security 2004. LNCS, vol. 3269. Berlin: Springer; 2004. p. 414–26.

[338] Moriai S, Shimoyama T, Kaneko T. Higher order differential attack using chosen higher order differences. In: Selected areas crypt. 1998. LNCS, vol. 1556. Berlin: Springer; 1999. p. 106–17.

[339] Murphy S, Robshaw MJ. Esential algebraic structures within the AES. In: Adv. in crypt.–Crypto 2002. LNCS, vol. 2442. Berlin: Springer; 2002. p. 1–16.

[340] Murphy S, Robshaw MJ. Comments on the security of the AES and the XSL technique, http://www.isg.rhul.ac.uk/~mrobshaw/rijndael/xslnote.pdf, September 26; 2002.

[341] Neumann PM. A lemma that is not Burnside's. Math Scientist 1979;4:133–41.

[342] NIST. Modes of operations for symmetric block ciphers (available at: http://csrc.nist.gov/CryptoToolkit/modes/).

[343] NIST. Request for AES proposals, http://csrc.nist.gov/CryptoToolkit/aes/pre-round1/aes_.9709.htm

[344] Nyberg K. Construction of bent functions and difference sets. In: Adv. in crypt.–Eurocrypt '90. LNCS, vol. 473. Berlin: Springer; 1991. p. 151–60.

[345] Nyberg K. Perfect nonlinear *S*-boxes. In: Adv. in crypt.–Eurocrypt '91. LNCS, vol. 547. Berlin: Springer; 1992. p. 378–86.

[346] Nyberg K. On the construction of highly nonlinear permutations. In: Adv. in crypt.–Eurocrypt '92. LNCS, vol. 658. Berlin: Springer; 1992. p. 92–8.

[347] Nyberg K. New bent mapping suitable for fast implementation. In: Fast software encryption–Cambridge 1993. LNCS, vol. 809. Berlin: Springer; 1994. p. 179–84.

[348] Nyberg K. *S*-boxes and round functions with controllable linearity and differential uniformity. In: Fast software encryption–Leuven 1994. LNCS, vol. 1008. Berlin: Springer; 1995. p. 111–30.

[349] O'Connor L. An upper bound on the number of functions satisfying the strict avalanche criterion. Inform Process Lett 1994;52:325–7.

[350] O'Connor L, Klapper A. Algebraic nonlinearity and its application to cryptography. J Cryptology 1994;7:213–27.

[351] O'Connor JJ, Robertson EF. George Boole, http://www.history.mcs.st-andrews.ac.uk/Biographies/Boole.html

[352] Olsen JD, Scholtz RA, Welch LR. Bent-function sequences. IEEE Trans Inform Theory 1982;28:858–64.

[353] Oswald E, Daemen J, Rijmen V. AES – The state of the art of Rijndael's security. Technical report (October 30, 2002), http://citeseerx.ist.psu.edu/viewdoc/summary?doi=10.1.1.3.164

[354] Park S-J, Lee S-J, Goh S-C. On the security of the Gollmann cascades. In: Adv. in crypt.–Crypto '95. LNCS, vol. 963. Berlin: Springer; 1995. p. 148–56.

[355] Park SM, Lee S, Sung SH, Kim K. Improving bounds for the number of correlation immune Boolean functions. Inform Process Lett 1997;61:209–12.

[356] Pasalic E, Johansson T, Maitra S, Sarkar P. New constructions of resilient and correlation immune Boolean functions achieving upper bounds on nonlinearity. In: Workshop on Coding and Cryptography – WCC2001. Electronic notes in discrete mathematics, vol. 6. Amsterdam: Elsevier Ltd; 2001.

[357] Pasalic E, Maitra S. Linear codes in generalized construction of resilient functions with very high nonlinearity. IEEE Trans Inform Theory 2002;48:2182–91.

[358] Patarin J. Hidden fields equations (HFE) and isomorphisms of polynomials (IP): two new families of asymmetric algorithms. In: Adv. in crypt.–Eurocrypt '96. LNCS, vol. 1070. Berlin: Springer; 1996. p. 33–48.

[359] Patel S, Sundaram GS. An efficient discrete log pseudo-random generator. In: Adv. in crypt.–Crypto '98. LNCS, vol. 1462. Berlin: Springer; 1998. p. 304–17.

[360] Patterson NJ, Wiedemann DH. The covering radius of the $(2^{15}, 16)$ Reed–Muller code is at least 16276. IEEE Trans Inform Theory 1983;29:354–6, see also the correction in IEEE Trans Inform Theory 36 (1990), 443.

[361] Penzhorn WT. Correlation atttacks on stream ciphers: computing low-weight parity checks based on error-correcting codes. In: Fast software encryption–Cambridge 1996. LNCS, vol. 1039. Berlin: Springer; 1997. p. 159–72.
[362] Peralta R. Simultaneous security of bits in the discrete log. In: Adv. in crypt.–Eurocrypt '85. LNCS, vol. 219. Berlin: Springer; 1986. p. 62–72.
[363] Pieprzyk JP. Nonlinearity of exponent permutations. In: Adv. in crypt.–Eurocrypt' 89. LNCS, vol. 434. Berlin: Springer; 1990. p. 80–92.
[364] Pieprzyk JP. On bent permutations. In: Mullen GL, Shiue PJ-S, editors. Finite Fields, Coding Theory and Adv. in Communications and Computing, vol. 141. New York: Marcel Decker; 1993. p. 173–81.
[365] Pieprzyk JP, Qu CX. Fast hashing and rotation-symmetric functions. J Universal Computer Science 1999;5:20–31.
[366] Piret G-F. Block ciphers: security proofs, cryptanalysis, design, and fault attacks. Ph.D. thesis. Université Catholique de Louvain, 2005.
[367] Pommerening K. Fourier analysis of Boolean maps – a tutorial, manuscript, 2005 (available at: http://www.staff.uni-mainz.de/pommeren/Kryptologie/Bitblock/A_Nonlin/Fourier.pdf).
[368] Preneel B. Analysis and design of cryptographic hash functions. Ph.D. thesis. Katholieke Universiteit Leuven, Belgium; 1993.
[369] Preneel B, Biryukov A, de Cannière C, Örs SB, Oswald E, van Rompay B et al. Final report of European project IST-1999 12324. New European Schemes for Signatures, Integrity, and Encryption, https://www.cosic.esat.kuleuven.be/nessie/
[370] Preneel B, Govaerts R, Vandewalle J. Boolean functions satisfying higher order propagation criteria. In: Adv. in crypt.–Eurocrypt '91. LNCS, vol. 547. Berlin: Springer; 1992. p. 141–52.
[371] Preneel B, Van Leekwijck W, Van Linden L, Govaerts R, Vandewalle J. Propagation characteristics of Boolean functions. In: Adv. in crypt.–Eurocrypt '90. LNCS, vol. 473. Berlin: Springer; 1991. p. 161–73.
[372] Prömel HJ, Schickinger T, Steger A. A note on triangle-free and bipartite graphs. Discrete Math 2002;257:531–40.
[373] Qu C, Seberry J, Pieprzyk JP. On the symmetric property of homogeneous Boolean functions. In: Proc. Australian conference on information security and privacy 1999. LNCS, vol. 1587. Berlin: Springer; 1999. p. 26–35.
[374] Qu C, Seberry J, Pieprzyk JP. Homogeneous bent functions. Discrete Appl Math 2000;102:133–9.
[375] Ralston A. De Bruijn sequences – a model example of the interaction of discrete mathematics and computer science. Math Mag 1982;55:131–43.
[376] Rao CR. Factorial experiments derivable from combinatorial arrangements of arrays. J Roy Statist Soc 1947;9:128–39.
[377] Ritter T. A variable size core for block cipher cryptography, cipher blocks of arbitrary size with good data diffusion (available at: www.ciphersbyritter.com/VSBCCORE.HTM).
[378] Rivest RL. The RC5 encryption algorithm. In: Fast software encryption–Leuven 1994. LNCS, vol. 1008. Berlin: Springer; 1995. p. 86–96.
[379] Rivest RL. Block encryption algorithm with data dependent rotation. US patent #5, 724, 428, issued on March 3, 1998.
[380] Rivest RL, Shamir A, Adleman L. A method for obtaining digital signatures and public-key cryptosystems. MIT Memo MIT/LCS/TM-82; 1977.
[381] Rivest RL, Shamir A, Adleman L. A method for obtaining digital signatures and public key cryptosystems. Comm ACM 1978;21:120–6.
[382] Rodier F. Sur la non-linéarité des fonctions booléennes. Acta Arithmetica 2004;115:1–22.
[383] Rodier F. Asymptotic nonlinearity of Boolean functions. Coding and cryptography–WCC 2003, INRIA, p. 397–405.
[384] Rosen K. Elementary number theory and its applications. 4th ed. Pearson Addison Wesley; 2000.
[385] Rothaus OS. On bent functions. J Combin Theory A 1976;20:300–5.
[386] Rotman J. An introduction to the theory of groups. Springer; 1995.

[387] Rueppel RA. Correlation immunity and the summation generator. In: Adv. in crypt.–Crypto '85. LNCS, vol. 218. Berlin: Springer; 1986. p. 260–72.
[388] Rueppel RA. Analysis and design of stream ciphers. Springer communications and control engineering series. Berlin: Springer; 1986.
[389] Rueppel RA. Stream ciphers. In: Contemporary cryptology: the science of information integrity. IEEE Press; 1992. p. 65–134, Chapter 2 in [427].
[390] Rueppel RA, Staffelbach OJ. Products of linear recurring sequences with maximum complexity. IEEE Trans Inform Theory 1987;33:124–31.
[391] Ruskey F, Sawada J. An efficient algorithm for generating necklaces with fixed density. SIAM J Comput 1999;29:671–84.
[392] Sağdiçoğlu S. Cryptological viewpoint of Boolean functions. Ph.D. thesis. The Middle East Technical University; 2003.
[393] Sakthivel G. Differential cryptanalysis of substitution permutation networks and Rijndael-like ciphers. Master's project report, Rochester Institute of Technology, Department of Computer Science; 2005.
[394] Sarkar P. A note on the spectral characterization of Boolean functions. Inform Process Lett 2000; 74:191–5.
[395] Sarkar P, Maitra S. Nonlinearity bounds and constructions of resilient Boolean functions. In: Adv. in crypt.–Crypto 2000. LNCS, vol. 1880. Berlin: Springer; 2000. p. 515–32.
[396] Sarkar P, Maitra S. Construction of nonlinear Boolean functions with important cryptographic properties. In: Adv. in crypt.–Eurocrypt 2000. LNCS, vol. 1807. Berlin: Springer; 2000. p. 485–506.
[397] Sarkar P, Maitra S. Cross-correlation analysis of cryptographically useful Boolean functions and *S*-boxes. Theory Comput Syst 2002;35:39–57.
[398] Sarkar P, Maitra S. Construction of nonlinear resilient Boolean functions using 'small' affine functions. IEEE Trans Inform Theory 2004;50:2185–93.
[399] Savard J. The Advanced Encryption Standard (Rijndael), http://www.quadibloc.com/crypto/co040401.htm
[400] Savicky P. On the bent Boolean functions that are symmetric. European J Combin 1994;15: 407–10.
[401] Schneeweiss WG. Boolean functions with engineering applications and computer programs. Springer; 1989.
[402] Schneider M. A note on the construction and upper bounds of correlation-immune functions. In: Cryptography and coding 6th IMA conference. LNCS, vol. 1355. Berlin: Springer; 1997. p. 295–306.
[403] Schneider M. On the construction and upper bounds of balanced and correlation-immune functions. Selected Areas Crypt 1997;73–87.
[404] Schneier B. Applied cryptography. 2nd ed. John Wiley & Sons; 1996.
[405] Schnorr CP. The multiplicative complexity of Boolean functions. In: Applied algebra, algebr. algorithms and error-correcting codes, Italy; 1988.
[406] Seberry J, Xia T, Pieprzyk JP. Construction of cubic homogeneous Boolean bent functions. Australas J Combin 2000;22:233–45.
[407] Seberry J, Zhang X-M. Highly nonlinear 0-1 balanced Boolean functions satisfying strict avalanche criterion. In: Adv. in crypt.–Auscrypt '92. LNCS, vol. 718. Berlin: Springer; 1993. p. 145–55.
[408] Seberry J, Zhang X-M, Zheng Y. On construction and nonlinearity of correlation immune functions. In: Adv. in crypt.–Eurocrypt '93. LNCS, vol. 765. Berlin: Springer; 1994. p. 181–99.
[409] Seberry J, Zhang X-M, Zheng Y. Improving the strict avalanche characteristics of cryptographic functions. Inform Process Lett 1994;50:37–41.
[410] Seberry J, Zhang X-M, Zheng Y. Nonlinearly balanced functions and their propagation characteristics. In: Adv. in crypt.–Crypto '93. LNCS, vol. 773. Berlin: Springer; 1994. p. 49–60.

[411] Seberry J, Zhang X-M, Zheng Y. Relationships among nonlinearity criteria. In: Adv. in crypt.–Eurocrypt '94. LNCS, vol. 950. Berlin: Springer; 1995. p. 389–98.

[412] Seberry J, Zhang X-M, Zheng Y. Nonlinearity and propagation characteristics of balanced Boolean functions. Inform Comput 1995;119:1–13.

[413] Segers AJM. Algebraic attacks from a Gröbner basis perspective. Master's Thesis. Technische Universiteit Eindhoven, Netherlands; 2004.

[414] Selmer ES. Linear recurrence relations over finite fields, 1965 Cambridge University lecture notes. Bergen, Norway: Dept. of Math., University of Bergen; 1966.

[415] Selmer ES. From the memoirs of a Norwegian cryptologist. In: Adv. in crypt.–Eurocrypt '93. LNCS, vol. 765. Berlin: Springer; 1994. p. 142–50.

[416] Shamir A, Patarin J, Courtois N, Klimov A. Efficient algorithms for solving overdefined systems of multivariate polynomial equations. In: Adv. in crypt.–Eurocrypt 2000. LNCS, vol. 1807. Berlin: Springer; 2000. p. 392–407.

[417] Shannon CE. A mathematical theory of communication. Bell System Tech J 1948;27:379–423, 623–56.

[418] Shannon CE. Communication theory of secrecy systems. Bell System Tech J 1949;28:656–715.

[419] Sloane NJA, Wyner AD, editors. Claude Elwood Shannon: collected papers. NY: IEEE Press; 1993 (with a profile of Shannon by Anthony Liversidge).

[420] Shoup V. A computational introduction to number theory and algebra. Cambridge University Press; 2005.

[421] Shrikhande SS. The uniqueness of the L2 association scheme. Ann Math Statist 1959;30:781–98.

[422] Siegenthaler T. Correlation immunity of nonlinear combining functions for cryptographic applications. IEEE Trans Inform Theory 1984;30:776–80.

[423] Siegenthaler T. Decrypting a class of stream ciphers using ciphertext only. IEEE Trans Comput 1985;34:81–5.

[424] Siegenthaler T. Cryptanalysts representation of nonlinearly filtered ML sequences. In: Adv. in crypt.–Eurocrypt '85. LNCS, vol. 219. Berlin: Springer; 1986. p. 103–10.

[425] Siegenthaler T. Design of combiners to prevent divide and conquer attacks. In: Adv. in crypt.–Crypto '85. LNCS, vol. 218. Berlin: Springer; 1986. p. 273–9.

[426] Simon MK, Omura JK, Scholtz RA, Levitt BK. Spread spectrum communications, vol. 1. Rockville, MD: Computer Science Press; 1985.

[427] Simmons GJ, editor. Contemporary cryptology: the science of information integrity. IEEE Press; 1992.

[428] Simpson L, Golić J, Dawson E. A probabilistic correlation attack on the shrinking generator. In: Information security and privacy 1998. LNCS, vol. 1438. Berlin: Springer; 1998. p. 147–58.

[429] Simpson L, Golić J, Salmasizadeh M, Dawson E. A fast correlation attack on multiplexer generators. Inform Process Lett 1999;70:89–93.

[430] Skiena S. Implementing discrete mathematics: combinatorics and graph theory with Mathematica. Reading, MA: Addison-Wesley; 1990.

[431] Sloane NJA. On single-deletion-correcting codes. In: Arasu KT, Seress Á, editors. Codes and designs-Ray-Chaudhuri festschrift. de Gruyter; 2002. p. 273–92.

[432] Song B, Seberry J. Further observations on the structure of the AES algorithm. In: Fast software encryption–Lund, Sweden 2003. LNCS, vol. 2887. Berlin: Springer; 2003. p. 223–34.

[433] Stallings W. Cryptography and network security. 4th ed. Prentice Hall; 2006.

[434] Stănică P. Chromos, Boolean functions and avalanche characteristics. Ph.D. thesis. State University of New York at Buffalo; 1998.

[435] Stănică P. Nonlinearity, local and global avalanche characteristics of balanced Boolean functions. Discrete Math 2002;248:181–93.

[436] Stănică P. Graph eigenvalues and Walsh spectrum of Boolean functions. In: Combinatorial number theory, de Gruyter. In: Proc. of the 'Integers Conference 2005' in celebration of the 70th birthday of Ron Graham, Carrollton, Georgia, 431–42. (Appeared also in Integers-J. Combinatorial Number Theory 7(2), Art. 32).

[437] Stănică P. On the nonexistence of homogeneous rotation symmetric bent Boolean functions of degree greater than two. In: Proceedings of NATO adv. stud. instit. – Boolean functions; 2008. p. 214–8.

[438] Stănică P, Maitra S. A constructive count of rotation symmetric functions. Inform Process Lett 2003;88:299–304.

[439] Stănică P, Maitra S. Rotation symmetric Boolean functions–count and cryptographic properties. Discrete Appl Math 2008;156:1567–80; preliminary version appeared in Electronic Notes in Discrete Math 15 (2003), 141–7.

[440] Stănică P, Maitra S, Clark J. Results on rotation symmetric bent and correlation immune Boolean functions. In: Fast software encryption–Delhi 2004. LNCS, vol. 3017. Berlin: Springer; 2004. p. 161–77.

[441] Stănică P, Sung SH. Improving the nonlinearity of certain balanced Boolean functions with good local and global avalanche characteristics. Inform Process Lett 2001;79:167–72.

[442] Stănică P, Sung SH. Boolean functions with five controllable cryptographic properties. Des Codes Cryptogr 2004;31:147–57.

[443] Steel A. Gröbner basis timings page, http://magma.maths.usyd.edu.au/users/allan/gb; 2004.

[444] Stinson DR. Resilient functions and large sets of orthogonal arrays. Congr Numer 1993;92: 105–10.

[445] Stinson DR. Cryptography, theory and practice. 3rd ed. Boca Raton: Chapman & Hall/CRC; 2006.

[446] Strassen V. Gaussian elimination is not optimal. Numer Math 1969;13:354–6.

[447] Tarannikov Y. On resilient Boolean functions with maximal possible nonlinearity. In: Indocrypt 2000. LNCS, vol. 1977. Berlin: Springer; 2000. p. 19–30.

[448] Tarannikov Y. Spectral analysis of high order correlation immune functions, preprint 2000.

[449] Tarannikov Y. New constructions of resilient Boolean functions with maximal nonlinearity. In: Fast software encryption–Yokohama 2001. LNCS, vol. 2355. Berlin: Springer; 2002. p. 66–77.

[450] Tarannikov Y, Korolev P, Botev A. Autocorrelation coefficients and correlation immunity of Boolean functions. In: Adv. in crypt.–Asiacrypt 2001. LNCS, vol. 2248. Berlin: Springer; 2001. p. 460–79.

[451] Toli I, Zanoni A. Looking inside AES and BES. In: Levy J-J, Mayr EW, Mitchell JC, editors. Exploring new frontiers of theoretical informatics. Proc. 18th IFIP world computer congress (TCS 2004), USA: Kluwer; 2004. p. 23–36.

[452] U.S. Senate, Select Committee on Intelligence. Unclassified summary: involvement of NSA in the development of the data encryption standard; 1978.

[453] Vaudenay S. On the need for multipermutations: cryptanalysis of MD4 and SAFER. In: Fast software encryption–Leuven 1994. LNCS, vol. 1008. Berlin: Springer; 1995. p. 286–97.

[454] Vazirani UV. Towards a strong communication complexity theory or generating quasi-random sequences from two communicating slightly random sources. In: Proc. 17th symposium on theory of computing; 1985. p. 366–78.

[455] Vazirani UV, Vazirani VV. Efficient and secure pseudo-random number generation. In: Proc. 25th IEEE symposium on foundations of computer science; 1984. p. 458–63.

[456] Vazirani UV, Vazirani VV. Efficient and secure pseudo-random number generation. In: Adv. in crypt.–Crypto '84. LNCS, vol. 196. Berlin: Springer; 1985. p. 193–202.

[457] Webster AF, Tavares SE. On the design of *S*-boxes. In: Adv. in crypt.–Crypto '85. LNCS, vol. 218. Berlin: Springer; 1986. p. 523–34.

[458] Weinmann R-P. Evaluating algebraic attacks on the AES. Diplomarbeit, Darmstadt University, Germany; 2004.

[459] Weisstein EW. Hadamard matrix. Mathworld – a Wolfram web resource. http://mathworld.wolfram.com/HadamardMatrix.html

[460] Wiener MJ. Efficient DES key search. Technical report TR-244 (School of Computer Science, Carleton University, Ottawa, Canada), May 1994. Presented at the Rump Session of Crypto' 93.

[461] Wiener MJ. Efficient DES key search – an update. CryptoBytes 1998;3(2):6–8.

[462] Wikipedia. http://en_wikipedia.org/wiki/Data_Encryption_Standard
[463] Wikipedia. http://en_wikipedia.org/wiki/Linear_feedback_shift_register
[464] Wilf HS. The eigenvalues of a graph and its chromatic number. J London Math Soc 1967;42: 330–2.
[465] Xia T, Seberry J, Pieprzyk JP, Charnes C. Homogeneous bent functions of degree n in $2n$ variables do not exist for $n > 3$. Discrete Appl Math 2004;142:127–32.
[466] Xiang Q. Recent progress in algebraic design theory. Finite Fields Appl (ten year anniversary edition) 2005;11:622–53.
[467] Xiao G-Z, Massey JL. A spectral characterization of correlation-immune combining functions. IEEE Trans Inform Theory 1988;34:569–71.
[468] Yang YX, Guo B. Further enumerating Boolean functions of cryptographic significance. J Cryptology 1995;8:115–22.
[469] Yao A. Theory and applications of trapdoor functions. In: Proc. 23rd IEEE symposium on foundations of computer science; 1982. p. 80–91.
[470] Yarlagadda R, Hershey JE. A note on the eigenvectors of Hadamard matrices of order 2^n. Linear Algebra & Appl 1982;45:43–53.
[471] Yarlagadda R, Hershey JE. Analysis and synthesis of bent sequences. IEE Proc E 1989;136:112–23.
[472] Youssef AM, Cusick TW, Stănică P, Tavares SE. New bounds on the number of functions satisfying the strict avalanche criterion. Selected Areas Crypt 1996;49–56.
[473] Youssef AM, Tavares SE. Comment on Bounds on the number of functions satisfying the strict avalanche criterion (1996). Inform Process Lett 1996;60:271–5.
[474] Yu NY, Gong G. Constructions of quadratic bent functions in polynomial forms. IEEE Trans Inform Theory 2006;52:3291–9.
[475] Zeng K, Huang M. On the linear syndrome method in cryptanalysis. In: Adv. in crypt.–Crypto '88. LNCS, vol. 403. Berlin: Springer; 1990. p. 469–78.
[476] Zeng K, Yang CH, Rao TRN. On the linear consistency test (LCT) in cryptanalysis with applications. In: Adv. in crypt.–Crypto '89. LNCS, vol. 435. Berlin: Springer; 1990. p. 164–74.
[477] Zeng K, Yang CH, Rao TRN. An improved linear syndrome algorithm in cryptanalysis with applications. In: Adv. in crypt.–Crypto '90. LNCS, vol. 537. Berlin: Springer; 1991. p. 34–47.
[478] Zenner E, Krause M, Lucks S. Improved cryptanalysis of the self-shrinking generator. In: Information security and privacy 2001. LNCS, vol. 2119. Berlin: Springer; 2001. p. 21–35.
[479] Zhang J-Z, You Z-S, Li. Z-L. Enumeration of binary orthogonal arrays of strength 1. Discrete Math 2001;239:191–8.
[480] Zhang X-M, Zheng Y. Cryptographically resilient functions. IEEE Trans Inform Theory 1997; 43:1740–7.
[481] Zhao Y, Li H. On bent functions with some symmetric properties. Discrete Appl Math 2006;154: 2537–43.
[482] Zheng Y, Zhang X-M. Improved upper bound on the nonlinearity of high order correlation immune functions. In: Selected areas crypt. 2000. LNCS, vol. 2012. Berlin: Springer; 2001. p. 262–74.
[483] Zheng Y, Zhang X-M. New results on correlation immune functions. In: Proc. 3rd inter. conf. on inform. security and cryptology 2000. LNCS, vol. 2015. Berlin: Springer; 2001. p. 49–63.
[484] Zheng Y, Zhang X-M. On relationships among avalanche, nonlinearity and correlation immunity. In: Adv. in crypt.–Asiacrypt 2000. LNCS, vol. 1976. Berlin: Springer; 2000. p. 470–82.
[485] Zheng Y, Zhang X-M. On plateaued functions. IEEE Trans Inform Theory 2001;47:1215–23.
[486] Zheng Y, Zhang X-M, Imai H. Restriction, terms and nonlinearity of Boolean functions. Theoret Comput Sci 1999;226:207–23.
[487] Zierler N. Linear recurring sequences. J Soc Industr Appl Math (SIAM) 1959;7:31–48.
[488] Zierler N, Mills WH. Products of linear recurring sequences. J Algebra 1973;27:147–57.
[489] Živković M. An algorithm for the initial state reconstruction of the clock controlled shift register. IEEE Trans Inform Theory 1991;37:1488–90.

Index

Symbols

A

B

C

Printed and bound by CPI Group (UK) Ltd, Croydon, CR0 4YY
11/06/2025
01899189-0016